2nd International Conference on Environmentally Sustainable Animal Industry (The 2nd ICESAI 2021)

E3S Web of Conferences Volume 335 (2022)

Malang, Indonesia
12 October 2021

Editors:

E. Widodo
A Jayanegara
Suyadi
V.M.A. Nurgiartiningsih
G. Ciptadi
M.H. Natsir
S. Wahjuningsih
O. Sjofjan
Marjuki

ISBN: 978-1-7138-4243-9

Printed from e-media with permission by:

Curran Associates, Inc.
57 Morehouse Lane
Red Hook, NY 12571

Some format issues inherent in the e-media version may also appear in this print version.

For additional information, please contact EDP Sciences – Web of Conferences
at the address below.

EDP Sciences – Web of Conferences
17, Avenue du Hoggar
Parc d'Activité de Courtabœuf
BP 112
F-91944 Les Ulis Cedex A
France

Phone: +33 (0) 1 69 18 75 75
Fax: +33 (0) 1 69 28 84 91

contact-edps@webofconferences.org

Additional copies of this publication are available from:

Curran Associates, Inc.
57 Morehouse Lane
Red Hook, NY 12571 USA
Phone: 845-758-0400
Fax: 845-758-2633
Email: curran@proceedings.com
Web: www.proceedings.com

TABLE OF CONTENTS

Sustainable animal genetic improvement

Julius H. J. van der Werf*

School of Environmental and Rural Science, University of New England, Armidale, NSW 2351, Australia

Abstract. Well designed and targeted animal breeding programs allow a sustainable genetic improvement of livestock with increases in animal productivity of 1-2% per annum. Over time, massive improvements have been achieved, e.g. in dairy and pig and poultry production systems, now resulting in higher production that requires much fewer input of resources. Although reproductive and genomic technologies contribute nowadays to increases in rates of genetic improvement, the key to successful breeding programs lies in a strong focus on simple and well-defined breeding objectives, effective investment in trait measurement, a clear understanding of the structure of the breeding program, and efficient systems for genetic evaluation, selection and mating of elite animals. The dissemination of the genetics of selected animals to the wider population also needs consideration, requiring commercial farmers or smallholders to have the means to have access to improved genetics as well as an understanding of the value of using improved bulls. Definition of the breeding objectives as well as evaluation of genetic merit needs to be based on the local environment. Compared with other interventions to improve productivity and stability, genetic improvement is a critical and cost-effective approach as genetic gains are permanent and cumulative.

1 Introduction

The power of long-term genetic improvement has been well documented in various livestock species. For example, in many countries the milk production per cow has more than doubled since the introduction of AI in the 1950ties, with an annual increase of about 2%. In the UK the average milk production per cow per year in the UK was 5151kg in 1990 whereas it was 7899 kg in 2014 (AHDB Dairy), which is a 53% productivity increase in 24 years. In the United States, the production per cow per lactation increased by 72% between 1985 and 2015 (USDA). The genetic trend between 1990 and 2000 was 1092 kg [1] implying that about 70% of the phenotypic increase was due to genetic selection. Likewise, in poultry production, the growth rate of broilers has increased by about 400% between 1957 and 2005 [2], with the majority (~80-90%) of this increase being due to genetic selection. These examples demonstrate that genetic selection implemented in effective breeding programs has the potential to drive large phenotypic changes in livestock resources. In first instance, genetic selection focused on traits that led to productivity improvements. A higher efficiency of

* Corresponding author: jvanderw@une.edu.au

animal production systems leads to producing more animal product from fewer resources and this is therefore an important contribution to sustainability. However other aspects such a genetic diversity, product quality and the quality of the animals' wellbeing have become additional considerations, leading to adjustments of breeding objectives and breeding strategies. Large phenotypic changes and their consequential effect on productivity are less pronounced in most developing countries. This paper explores briefly the factors that drive the success of breeding programs, and how they might be applied to breeding programs, in particular those in developing countries.

2 Genetic improvement in developing countries

While genetic improvement programs have been very successful in developed countries it has been difficult to establish programs with significant genetic improvement in developing countries. There are several reasons for this. Different socio-economic factors play a main role, as pointed out by Rege *et al.* [3], who both provided an analysis of the role of genetic improvement in African livestock systems. Mostly, individual farmers are smallholders and the aim of the enterprises is focused on food security and overall livelihoods, besides income. Marshall *et al.* [4] described the various roles of livestock holding as well as the many challenges within the production systems and genetic improvement of productivity is not highest on the priority list. Besides, the incentives for increased production are often hindered by the lack of access to markets. Therefore, it is acknowledged that implementing breeding programs in developing countries should be part of a broader approach where infrastructure and market development are important aspects. Nevertheless, with increased population growth, urbanisation and development, the demand for animal products is expected to increase significantly and it is well recognized that genetic improvement of animal resources can play an important role in a development strategy.

3 Between versus within breed selection

One difference between the current situation in developing countries and the state of animal resources in developing countries a few decades ago, is the availability of improved breeds and strain of livestock that exist. A local breeding program is now faced with the choice of improving a local breed versus upgrading by making use of improved breeds elsewhere. In the past an example of the latter has been the upgrading of many local cattle breeds with Holstein Friesian semen importation, initially mainly from the United States. This phenomenon has become known as the Hosteinization' of many local cattle breeds. This process has had a large effect of the rate of genetic improvement of many national dairy cattle populations, and at the same time it has greatly decreased the genetic diversity, at least in black and white dairy cattle. There are several reasons why improving a local breed is preferably over importation of exogenous genetics. These include the maintenance of between breed genetic variation globally and local breeds being more adapted to the local environment, both the natural and the socio-economic environment. On the other hand, genetic improvement is about utilizing genetic variation, and this can be utilized both within and between breeds. In other words, it maybe more rational to upgrade a local population, the more so if its production potential falls behind that of alternative breeds. However, a genetic improvement program that uses exogenous breeds would only be successful if it was based on performance data, where progeny of imported bulls are compared with progeny of the local breed, with performance measured in the environment in which we aim to develop improved genetic resources for. Even the creation of new composites based on diverse local breeds, which all are more likely to be more adapted to the local environment, could be a strategy-based on that same principle, especially when indigenous populations are small. In

Australia, the meat sheep industry has made massive genetic gain over the last two decades, with lamb carcass weights improving from 18 to 23 kg. Much of that improvement was based on the introduction of LAMPLAN, a genetic evaluating system that allowed selection of sires across breeds. As a result, breeders are not too worried about 'pure-breeding' and more opportunistically use sires that have a mixed breed background. The resulting genetic gain is likely higher than what could have been achieved with within breed selection alone, simply because selection was based on genetic variation across breeds, which is generally higher than within breed variation. Rege *et al.* [3] concluded that breeding strategies for resource poor smallholders should in first instance focus on using appropriate genotypes from the available range of global breed resources. A breeding system should be set up that allows those genotypes to express their potential. My interpretation is that one should be open to what genotypes should be used in local production systems, but that the choice of genotypes should be based on data.

4 Breeding objectives

The definition of the breeding objective is the first and probably most important step to be taken in any activity with the aim of genetic improvement. Defining breeding objective involves simply to define the traits (phenotypes) that need to be improved and the relative weight for each trait, expressing the value of improvement of each trait relative to each other. Defining the value of improvement requires valuation. If this is according to an economic framework, then economic values can be derived from a financial cost-benefit model. However, not all genetic change can be valued in financial terms. Relative weights of traits in a breeding objective can also be derived from an evaluation of the required genetic changes, using a selection index framework [5].

Improving the wrong traits could be equivalent or even worse than no improvement at all. If many breeding animals will be chosen for reasons irrelevant to the breeding objective, then selected parents will be less superior with regard to the breeding objective. It is important in the selection process that the selection criterion is clear, and whether the selection is efficient in relation to the breeding objective. For example, selection based on body measures is less efficient than selection for body weight itself, when the breeding objective is to increase body weight.

Many practical breeding programs suffer from the fact that the objectives are not properly defined. Selection decisions are often influenced by attention for characteristics that are not formally defined in the objective. Furthermore, the outcome for breeding programs is noticed many years after selection decisions are made. Hence objectives have to be designed for future circumstances. These are not always easy to predict in advance.

As pointed out earlier, significant changes in phenotypes have been achieved in many countries by selection on production traits. However, after some time it was noted that these genetic improvements led to negative correlated responses in other traits, often related to fitness and longevity, and this has led to adjustments of breeding objectives, including more traits related to product quality and animal fitness. For example, there is a large body of literature to demonstrate that fertility has decreased in Holstein dairy cattle [6,7]. A more detailed review across species on the effects of one-sided selection for production traits was published by Rauw *et al.* [8]. Adjustment to breeding objectives were often implemented by increasing the selection emphasis on traits that showed undesirable genetic trends as a result of one-sided selection for production traits. The realisation that a more balanced breeding objective is more sustainable in the long term is often the result of observing undesirable correlated responses with selection for production traits only. These corelated responses are not only in fitness traits. Australian research showed selection for lean meat growth in lambs has undesirable effects on the eating quality of lamb meat [9] and breeding programs were adjusted to address this problem [10]. These adjustments affect the direction of the genetic improvement, but overall it has become clear that sustainable large genetic improvement are possible over time.

Breeding objectives are critical in the development of genetic improvement programs for environmentally sustainable animal industries. These objectives should be related to development goals of such industries. Production efficiency is a main contribution to sustainability, both economically and environmentally. It is important to be able to generate the required outputs, mainly food, while minimizing the resources needed. For example, the selection for growth in broilers has led to an efficiency of growth such that, compared to half a century ago, about one third of the same feed is now required to obtain the same growth. Traits related to animal fitness and health have become more important with the sharp increase in productivity, and improvement of such traits contribute to both productivity as well as animal wellbeing, the latter becoming increasing important to the social license for animal production. In the last decade, traits directly related to enteric emission of greenhouse gasses from livestock have become a new focus and there is sufficient genetic variation between animals in methane production to allow selection for this trait. Lower methane production is closely related to feed efficiency, making it economically attractive to include these traits in selection indexes [11], and more so when the carbon price becomes a firmer tool in greenhouse gas mitigating policies.

5 Achieving genetic improvement

Breeding goal definition is the first step to be made in designing animal breeding structures. Then, a second step is to implement a structure of gathering information, a recording system, and a genetic evaluation to identify those animals that have the highest breeding value for traits in the breeding goal. After animals are ranked, a mating strategy is required. A third step is therefore to make a well-organised structure for the use of animals with highest genetic merit. In principle, the best males should be mated to the best females in order to obtain the best offspring. Therefore, a tiered structured is usually desirable where the nucleus matings are based on 'elite matings' between the males and females with the highest merit in a population (Fig. 1). The resulting progeny from these matings can become selection candidates for the future generation of nucleus animals, and if they are chosen through selection we can expect a genetic improvement with each generation. The nucleus born animals that are not selected as nucleus parents can be used for the dissemination of superior genotypes through the commercial population (the 'breeding bulls"). Dissemination could be via artificial insemination (AI), in which case a relatively small nucleus could be used to serve a large population of commercial females. However, with natural mating, an additional 'multiplier' tier is required Since the genetic mean of the nucleus increases each generation, the genetic mean of the animals used for dissemination will increase as well, and as a result, the commercial population will be improved Note that recording of phenotypes to facilitate selection is only required in the nucleus, yet the wider commercial population will benefit. In fact, the rate of genetic gain in each tier will be the same as that in the nucleus, there is simply a genetic lag, typically equal to two generations between tiers if dissemination is only via males.

The breeding nucleus can be an easy identifiable and be physically a separate operation, but it can also be dispersed with multiple breeders using the semen of top sires to service their highest merit cows. The nucleus can also be open to influx of genetic material from other populations or breeds. For the purpose of maintaining sufficient genetic diversity, the size of the nucleus needs to be sufficient as the rate of inbreeding depends on the effective population size. This is largely driven by the number of males that is used each generation, because breeding schemes require usually fewer males than females. A ballpark figure is to use at least 10 males "unrelated" males each generation [12] but more sophisticated methods are now available to optimally balance the need for genetic gain while constraining the rate of inbreeding. These methods are referred to as 'optimal contribution selection', e.g. as implemented in the MateSel software [13].

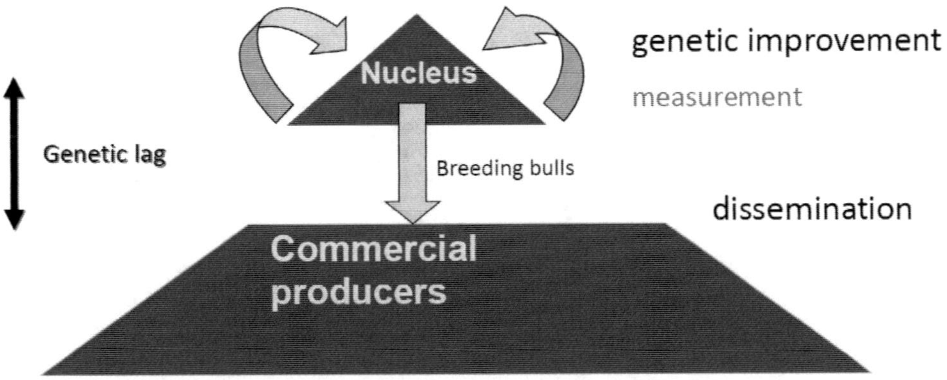

Fig. 1. Diagram of a two-tier breeding structure

6 Success factors in breeding programs

The rate of genetic gain per year (R_{year}) can be summarized by one simple formula from Rendel and Robertson [14].

$$R_{year} = \frac{S_m+S_f}{L_m+L_f} = \frac{i_m r_m+i_f r_f}{L_m+L_f}\sigma_A \qquad (1)$$

where the selection differential (S) of parents is determined by the selection intensity (i), the selection accuracy (r) and the genetic standard deviation (σ_A): S= i.r.σ_A, where the subscripts m and f refer to male and female selection, and the aim is to maximise this selection differential and to minimize the generation interval (L). The formula says in simple words: "mate the very best males with the very best females and do that as soon as possible". There are a few balances that need to be struck. For example, younger animals have usually lower accuracy of breeding value and it is tempting to progeny test sires in order to have highly accurate estimates of breeding values. However, progeny testing requires several years of waiting time before selected bull can be used, therefore drastically increasing the generation interval. The generation interval can be optimized by balancing the accuracy of selection and the generation interval. Also, selection intensity and generation interval are interdependent, as fewer replacement breeding animals are required every year (i.e. higher selection accuracy) if breeding animals are kept longer. Fortunately, it has been shown that BLUP selection optimises the age structure, and therefore the generation interval, as it allows direct comparison of genetic merit between old and young breeding animals [15].

Although understanding and implementing an efficient breeding structure is important. The key driver of genetic selection is the information that is available for selection. The information that can be used is phenotypic measurement, preferably on breeding animals, and appropriate data surrounding these measurements, e.g. age and management group, to allow unbiased genetic evaluation. Pedigree information is important to allow the use of information from relatives as well as to link data in different management groups (e.g. through link sires) and data from different generations or age classes. Nowadays, pedigree information can be substituted by genetic information, allowing a more detailed and wider linkage of animals to data. Multiple trait evaluation allows estimation of breeding value based on correlated traits. However, with antagonistic correlations, it is important to measure the key traits in the breeding objective in order to achieve a balanced genetic gain. For example, selecting for higher fertility based on information on body weight only might increase genetic gains infertility, but t will increase the response to body weight even more.

Recognizing the breeding structure in a population is important as well as understanding which decisions are made by which players. Who determines the breeding objective? Who decides on a trait recording scheme? Who makes selection decisions? How does the commercial farmer have access to the improved genetics that flows from the nucleus?

It should be noted that the rate of gain depends on the species, as it is a function of reproductive rates and generation intervals. But within species there is a large variation in genetic gain achieved and this is related to how well the breeding program is run. Breeding programs with a centralised nucleus and clear objectives related to measured traits are relatively easy as there are few decision makers and trait measurement and genetic evaluation can be approached rationally. With a dispersed nucleus, it is more challenging to collect quality data in well-recorded and well-linked management groups. Moreover decisions about breeding animals are made by many individual breeders, it is more challenging to have consensus on selection criteria, especially where more traditional views about animal selection are dominant, i.e. less emphasis on objective measurement of breeding objective traits related to productivity and sustainability and more emphasis on vaguely related criteria related to animal appearance. Successful systems also require an understanding and agreement of selection criteria at multiple tier levels, as buyers of bulls or semen need to have a clear understanding on the differences in merit between the animals on offer.

7 Use of new technologies

Animal breeding has been revolutionized in the last decade by the introduction of genomic technologies. Knowledge of animal genomes and the development of tools such as chips that allow fast and cheap genotyping for thousands of Single Nucleotide Polymorphisms (SNP) have facilitated the introduction of 'genomic selection'. This information is essentially used to link phenotype information from many more animals (compared to pedigree information) to a target animals of which we want to estimate a breeding value. Therefore, we can now estimate breeding values much more accurate at a younger age and for traits that can not be easily measured on breeding animals. Genomic selection mostly affects breeding programs where key objective traits are hard to measure, e.g. in dairy cattle where nearly all traits are sex-limited, and within breeding programs there can be a shift of selection emphasis to harder to measure traits, e.g. more on carcass quality at the expense of just selecting for growth. Moreover, new traits can now be selected for by forming reference populations where traits such as feed intake, fertility or health traits can be efficiently measured in designated cohort of animals. Compared with traditional selection based on pedigree, it is now less important that the animals whose phenotypes is measured are directly related to the selection candidates. Genomic selection has been shown to be able to nearly doubling the rates of genetic gain in dairy cattle programs [16].

Reproductive technologies can also have a substantial impact on the rate of genetic improvement. Artificial Insemination is a key technology for effectively disseminating superior genetic stock from the nucleus, and it can conveniently link data in different locations, making it easier to important genetic material from different locations and populations. AI technology has been long used for progeny testing, although that role has become less important in the genomic era. Embryo technology allows making more intensive use of the best females, allowing multiple progeny per year rather than just one, as in ruminants. This can increase selection intensity of females and allows importation of high genetic merit stock from other locations. Sexed semen is now successfully used to determine the sex in targeted matings, e.g. sons of top sires. Sexed semen has great potential in the commercial tier, where more males can be produced for more efficient meat production. The effect of sexed semen on rates of genetic gain in the nucleus however is limited.

8 Conclusion

Animal breeding programs can be extremely powerful in improving the genetic resource in animal production systems, therefore contributing to long term sustainability of the sector. It is challenging to set up effective breeding programs at a national level, with only limited success in many developing countries. The principles and success factors of genetic improvement programs are summarised in this paper. Defining clear breeding objectives relevant to a long-term sustainability goals is a key factor, where improving productivity is still a major component. Breeding goals can be further finetuned, including more emphasis on traits related to animal health and their efficiency and environmental impact. New technologies such as genomics and reproductive technologies can be useful, but only where there is already an existing breeding structure of achieving genetic improvement successfully.

References

1. P. VanRaden. J. Dairy Sci. **87**:3125–3131 (2004).
2. M.J. Zuidhof, B. L. Schneider, V. L. Carney, D. R. Korver, and F. E. Robinson. Poultry Sci., **93**: 2970–2982 (2014).
3. J.E.O. Rege, K. Marshall, A. Notenbaert, J.M. Ojango and A.M. Okeyo. Livestock Sci., **136**: 15-28 (2011).
4. K. Marshall, J.P. Gibson, O. Mwai, J.M. Mwacharo, A. Haile, T. Getachew, R. Mrode and S.J. Kemp. Frontiers in Genetics, **10**: 297 (2019).
5. L.N. Hazel, 1943. Genetics, **28**: 476-490 (1943).
6. M.D. Royal, A.O. Darwash, A.P.F. Flint, R.Webb, J.A.Woolliams, J.A. and G.E. Lamming. Animal Sci., **70**: 487-501 (2000).
7. M.C. Lucy. Journal of Dairy Science, **84** : 1277-1293 (2001).
8. W.M. Rauw, E. Kanis, E.N. Noordhuizen-Stassen and F.J. Grommers, F.J., Livestock Production Science, **56**: 15-33 (1998)
9. S.I. Mortimer, N.M. Fogarty J.H. van der Werf, D.J. Brown, A.A. Swan, et.al. J. Animal Sci., **96**: 3582-3598 (2018).
10. A.A. Swan, T. Pleasants and D. Pethick. Proc. 21st Conf. Assoc. Adv. Anim. Breed. Genet., Lorne, Australia. p. 29-32 (2015).
11. C.I.V. Manzanilla-Pech, D.M. Gordo, G.F. Difford, J.E. Pryce, F. Schenkel, S. Wegmann, F. Miglior, T.C. Chud, P.J. Moate, S.R.O. Williams, C.M. Richardson, P. Stothard and J. Lassen. J. Dairy Sci. **104**:8983–9001 (2021).
12. M.G. Goddard and C. Smith. J. Dairy Sci, **73**: 1113-1122 (1990).
13. B.P. Kinghorn. Genetics Selection Evolution, **43**: 1-9 (2011).
14. J.M. Rendel, and A. Robertson. J. Genet. **50**: 1–8 (1950).
15. J.W. James, J. Animal Breed. Genet.**104**:23 – 27. (1987).
16. L.R. Schaeffer. J. Animal Breed. Genet **123**: 218-223 (2006).

Sustainable feed resources for optimizing broiler performance in tropical countries

Anjas Asmara Samsudin[1*], *Mamat Hamidi* Kamalludin[1], and *Abdul Razak* Alimon[2]

[1]Department of Animal Science, Faculty of Agriculture, Universiti Putra Malaysia, 43400 Serdang, Selangor, Malaysia
[2]Faculty of Animal Science, Universitas Gadjah Mada, Indonesia

Abstract. Soybean meal and grain corn are feedstuffs that normally used as the main ingredients in poultry feeds as a source of protein and energy in tropical countries. The availability of these feedstuffs for broiler nutrition nowadays is becoming more competitive. Thus, food security, especially in the developing countries, would be susceptible. Currently, efforts are being made to use alternative feedstuffs as a sustainable feed resources to substitute portion of soybean meal and grain corn in poultry diets due to their high price since both of the raw material were imported. In Malaysia, usage of PKC is common in ruminant diets, but limited in the non-ruminant diets especially in poultry diets due to the high fiber content of PKC. Several works have been conducted to increase the nutritional contents of PKC as one of the measures to reduce and/or eliminate the constraints of utilizing PKC in poultry diets. The method used to achieve this target is either through physical, chemical, biological or combination of these treatments. This paper discusses the use of PKC as the sustainable feed resources that can be incorporated in the poultry feeds. Hence, emphasis should be given to improve nutritional values of PKC in order to reduce feeding cost of poultry.

1 Introduction

Under tropical environment, poultry production especially broiler is encumbered by several constraints which preventing them to achieve their full genetic potential. The shortage and high feed price, limited availability of raw material such as grain corns and soybean, has been a big challenge to this sector, since feed still represents a major operational cost in producing the birds. The consumption of poultry products, appears to increase in the tropical countries since it is considered as one of the cheaper protein source. This has demanded the grains to be imported as the main poultry feedstuffs which causing an increased in production cost since the raw materials (corn grain and soy bean) are not suitable to grow in the tropic. The increase of world population year by year would definitely create a competition between human and the poultry industry since both of these group used the grains are the main feed ingredients.

*Corresponding author: anjas@upm.edu.my

In addition, there is an increasing demand to produce a biofuel from plants, especially from corn grain. This further poses a severe food security risk, especially in the tropical region. Rapid development of the poultry industry in tropical countries has created a situation of overdependence on the conventional feedstuffs. Currently, efforts are being made to use alternative energy and protein sources to substitute soybean and corn in monogastric industry. The soybean meal is usually used as a main protein source in poultry diet ingredient. Due to price of this feedstuff is quite expensive and unaffordable, some poultry farmers have used a non-traditional feed ingredients that can substitute a certain amount of soybean meal in the diets. Those ingredients are insect meal (e.g. black soldier fly larvae and earthworm), algae, azolla and single-cell protein.

In many developing tropical countries, a huge amount of alternative feedstuffs, considered as agricultural by-products such as rice bran, copra meal and palm kernel cake (PKC) is widely available. For instance in Malaysia, the used of PKC as animal feed has been widely practiced especially in ruminant industry. However, it's used as part of poultry diet ingredient has some limitations. It contained a non-starch polysaccharides (NSPs) such as mannan and xylan as well as secondary metabolite that limits the intake and digestibility of the feeds in the broiler [1].

1.1 Palm oil industry in Malaysia

Malaysia have a vast agriculture land that is planted with palm oil tree and this has produced an abundant amount of PKC, which is considered an agricultural industrial waste derived from the extraction process of oil from palm fruits. An increased in the oil palm planted area has been recorded from 5.23 million ha in 2013 to 5.86 million ha in 2020 [2]. This is mainly due to an increase of planted area in Sarawak and Sabah. However, Sarawak is still the largest oil palm planted area (1.58 million ha), followed by Sabah (1.54 million ha) (Table 1) [2].

Table 1. Oil palm planted area in Malaysia [2]

State	Mature (Ha)	%	Immature (Ha)	%	Total (Ha)	%
Johor	688,291	92.9	52,537	7.1	740,828	12.6
Kedah	80,210	89.3	9,572	10.7	89,782	1.5
Kelantan	131,768	78.6	35,831	21.4	167,599	2.9
Melaka	51,672	91.7	4,689	8.3	56,361	1.0
Negeri Sembilan	173,490	91.1	16,972	8.9	190,462	3.2
Pahang	702,163	89.8	80,084	10.2	782,247	13.3
Perak	362,557	90.0	39,211	10.0	391,768	6.7
Perlis	689	99.3	5	0.7	694	0.0
Pulau Pinang	12,540	97.7	289	2.3	12,829	0.2
Selangor	113,911	90.0	12,614	10.0	126,525	2.2
Terengganu	148,244	83.0	30,384	17.0	178,628	3.0
Peninsular Malaysia	2,455,535	89.7	282,188	10.3	2,737,723	46.7
Sabah	1,344,608	87.1	198,446	12.9	1,543,054	26.3
Sarawak	1,431,600	90.3	152,920	9.7	1,584,520	27.0
Malaysia	5,231,743	89.2	633,554	10.8	5,865,297	100

Apart producing palm oil, the plantations of oil palm industry in Malaysia also produce a number of useful by-products such as PKC, palm kernel shells (PKS), empty fruit bunches, (EFB), oil palm fronds (OPF), palm oil mill effluent (POME; also called sludge and decanter cake), oil palm trunks (OPT), and palm press fibre (PPF) throughout the year and this guarantees their availability as livestock feed ingredients.

Although large quantities of PKC produced from this industry are readily available as feed for livestock, but the use of PKC is usually limited to the ruminant sector, due to coarse texture, high fibre content, and presence of a non-starch polysaccharides (NSP) such as 58% mannan, 12% cellulose and 4% xylan [3]. Monogastric animals such as poultry and swine have limitation in utilizing this feedstuff because of the anatomical structure of their gastrointestinal system that does not have the capability to produce enzymes that can efficiently digest fibrous feed material. The digestion of NSP fractions in the PKC especially mannans that is hard and crystalline tends to be more variable due to lack of digestive enzymes and their predilection to create a viscous environment in the intestinal lumen [4, 5].

The palm kernel cake supplies both energy and protein, but it is looked upon more as a protein source. The crude protein (CP) content of PKC varies between 10.0 to 19.8% [6,7]. Furthermore, PKC also contains high crude fibre (CF) ranging between 13 to 20% (Table 2) [8] and this contributes to the low digestibility in poultry.

Table 2. Nutrient composition of PKC [8]

Ingredients	Content (%)
Dry matter	88.0-94.5
Crude fibre (CF)	13.0-20.0
Crude protein (CP)	14.5-19.6
Ash	3.0-12.0
Ether extract	5.0-8.0
Calcium	0.2-0.3
Phosphorus	0.48-0.7
ME, MJ/Kg	6.50-7.50
Amino acid, g/16 g N	
Lysine	2.68
Methionine	1.75

Several studies have reported that supplementation of exogenous enzymes such as mannanase, cellulase and xylanase on diets containing PKC could improve its nutritive value making it more available for broiler [3,7].

1.2 PKC treatment to improve the digestibility

There are lots of treatment available to breakdown the cellulose chain in PKC in order to make it more digestible. The mechanical, chemical and biological treatment of PKC appears to improve the nutritive values of PKC.

The choices of PKC treatment has been considered an important issue faced by nutritionists, and it has limited the manipulation of PKC to be used in the feed formulation. It has been reported that the maximum level of PKC that can be incorporated in the broiler diet is around 10%. However, the treatment of the PKC via solid-state fermentation by cellulolytic bacteria may increase the inclusion of PKC in the broiler diet up to 15%.

A study by [9] on the PKC treated by enzyme produced by cellulolytic bacteria (*Paenibacillus curdlanolyticus* DSMZ 10248) via solid state fermentation and fed to broiler has found out that the birds fed with 1% of enzyme had significantly increase the feed intake and growth performance. In other study, the apparent metabolizable energy (AME) and ileal crude protein digestibility in broiler fed with 25% of enzyme treated PKC in the diet were significantly increased than control group [10]. Supplementation of NSP (e.g. xylanase) at concentration of 200U/kg in the broiler diet incorporated with 20% of PKC has significantly increase the growth performance in the treated group than control [7]. Extrusion is another way of physical treatment on the PKC that can increased the nutritive value and broiler performance. A study by [11] reported that extrusion process on the PKC has increased the AME and CP digestibility by 6% and 32% respectively in broiler chicken fed with a basal diet supplemented with 20% PKC.

2 Conclusion

In summary, the use of alternative feedstuffs in poultry industry has received many attention nowadays because of their readily availability, nutritive value and a cheap source of protein and energy. Furthermore, the used of these non-conventional feedstuffs are not competing with humans as a source of food. At the same time, the used of PKC in the broiler diet can be considered as one of the promising sustainable feed resources as it can substitute certain portions of soybean and corn which will reduce the cost of broiler production.

References

1. A. Sharmila, A. R. Alimon, A. Kassim, M. N. Hezmee, A. A. Samsudin, Malaysian J. Anim. Sci. **17** (2014)
2. MPOB. https://mpob.gov.my/data-set
3. C. H. Chong, I. Zulkifli, R. Blair, Asian-Australas. J. Anim. Sci. **21** (2008)
4. M. Choct, G. Anisson, Br. Poult. Sci. **33** (1992)
5. D. Józefiak, A. Rutkowski, S. A. Martin, Anim. Feed Sci. Technol. **113** (2004)
6. M. Ramin, A. R. Alimon, M. Ivan, Livest. Res. Rural Dev. **22**, 82 (2010)
7. A. Sharmila, A. Kassim, M. N. Hezmee, A. A. Samsudin, J. Anim. Poult. Sci. **3** (2014)
8. A. R. Alimon, Palm Oil Developments **40** (2004)
9. H. U. Zamani, T. C. Loh, H. L. Foo, A. A. Samsudin, M. I. Alshelmani, Int. J. Microbiol. Biotechnol. **2** (2017)
10. A. H. Hakim, I. Zulkifli, A. Soleimani Farjam, E. A. Awad, N. Abdullah, W. L. Chen, Ital. J. Anim. Sci. **19** (2020)

The effect of storage period on calcium and magnesium level and antimicrobial activity of goat's milk kefir

Lilik Eka Radiati[1,*], *Hariana Tri* Juliyanti[2], and *Cindy Heryanti Kusuma* Wardhani[3]

[1]Department of Animal Product, Faculty of Animal Science, Universitas Brawijaya, Malang 65145, Indonesia
[2]Undergraduate Program, Faculty of Animal Science, Universitas Brawijaya, Malang 65145, Indonesia
[3]Master Program, Faculty of Animal Science, Universitas Brawijaya, Malang 65145, Indonesia

Abstract. The purpose of this reasearch was to determine the effect of storage period goat's milk kefir (GMK) on total LAB, total mineral levels included Ca and Mg and antimicrobial activity to pathogenic microbes. The materials used was GMK. This research was conducted using Completely Randomized Design with 4 treatments and 4 replications. The treatments were storage period that consist of D0 (0 day), D1 (7 days), D2 (14 days), and D3 (21 days). Total calculation of LAB used TPC (Total Plate Count). Samples were diluted from 10^{-1} to 10^{-8}, then incoculated PCA medium by using pour plate method. Petri dish was incubated at 37°C for 48 hours. Calcium and magnesium levels were calculed base AAS method. The result of this study was analyzed using analysis of variance (ANOVA), then followed by Duncan's Multiple Range Test (DMRT). The storage time treatment has a significant effect on the total microbial count, mineral levels (Ca and Mg) and inhibition zone of *Staphylococcus aureus*, *Escherichia coli* and *Salmonella enterica serovar typhi* on goat's milk kefir. Average clear zone *Staphylococcus aureus* 2.60 – 3.18 mm. Average inhibition zone *Escherichia coli* 2.09 – 3.32 mm. Average clear zone *Salmonella enterica serovar typhi* 1.97 – 2.72 mm.

1 Introduction

Along with increasing human awareness of the importance of healthy living, there is also an increase in research and marketing of food and beverage products that have the potential to maintain a healthy body. Nutritious food and beverage products are better known as functional foods. One of the functional food products that are currently popular in the community is fermented milk [1]. Reported that kefir is a fermented milk beverage product by kefir grain that contain of lactic acid bacteria and yeast [2]. This lactic acid bacteria will cause a sour kefir taste. Kefir grains are a mass consisting of various kinds of bacteria and yeast arranged in a complex protein and carbohydrate matrix [3]. [4] argue that based on research in the world of health, kefir has many benefits, including inhibit tumor growth more effectively than yogurt, protect digestion from pathogenic bacteria attack, maintain

* Corresponding author: lilik.eka@ub.ac.id

metabolism and human immune function, and maintain cholesterol levels in the body blood. Kefir has long being considered beneficial to human health. For example, kefir enhance immune system, digestive health, treat the metabolic disorders, hypertension, ischemic heart disease, and allergies. However, it has antimicrobial, antitumor, antiviral, antimutagenic, antiinflammatory and antioxidant effects [5]

Kefir consists of 4% fat, 12% ash, 46% exopolysaccharide, 34% protein, vitamins B and K, tryptophan, several minerals such as Ca, P and Mg [6]. Minerals is nutrient that the body needs in small amounts, but has many benefits that are very important for health. Minerals can be found in dairy products, for example in kefir. Kefir is a good source of Ca, Co, Cu, Fe, Mg, Mo and Zn [6]. According to [7] the process of making kefir itself has an influence on the vitamin and mineral content in it. The content of several minerals such as phosphorus and selenium decreased significantly after the fermentation process [5]. Meanwhile, the manganese content increases after the fermentation process. This occurs because some minerals are important components for the growth and metabolic activity of most organisms, including Lactic Acid Bacteria (LAB) [8].

Kefir has a water content of 89.5%, 1.5% fat, 3.5% protein, 0.6% ash, 4.55% lactose, and a pH of 4.6% [9]. Fermentation is an aerobic and anaerobic process that produces various kinds of primary and secondary metabolites. Such as lactic acid bacteria which have the role of producing anti-microbial substances that can inhibit the growth of panthogens and extend the shelf life of products. Long storage period means extending the fermentation process at cold temperatures. Antimicrobial compounds are chemical or biological compounds that can inhibit the growth and activity of pathogenic microbes [11]. Many pathogenic microbes contaminate food and cause disease, pathogenic microbes also cause harm because they have the ability to infect, cause disease and damage the quality of foodstuffs [12]. Pathogenic microbes are one of the causes of disease in humans and other living things, for example *Staphylococcus aureus, Escherichia coli*, *Salmonella enterica serovar typhi* and others [13]

Microbial activity in the fermentation process and storage period can produce antimicrobial compounds [14]. Antimicrobial compounds are chemical or biological compounds that can inhibit the growth and activity of pathogenic bacteria [15]. Pathogenic bacteria that contaminate food and cause disease, for example *Staphylococcus aureus, Escherichia coli* and *Salmonella enterica serovar typhi.* There is limited research on the shelf life of fermented products, especially on antimicrobial activity. So that the purpose of this study was to determine the effect of storage period for goat milk kefir on the antimicrobial activity of pathogenic microbes.

2 Materials and Methods

This research was conducted at Livestock Product Technology Laboratory of Animal Science Faculty. Analysis of antimicrobial activity test results was conducted in Biomedical Laboratory, total microbial count was conducted in Laboratory of Animal Product Technology, Faculty of Animal Science and the mineral's level test was conducted in Chemical Laboratory of Brawijaya University. The material was fresh GMK from the farm of Mrs. Yuyun, Sumberejo, Batu. The equipment used in the manufacture of GMKwas container as fermentor, filter stirrers, measuring cups, analytical scales and spoons.

These study was a laboratory experiment using a completely randomized design (CRD) with 4 treatments and 4 replications. The treatments consisted of of D0 (0 day), D1 (7 days), D2 (14 days), and D3 (21 days). Data were analyzed using analysis of variance (ANOVA), followed by the Duncan Multiple Range Test (DMRT).

Analysis of mineral's level using AAS (Atomic Absorption Spectrophotometry), total microbial analysis using TPC (Total Plate Count) pour plate method, and analysis of antibacterial activity using the disc diffusion method. Antimicrobial activity testing was carried out on Gram Positive Bacteria (*Staphylococcus aureus)* and Gram Negative Bacteria (*Escherichia coli* and *Salmonella enterica serovar typhi*.

2.1 Total microbial count test

A sample of 1 ml of kefir was prepared and then transferred with a sterile pipette into a solution of 9 ml of distilled water to obtain a 10^{-2} dilution. Do the same thing until the desired dilution is 10^{-8}. The media used was PCA (Plant Count Agar) media. A suspension (culture medium) was diluted and then inoculated on an empty petri dish. Pour the media so that it is still liquid then mix the media with the sample by turning the petri dish following the figure eight pattern [16]. The samples were then incubated at 37°C for 2 days. After that, the results of colony growth on agar media and the number of microbes can be calculated using a colony counter [17].

2.2 Test of mineral levels

Analysis of Ca and Mg using Atomic Absorption Spectrophotometry. Sample preparation was carried out by wet digestion method. In this digestion process, 8 mL of 65% HNO_3 and 2 mL of 30% H_2O_2 were used. The addition of HNO_3 functions as a destructor, H_2O_2 acts as an oxidizer to accelerate the oxidation process. After digestion, the sampe solution was added with 0.1 N HNO_3 then the sample solution was filtered using Whatman paper number 1 to produce a clear and free solution from particles that could interfere with the measurement process at AAS [18].

2.3 Test of antimicrobial on *Staphylococcus aureus, Escherichia coli* and *Salmonella enterica serovar typhi*

The antimicrobial test media used by Salmonella typhi is Media Mueller Hinton Agar (MHA). Put the microbial culture test into the petri dish. Inserted dipped disc paper in the sample. Incubated at 37°C for 24 hours. After 24 hours, observed areas of clear zones formed and measured in diameter.

2.4 Data analysis

In this study the data are presented as mean with standard deviation. Significance between the mean values was determined statistically by ANOVA in 4 treatments with 4 replications. Results are reported as mean ± standard deviation.

3 Results and discussion

3.1 Analysis of Total Microbial Count

The results of statistical analysis of total microbial showed that tthe storage period showed significant effect ($P < 0.01$) on the total value of kefir microbes. The results showed that the total number of microbes in D0 treatment (storage time on day 0) had the highest number of microbes, while D3 (storage period on day 21st) showed the lowest total number of microbes.

It is suspected because the ability of kefir microbes to survive has decreased due to the unavailability of substrates used for metabolism and survival by kefir microbes. [19] explains that the death phase occurs after the log phase and occurs quickly. This is because the nutrient sources found in the growth media have run out. The death phase is caused because cell growth begins to stop and the bacteria have used up their reserve energy (ATP) for respiration, so that many bacterial cells die. There was a decrease in the amount of LAB at the incubation period of 48 to 72 hours, this was due to the competition between LAB and yeast for nutrition. Then the accumulation of alcohol produced by yeast causes damage to the bacterial cell wall [20]. Decreasing bacterial population in D3 treatment causes kefir pH to decrease and acidity increases, so that lactic acid bacteria that cannot tolerate too high acidity will die, so that the population of lactic acid bacteria in D3 treatment decreases [21].

Table 1. Mean of total microbial count

Storage period	Colony average (CFU/mL)	Log average (CFU/mL)
D0	$2.0 \times 10^{9\ a}$	8.89 ± 1.96
D1	$1.2 \times 10^{8\ ab}$	7.78 ± 0.47
D2	$1.7 \times 10^{7\ abc}$	7.12 ± 0.55
D3	$1.1 \times 10^{7\ c}$	6.78 ± 0.11

3.2 Analysis of Ca (Calcium) Levels

The results of statistical analysis of Calcium levels showed that the duration of the storage process had a significant effect ($P < 0.05$) to Calcium levels of kefir.

Table 2. Mean of Ca levels

Storage Period	Mean of Ca levels (mg/kg)
D0	191.62 ± 26.83^{a}
D1	$214.75 \pm 21.40a^{b}$
D2	$228.03 \pm 22.19a^{bc}$
D3	248.34 ± 13.86^{c}

It is suspected because Ca is one of the factors needed in cell metabolism. At the period of storage on Day 0 (D0) the total number of kefir microbes reaches the highest value, so the need for calcium is also a lot, so that a lot of calcium is used and possibly only a little left. Meanwhile, at D1 (7th day of storage), the total kefir microbes had started to decline, but the amount of calcium increased slightly from the amount of D0. It is suspected that some of the dead microbes still have some residual calcium, so that when these microbes die, they still leave the remaining calcium in their bodies and accumulate so that they increase the amount of calcium levels. Quantitatively, the calcium increased 22.69% was compaired with control. Calcium and inorganic phosphate is the main components of the kefir grain culture wall as calciset cell wall component (mannose oligosaccharide, β-glucan), directly the composition of the cell did effect on kefir calcium, showed that during storage period the grain kefir culture synthesize calcium or calcium carbonate form as well as the culture growth.

Milk was the dietary source of (33.2%) calcium, (60.4%), vitamin D and (18.8 %) potassium [23]. The calcium was investigated as nutrition has been performed on elemental calcium and relatively little on calcium as supplied by dairy food. Dairy products as a products that hight calcium and are responsible for about 76% of the calcium available in diet [24]. In this study milk provided 22.5% of calcium intake and there were a larger number of smaller contributors to calcium intake than seen in previous of studies, primarily due to fortification of various products (e.g., fruit juice, RTEC). [25].

3.3 Analysis of Mg (Magnesium) Levels

The results of statistical analysis of Magnesium levels showed that the duration of the storage process had significant effect ($P < 0.01$) to Magnesium levels of kefir. Magnesium kefir levels increase with the length of the storage process. The longer the kefir is stored in the refrigerator temperature, the higher the Mg content in the kefir. The results of the analysis of variance carried out on magnesium kefir levels showed that the storage period had a very significant effect ($P < 0.01$) on the levels of Magnesium kefir.

Table 3. Mean of Mg Levels

Storage Period	Mean of Mg Levels (mg/kg)
D0	11.68 ± 0.73^{a}
D1	$13.74 \pm 0.31a^{b}$
D2	15.44 ± 0.45^{c}
D3	16.62 ± 0.93^{d}

It is suspected because magnesium is one of the substances needed for cellular metabolism. This is because the function of the minerals themselves is for metabolism and for cell growth. This is in accordance with the opinion of [7] this occurs because some minerals are an important component for the growth and metabolic activity of most organisms including LAB. The bacteria in kefir also need magnesium to help in this metabolic process. Therefore, it is suspected that on the 0th day of storage, the magnesium content in kefir has the lowest value because at that time the mineral, in this case magnesium, is needed by microbes for the overhaul of the substrate. In this case, the total number of microbes at the period of storage on the 0th day also has the highest value, so that the magnesium needed is also a lot, eventually the levels are reduced a lot.

The result has accordance with the opinion of goats' milk contains a similar magnesium concentration to cows' milk (mean, 122 mg/L; range 110–144 mg/L [26]. Magnesium is essential micromineral of biosynthetic processes. In both kefir samples produced from the goat milk, the concentration of Mg was found to increase when compared with the levels of Mg in milk samples (Tables 2 and 3). Furthermore, the concentration of Mg in kefir produced from goat milk was higher than fresh milk [27]. [28] Reported that increasing magnesium, calcium and phosphorus contents by 7.2, 12.4 and 9.6% in line with a similar increase of milk protein content. Finally, the essential nutrients, calcium, magnesium, and phosphorus was higher acoordance with aging period, minerals value has improved in kefir.

3.4 Analysis of antimicrobial activity in *Staphylococcus aureus, Escherichia coli and Salmonella enteritica serovar typhi*

The results showed that the effect of storage time treatment gave very significant difference ($P < 0.01$) effect on the bland zones of *Staphylococcus aureus and* the effect of storage period gave significant difference ($P < 0.05$) effect on the clear zones of *Escherichia coli* and *Salmonella enteritica serovar typhi*.

Table 4. Mean of inhibition zone for antimicrobial activity in inhibition *Staphylococcus aureus, Escherichia coli* and *Salmonella enteritica serovar typhi*

Treatments	Inhibition zone (mm) *Staphylococcus aureus*	Inhibition zone (mm) *Escherichia coli*	Inhibition zone (mm) *Salmonella enteritica serovar typhi*
D0	2.23 ± 1.87^{a}	3.32 ± 0.74^{b}	2.72 ± 0.21^{b}
D1	3.18 ± 0.33^{b}	2.70 ± 0.26^{ab}	2.37 ± 0.09^{ab}
D2	3.14 ± 0.42^{b}	2.37 ± 0.47^{a}	2.01 ± 0.14^{a}
D3	2.60 ± 0.10^{ab}	2.09 ± 0.19^{a}	1.97 ± 0.64^{a}

The results showed that the effect of storage period gave very significant difference ($P < 0.01$) effect on the inhibition zones of *Staphylococcus aureus.* Increase or decrease in the average of the inhibition zone along with the increase or decrease in the total lactic acid bacteria. D0 (0 days) lactic acid bacteria are still in the adaptation phase to the storage environment and cells are still able to carry out positive transport in the form of nutrients for growth. D1 (7 days) lactic acid bacteria enter the logarithmic phase which is the phase when bacteria divide rapidly indicated by the population of lactic acid bacteria begins to increase [29] and produce the bioactive compound as a metabolite such as peptide and organic. This compound have antibacterial activity which generated by utilizing lactose and protein of goat's milk. D1 (7 days) and D2 (14 days) enter the stationary phase of the growth of bacterial cell populations that tend to be constant. This is because in this phase (stationary) it is suspected that BAL produces the highest secondary metabolites for self-defense. D3 (21 days) there is a decrease in the average inhibition zone as the population of lactic acid bacteria enters the phase leading to the death of the growth curve. The phase to death is the phase when the medium nutrients and the number of bacterial populations decreases, which is followed by change the bioactive component.

The results showed that the effect of storage Period treatment gave significant difference ($P < 0.05$) effect on the clear zones of *Escherichia coli* and *Salmonella enteritica serovar typhi*. The decrease in the average inhibition zone in *Escherichia coli* and *Salmonella enteritica serovar typhi* is thought to be due to the long shelf life affecting antimicrobial activity, because the longer the shelf life, the more passive lactic acid bacteria and the fewer the number, so that it has the ability to break down the substrate the smaller. At a shelf life of 0 days has the ability to inhibit the growth of *Escherichia coli* and *Salmonella enteritica serovar typhi* highest, then decreased in shelf life 7 days, 14 days and 21 days. At a length of 21 days, it is suspected that the substrate has started to run out (Fardiaz, 1987). This is according to research [29], that antimicrobial activity is produced in the decay phase. According to [30], the criteria of antimicrobial power strength are divided into four groups, namely as follows: the diameter of the inhibition zone of 5 mm or less is categorized as weak, the 5 mminhibition zone is categorized as moderate, the 10 mm bland zone is categorized as strong, and the clear zone of 20 mm or more is categorized as very strong. The results of the study when compared to the standard fall into the category of weak antimicrobial activity.

4 Conclusion

Based on the research results, it can be concluded that the storage period treatment has a significant effect on the total microbial count, mineral levels and inhibition zone of *Staphylococcus aureu*, *Escherichia coli* and *Salmonella enteritica serovar typhi*. Storage period of 7 days and 14 days give the best results against the average inhibition zone in *Staphylococcus aureus*. Storage period of 0 days gives the best results to the average inhibition zone in *Escherichia coli* and *Salmonella enteritica serovar typhi* but the effect does not show any difference to the length in storage period of 7 days.

The reseach was funded by the Profesor Research Grant, Faculty of Animal Science Universitas Brawijaya.

References

1. R. C. Witthuhn, T. Schoeman, T. J. Britz, Int. Dairy J. **15**, 383–389 (2005)
2. G. L. Garrote, A. G. Abraham, G. L. De Antoni, J. Dairy Res. **68**, 639–652 (2001)
3. J. B. Prajapati, B. M. Nair, *The History of Fermented Foods* (2008)
4. M. Fatkhul Mubin *et al.*, J. Pangan dan Agroindustri **4**, 291–301 (2016)
5. L. Yilmaz-Ersan, T. Ozcan, A. Akpinar-Bayizit, and S. Sahin, Int. J. Chem. Eng. Appl. **7**, 22–26 (2016)
6. D. D. Rosa, M. M. S. Dias, Ł. M. Grześkowiak, S. A. Reis, L. L. Conceição, M. D. C. G. Peluzio, Nutr. Res. Rev. 30, 82–96 (2017)
7. H. Sa. a and I. S. a, Food Nutr. Sci. **2013**, 73–87 (2013)
8. X. Shi, H. Chen, Y. Li, J. Huang, Y. He, Acta Univ. Cibiniensis. Ser. E Food Technol. **22**, 43–50 (2018)
9. A. Bakar, S. Usmiati, Teknologi Pengolahan Susu **4** (2009)
10. Y. D. Putri, N. A. Setiani, S. Warya, Curr. Res. Biosci. Biotechnol. **2**, 101–104 (2020)
11. Sulmiyati, N. S. Said, D. U. Fahrodi, R. Malaka, F. Maruddin, Trop. Anim. Sci. J. **42**, 152–158 (2019)
12. U. Haroen, A. Budiansyah, Noperdiman, Harnita, Jusalia, Bulletin of Animal Science **43**, 109–117 (2019)
13. S. Juariah, D. Suryanto, I. Jamilah, Berk. Perikan. Terubuk **42**, 37–50 (2014)
14. M. J. Chen, J. R. Liu, C. W. Lin, Y. T. Yeh, Asian-Australas. J. Anim. Sci. 18, 711–715 (2005)
15. Z. Oner, A. Karahan, M. L. Cakmakci, Gida J. Food **35**, 177–182 (2010)
16. L. Tratnik, R. Božanić, Z. Herceg, I. Drgalić, Int. J. Dairy Technol. **59**, 40–46 (2006)
17. M. Yunita, Y. Hendrawan, R. Yulianingsih, J. Keteknikan Pertan. Trop. dan Biosist. 3, 237–248 (2015)
18. N. N. Susanti, Y. Sukmawardani, I. Musfiroh, Indones. J. Pharm. Sci. Technol. **3**, 26–30 (2016)
19. N. Y. Respati, E. Yulianti, A. Rakhmawati, J. Prodi Biol. **6**, 423–430 (2017)
20. R. La Sinurat, C. N. Ekowati, S. Sumardi, S. Farisi, J. Ilm. Peternak. Terpadu **6**, 111 (2019)
21. N. L. P. Sriyani, M. Hartawan, D. A. N. I. G. Suranjaya, Maj. Ilm. Peternak. 95–99 (2015)
22. F. Gaucheron, Reprod. Nutr. Dev. **45**, 473–483 (2005)
23. D. R. Keast, V. L. Fulgoni, T. A. Nicklas, C. E. O'Neil, Nutrients **5**, 283–301 (2013)
24. M. H. Tunick, J. Dairy Sci. **70**, 2429–2438 (1987)
25. C. E. O'Neil, D. R. Keast, V. L. Fulgoni, T. A. Nicklas, Nutrients **4**, 2097–2120 (2012)

26. H. E. Oh, H. C. Deeth, Int. Dairy J. **71**, 89–97 (2017)
27. G. Turker, B. Kizilkaya, N. Cevik, J. Food Agric. Environ. **11**, 62–65 (2013)
28. E. Bijl, H. J. F. van Valenberg, T. Huppertz, A. C. M. van Hooijdonk, J. Dairy Sci. **96**, 5455–5464 (2013)
29. H. A. El Enshasy, A. F. El Baz, E. M. Ammar, Curr. Res. Educ. Top. Trends Appl. Microbiol. A. Méndez-Vilas, Formatax, 315–321 (2007)
30. T. Yulinery, N. Nurhidayat, J. Teknol. Lingkung. **13**, 109 (2016)

Poultry biodiversity for alternative farming systems development

Edoardo Fiorilla[1], *Alice* Cartoni Mancinelli[2], *Marco* Birolo[3], *Cesare* Castellini[2], *Dominga* Soglia[1], *Elisabetta* Macchi[1], *Silvia* Mioletti[1], *Barbara* Miniscalco[1], *Stefano* Sartore[1], *Riccardo* Franco[1], *Stefano* Nurisso[1], *Maristella* Baruchello[4], *Martina* Tarantola[1], *Paola* Sacchi[1], *Achille* Schiavone[1], and *Cecilia* Mugnai[1*]

[1]Department of Veterinary Sciences, University of Turin, Largo Braccini 2, 10095, Grugliasco, Turin, Italy
[2]Department of Agricultural, Environmental and Food Science, University of Perugia, Borgo XX Giugno 74, 06124, Perugia, Italy
[3]Department of Agronomy, Food, Natural resources, Animals and Environment, University of Padova, Viale dell'università, 16, 35020, Legnaro, Padova, Italy
[4] Veneto Agricoltura, Agenzia Veneta per l'innovazione nel Settore primario, Viale dell'Università 14, 35020 Legnaro, Padova, Italy

Abstract. Poultry biodiversity represents a key factor to improve poultry resilience and promote sustainable and low input farming systems. The EU and member states promote protection of livestock biodiversity and the development of alternative farming through funding projects such as "Local Chicken Breeds in Alternative Production Chain: Welfare, Quality and Sustainability" (funded by the Italian Ministry of Research and University). The aim of the present research was to identify among five different poultry genotypes Bionda Piemontese (BP), Robusta Maculata (RM), RM x Sasso (RMxS), BP x Sasso (BPxS) and a commercial hybrid (Ross 308) the best suitable breed in terms of productivity and welfare for alternative housing system. A total of 300 (60 x genotype), 21 days old male birds were randomly allotted in two housing systems: 1) standard intensive farming (controlled environment, 33 kg/m^2 and standard diet) and 2) free-range ("natural" environmental conditions, 21 kg/m^2, access to outdoor area and low-input diet). Slaughtering was performed at 81 days of age. During the trial, the productive performance and behaviour of the animals were evaluated. The housing system, the genotype and their interaction significantly affected many of the studied variables, showing broiler not the ideal genotype for extensive farming system, which is more suited for low/medium performance strains.

1 Introduction

Humans are in continuous expansion, as FAO states, by 2050 global population will reach 10 billion people and, with this constant population increase, the consumption of food from

* Corresponding author: cecilia.mugnai@unito.it

animal origin will increase dramatically. It will be necessary to increase the number of farmed animals and implement new ways of farming.

The sustainability of intensive farming system is considered low, and the situation will worsen every year. The main phenomena related to intensive farming are deforestation, greenhouse gasses emissions, excessive consumption and pollution of land and water [1]. Another side effect of this process is the reduction of genetic variability and the consequent vulnerability of animals to environmental stress [2].

Poultry farming has a better environmental impact than other animal production chains thanks to the high efficiency in converting feed into meat or egg. The shorter production cycle and the strong genetic selection carried out to increase production performance represent an advantage over ruminant or swine farming. This is particularly evident in the poultry meat production, in fact, modern broilers reach slaughtering weights in short cycles of about 40 days with a high percentage of meat yield [3]. Unfortunately, in addition to all the benefits listed above these high-performance strains (HPS) show welfare and health issues, skeletal imbalances, metabolic disorders, and muscle abnormalities, which affect the appearance of the meat, nutritional traits, and acceptance by consumers [4, 5].

These are some of the main reasons why new sustainable and alternative farming systems need to be identified. However, poultry meat and egg production rely on these HPSs presenting major animal health and welfare concerns [6]. As a result, there is a growing demand in developed countries for poultry meat and eggs from welfare-friendly farming systems [7]. Alternative poultry production is more expensive than intensive, but supports biodiversity, animal welfare, local economies and the multifunctionality of farms, providing eggs and meat to which a growing part of consumers attributes high ethical value, quality and taste [8].

Alternative systems (organic, biodynamics, free-range) require birds adapted to less controlled environment, high foraging aptitude, active immune response, and thermo-tolerance. The response of chickens to alternative systems and to different climatic conditions have not been sufficiently investigated and only few commercial breeds are available for these rearing systems, HPS, unlike Local Breeds (LBs), are highly unsuited to this purpose [9].

Nowadays, the preservation of poultry biodiversity is a key objective in all developed countries [10]. The possibility of improving LBs originates from the balance between the possible benefits (good health and welfare, resistance and resilience to heath stress, lower dietary requirements, reduced veterinary cares) and other unfavourable aspects (low performance, low meat yield) [11]. Accordingly, an essential step is the improvement of production efficiency in LBs. Crossbreeding is the main tool used in poultry, which normally involves a cross between HPS and LBs, with the aim of combining the production capacity of the former with the latter adaptability to natural environment. Cross breeding also maximizes the expression of heterosis and normally improves fitness characteristics [12].

Moreover, since the current HPS are specialized in either meat or egg production and egg production requires only females, the male of egg-type strain, due to their slow-growth rate, are killed at 1 day-old. Ethical reasons exist against this practice and the use of dual-purpose breeds could be a solution to this issue: the males and the females of local breeds could be farmed for meat and eggs, respectively. An additional reason to safeguard LBs derives from their ability to produce meat and eggs in alternative systems, with low-input diets and with outdoor run that produce meat and eggs with higher welfare standards [13]. These aspects meet the attitude of consumer in developed countries where people can make an opinionated decision about their meal taking in considerations animal welfare, sustainability, nutritional, sensory, and ethical factors. It is obvious that, in most countries where poverty and hunger is widespread, HPS still represents a good opportunity to provide quite cheap high value food [14, 15]. Worldwide, conventional farming systems account for 67% of poultry meat production and 50% of eggs; an increase in local poultry farming is therefore feasible if supported by a productive and economic perspective [16]. The conservation of native breeds is also an important component of poultry biodiversity. The Food and Agriculture Organization of the United Nations (FAO) stated that 55% of all local poultry breeds are found in Europe and the Caucasus regions [16]. In Italy 22 breeds have been included in the Indigenous Poultry Register and most of them are included in the FAO Domestic Animal Diversity Information System (DAD-IS) database [17].

In European countries, the development of sustainable farming systems and the protection of biodiversity (animal and vegetal) are encouraged, especially in the more rural areas and the collaboration between associations, farmers, and scientific partners such as universities is incentivized. For this reason, the Italian Ministry of Research and University funded the project PRIN "*USE OF LOCAL CHICKEN BREEDS IN ALTERNATIVE PRODUCTION CHAIN: WELFARE, QUALITY AND SUSTAINABILITY*" with the aim to promote the use of local chicken breeds in alternative production chains through the development of crossbreeds to increase the production efficiency. This fact represents an opportunity to mobilize some resources on the breeding of local breeds and products that generate work and quality products obtained from animals that are present in the Italian territory [18]. In particular, in Italy there are some associations such as the Consortium for the Bionda Piemontese breed and Veneto Agriculture [19], partners of the above-mentioned project, that have been working for years for Italian poultry breeds conservation and to provide assistance and birds rural farmers.

The Bionda Piemontese (BP), is a native breed of the Piedmont region (north-west, Italy) is characterized by a blond plumage and a black tail. Currently they are bred mainly for meat production and are slaughtered at around 24 weeks of age, although they are considered dual-purpose breeds. In 2013, the population size was 16,000 birds. The annual laying production is concentrated in the spring-summer period and is around 180-200 units [20]. The breeders of BP and RM are included in the Consortium for the conservation of the breed and in the biodiversity conservation project of Italian poultry breeds and sponsored by the Ministry of Policies Agricultural, Food and Forestry (https://www.pollitaliani.it/en/). Knowledge of growth performance is therefore considered a requirement and a fundamental contribution to the improvement and conservation of local poultry breeds.

The Robusta Maculata (RM) is a native breed of the Veneto region (north-east, Italy) characterized by a silver-white plumage edged and irregularly spotted with black. It is a dual-purpose breed with a good aptitude for egg production and medium precocity. The Robusta Maculata is a rustic breed with good grazing aptitude and able to adapt to different agricultural environments. It adapts well to both extensive breeding and organic farming. Roosters reach a weight of 3.8-4.4 kg while females weigh an average of 2.8-3.3 kg. It is quite early and lays eggs with a pink to brown colour. Production is around 140-160 eggs per cycle with a weight of 55-60 g. The recommended slaughtering age is 18-22 weeks [21].

The sustainable use of genetic resources in extensive and organic systems is an alternative practice to conventional farming [22]. BP and RM perform well in these alternative systems and retain adaptation to low input farming systems that can positively affect welfare and quality; the products obtained are officially recognized as traditional and usually sold as whole carcasses and/or processed meat products [23, 24]. These properties contribute to the value of this production and support the proposal of a retention situation for this breed.

In the present research, preliminary results of the growth performance and behavioural patterns of BP and RM chickens and their F1, crossed with a higher productivity breed (Sasso), were evaluated and compared with the Ross 308 chickens in two housing systems (conventional and free-range), to identify the breed better adapted for a particular system, offering the greatest benefits in terms of productivity and welfare.

2 Materials and methods

2.1 Birds and housing systems

The present trial was performed in the poultry farm of the Department of Veterinary Sciences of University of Turin (Italy), from March to July 2021 and the experimental protocol was approved by the Bioethical Committee of the University of Turin (Italy) (Prot. ID: 251833). 300 chicks (60 x genotype) of five different poultry genotypes, Robusta Maculata (RM), Bionda Piemontese (BP), RM x Sasso (RMxS), BP x Sasso (BPxS) and commercial hybrid (Ross 308) were hatched by a local farmer [25]. Chicks were vaccinated against Marek and Newcastle diseases and sexing was done based on visual analysis of cloaca at hatching.

Chicks were reared for the first 20 days in the brood divided into five pens, one for each genotype. Each pen was 1 m wide and 2 m long and was covered with wood shavings (20 cm deep) as litter equipped with a waterproof floor and walls, environmentally controlled with temperature and relative humidity (RH) ranging from 32 to 20 °C and from 70 to 65% RH, respectively.

At 21 days until slaughter, birds were randomly divided into two different rearing systems: conventional (33 kg of meat per m^2) and free-range (21 kg of meat per m^2). For each genotype and system, three replicates were made (n = 10 cockerels/5 genotypes /2 housing systems/3 replicates).

In the conventional system, the lighting schedule was 16 h light, 8 h darkness for the whole trial. Environmental conditions in the poultry-house, temperature and relative humidity, were set according to the Ross guidelines [26]. Whereas for free-range system no lightning schedule was set, and birds were exposed to natural temperature and photoperiod. The animals were left free to stay either outside or inside at any time of the day, outdoor access was closed for the night (20.00-06.00) till May and then left open over 24 h till the end of experimentation.

The whole facility, including the outdoor areas, was protected from wild birds and predators with fences. The bird's health status and mortality and the environmental parameters were daily checked during the whole experimental period.

Slaughtering was performed in a commercial slaughterhouse at 81 days old, accordingly to EU recommendation for organic poultry production [27].

2.2 Diets

Three different feeds were used: a starter (administered from 0 to 21 days of age to all birds) and two grower/finisher diets (a standard and a low input diet). The standard diet was formulated to meet the recommended energy and protein levels for Ross and administered from 21 days of age until slaughter to the birds reared in the conventional system. The low input diet was formulated with the aim of achieving a reduced content (-10% on average) of protein and energy. In the low input diet, the imported soybean meal was replaced by soybean meal and faba bean, GMOs free and from Italian cultivations. This diet was administered from 21 days of age until slaughter to the birds reared of free-range group. The ingredients and chemical composition of the experimental diets are reported in Table 1. The diets were analyzed to determine the contents of dry matter (934.01), ash (967.05), and crude protein (2001.11) by AOAC [29] methods. Ether extract was analyzed after acid hydrolysis [30]

Table 1. Formulation (%), chemical composition (% as fed) and energetic value (MJ kg-1) of experimental diets

Housing System / **Diet**	**Starter**	**Industrial Standard**	**Free-range Low input**
Ingredients			
Corn	55	57	58
Soybean oil	2	3	2
Soybean meal*	-	-	20
Soybean meal**	25	15	-
Whole Soy	14	21	-
Faba bean	-	-	16
Calcium Carbonate	1.8	1.8	1.8
Monocalcium phosphate	1.3	1.3	1.3
Sodium chloride	0.4	0.4	0.4
Vitamin-mineral premix⁺	0.5	0.5	0.5
Chemical composition			
Dry matter	88.53	87.48	87.37
Crude protein	20.99	18.01	16.26
Ether extract	6.48	8.49	4.74
Crude fibre	3.17	1.72	2.68
Ash	6.15	6.02	6.06
Lysine§	1.20	1.07	0.95
Met. + Cist. §	0.71	0.66	0.53
Calcium§	1.01	1.00	0.99
Phosphorus§	0.70	0.68	0.68
Ca/P§	1.44	1.46	1.46
Metabolizable Energy§	3089	3252	2921

*: National Soybean meal GMOs free, 44.0% CP. **: Imported Soybean meal GMOs, 49.0% CP. ⁺ Amounts per kg: Vit. A 12,500 IU; Vit. D3 5000 IU; Vit. B1 2.5 mg; Vit. B2 9 mg; Vit. B6 1.25 mg; Vit. B12 0.01 mg; α-tocopherol acetate 75 mg; Biotin 0.15 mg; Vit. K 4.0 mg; Niacin 60 mg; Folic acid 2.00 mg; Pantothenic acid 10 mg; Choline chloride 600 mg; Mn 100 mg; Fe 80 mg; Zn 80 mg; Cu 8.0 mg; I 2.0 mg; Se 0.2 mg. § Estimated values Feedipedia, [28].

2.3 Productive performance

The individual live weight (LW) and feed intake (FI; at a pen level) of the chickens were weekly recorded using an electronic scale (KERN PLE- N v. 2.2). Daily weigh gain and feed conversion ratio (FCR) were also calculated. To calculate mortality death birds were also daily recorded and causes of death evaluated by necropsy.

2.4 Behavioural analysis

During the week prior to slaughtering (from 73 to 80d), behavioural observations using a digital video camera (Sony 4K, FDR-AX43) were performed. The videos were analysed afterwards by two trained observers, using the continuous sampling method [31] for three minutes for each replication, during 2 periods (morning, starting at 09.00 am and afternoon, from 05.00 pm), of 1.5 h each.

The behavioural observations (Table 2) were divided into 6 macro-categories: feeding (scratching, feeding from feeders and drinkers, grass/object pecking); locomotor (walking, running, flapping wings while running); social (menacing each other without fighting, fighting, gentle pecking, allo-preening); resting (crouching, standing still); comfort (sand bath, self-preening); stretching (leg stretching, wing stretching).

The behaviour was recorded on a purpose-designed table (based on the major behaviours in the broiler ethogram chosen from preliminary observations), and their respective frequencies were calculated as a percentage of the total observed behaviours. As no significant differences were found between periods, all data were pooled.

Table 2. Designed table of macro-categories and corresponding behaviours used for behavioural analysis

Macro-categories	**Behaviour**
Feeding	*Scratching*
	Feeding
	Grass/objects pecking
Locomotor	*Walking*
	Running
	Wing flap running
Social	*Menacing*
	Fighting
	Gentle Pecking *Allo-Preening*
Resting	*Crouching*
	Standing
Comfort	*Sand bath*
	Self-Preening
Stretching	*Leg stretching*
	Wing stretching

2.5 Statistical analysis

A linear model [32] was used to evaluate the effect of genotype and housing system. For the final live weight, the repeated effect of chicks was included as random in the model. The Gompertz coefficients have been evaluated for the different genetic groups with a non-linear model (proc NL) using the following equation:

$$LW_t = b1 * \exp\{-\exp[-b3*(\text{time in wk} - b2)]\}, \quad (1)$$

LW_t was the live weight of any chicken at a given wk,
$b1$ = adult weight (asymptote),
$b2$ = the slope at flex point
$b3$ = the point of flex.

Goodness of fit of the model to the data was determined on each breed and housing system separately using adjusted coefficient of determination (R^2).

Difference between groups were assessed by ANOVA test with a Bonferroni multiple t-test. Differences with at least a $P \leq 0.05$ value were considered statistically significant. Nonparametric tests were performed on the behavioural categories and mortality rate and significance was evaluated by χ^2 value and set at $P \leq 0.05$.

3 Results and discussion

3.1 Growth performance

The data relating to the productive performance are shown in Table 3. All parameters (LW, FI, FCR, mortality) were affected by genotype ($P < 0.05$) and housing system ($P < 0.005$). Furthermore, the interaction housing system x genotype effects resulted significant ($P < 0.005$).

As expected, the final LW, the FI and the FCR differed between genotypes with higher values for Ross 308 compared to LBs and their crossbred, in both conventional and free-range systems. Mortality rate showed interesting results. Ross mortality was significantly ($P < 0.005$) higher in both housing systems with higher percentage in the free-range system compared to the conventional.

As far as Ross growth performance is concerned, the comparison between the two rearing systems shows how this HPS performs less in a free-range system. On the other hand, all LBs and F1 crossed birds considered performed better in free-range rearing systems.

Previous research [33, 34] showed that the strong genetic selection for meat yield of HPS, provide increasingly better growth performances and yields, but the obvious consequence is that these animals are more susceptible to environmental variations, reducing their adaptability to uncontrolled environments and making them inappropriate for any type of farming other than the conventional intensive one.

Table 3. Productive performance of each genotype in the two rearing systems

Housing System	Genotype	LW (g)	FI (g/d)	FCR	Mortality (%)
Conventional	BP	1.500	85.11	4.59	1.2
	BPxS	1.982	87.81	3.55	0.9
	RM	1.588	81.61	4.13	1.4
	RMxS	2.035	90.61	3.58	1.7
	Ross	4.920	175.15	2.88	3.0
Free-range	BP	1.660	99.88	4.87	0.0
	BPxS	2.052	102.38	4.04	0.0
	RM	1.620	96.58	4.89	0.5
	RMxS	2.380	105.39	3.58	0.0
	Ross	4.185	191.55	3.69	3.9
Significance of effect					
Housing		**	**	**	**
Genotype		*	*	*	*
H x G		**	**	**	**

* = P < 0.05. ** = P < 0.005 ***. H = housing system. G = genotype. H x G interaction.

LBs, on the other hand, showed lower to none mortality rate in the free-range system, thus implying a better resilience of these animals and a more appropriate fit to this rearing system [35].

Causes of Ross mortality were related to fast growth and were mainly Sudden Death Syndrome (SDS) (65%) and ascites (35%) in conventional system, whereas for free-range mainly ascites (80%). This lower incidence of SDS in the free-range system (20%) was probably a consequence of the lowering of the growing rhythm due to the housing system (diet + uncontrolled environmental condition + lower density + higher locomotor activity) [36, 37].

Another interesting finding is that for the F1 crossbred (BPxS, RMxS) showed adequate performance in both environments, conjugating the strength and resistance of LBs with higher yields [38, 39]. The FI was lower in free-range housing system of LBs and their crossbred, probably due to the grass ingestion (as confirmed by behavioural analysis), among other things, positively conditioning the other production parameters such as LW, FCR and mortality, too. These results are confirmed also by previous research [40] in different genotypes (mainly HPS vs LBs) reared in organic system, where grass ingestion improved LBs health and welfare and products (meat and egg) quality.

The crossing of pure breeds with HPS [41] had a positive impact in LW of LBs genotypes, in both rearing systems, this result can be deeply analysed observing the LW Gompertz curves (fig. 2, 3). In fact, as previous studies confirm [42], BP has a faster growing rate in the first part of its life cycle (4-7 weeks), while RM in the second part of its life cycle (8-13 weeks). In the Gompertz model, parameter b3, which was related to the shape of the growth curve, showed positive correlation with BW and in RM free-range birds showed the highest value (11.75, $R^2 = 0.99$). Crossing LBs with HPS bird line, improved RM growth in the first phase while in the second sustained the BP growth allowing the curves of BPxS with RMxS to be comparable since week 1 in both systems (b3 = 9.61 and 10 for BPxS and RMxs, respectively). The relative growth rate b2, related to the slope of the curve, was the highest in conventional Ross (0.27, $R^2 = 0.98$). Accordingly, the higher b2 value corresponded to a shorter time to reach maturity.

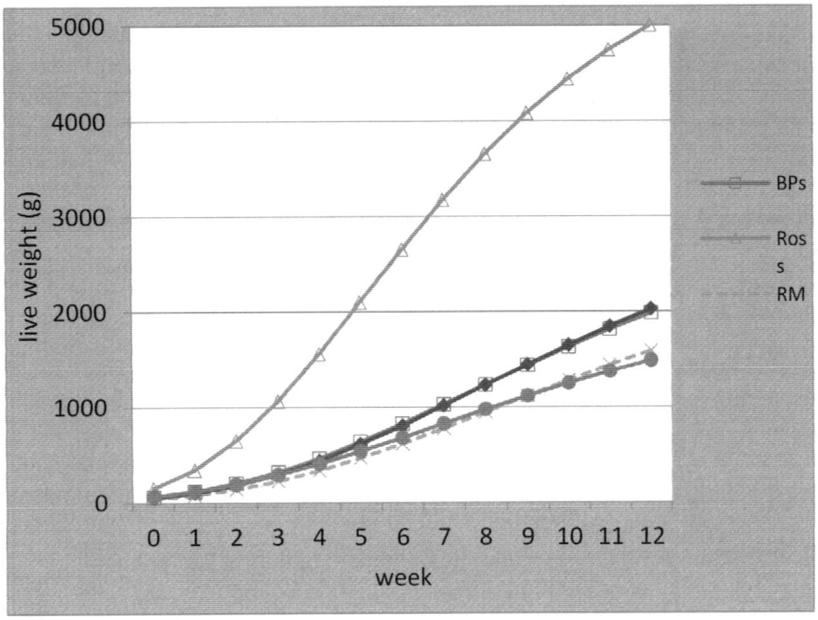

Fig. 1. Gompertz distribution of LW from 0 to 12 weeks of conventional system.

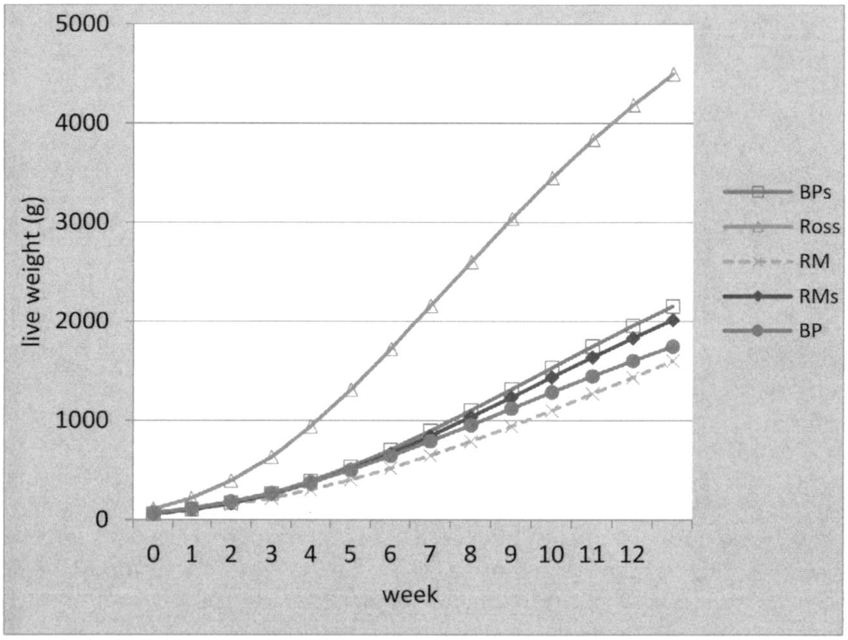

Fig. 2. Gompertz distribution of LW from 0 to 12 weeks in free-range system.

3.2 Behavioural analysis

Analysis of animal behaviours in farming is important from a multifactorial point of view; results obtained from more traditional observed point (i.e. performance) in poultry farming, can be put in correlation and explained observing the behaviour that animals show [43].

The results obtained are present in Table 4; macro-categories like feeding, locomotor, resting, and comfort showed strong significance ($P < 0.005$) for housing system and genotype, the interaction between housing and genotype was significant ($P < 0.05$) for feeding and rest and showed a strong significance ($P < 0.005$) only for locomotor activities.

To deeply understand differences founds in birds feeding activities, we must analyse the behavioural patterns that constitute this macro-category. In fact, feeding was constituted by scratching, feeding from feeders and drinkers, and grass/object pecking. In free-range birds, feeding was mainly constituted by grass pecking in all genotypes, while in conventional, this behaviour was not allowed, and it was substituted by pecking objects. Regarding feeding from feeders, the conventional birds performed the greater percentage (data not shown). Respect to genotype Ross showed in both systems the highest percentage of feeding from feeders according to due to genetic selection [44].

As expected, in both rearing systems, Ross showed the lowest locomotor activities; particularly interesting is that in both housing systems there was no significant difference, these animals even if given the opportunity of an outside area where they could move freely, spent, like in the conventional housing system, a lot of their daily time resting and feeding at feeders. On the contrary free-range LBs and their crossbred took advantage of the outdoor area, showing more locomotor and less resting feeding at feeders activities compared to conventional farmed animals, these results are confirming previous studies [40] showing how strong genetic selections have impacted the life of this animals.

Since active behaviours (i.e., running and walking) represent an energy cost for the animal, it is plausible that the strong genetic selection carried out to improve productive performance modifies the behaviour of these animals, forcing them to direct as much energy as possible to growth, by limiting some behaviours wasteful in terms of energy [45]. This confirms what has been observed, in both housing systems, regarding Ross, which spent most of their energy on muscle growth, while in LBs a greater percentage of available resources have been used for movement, food search and performing behaviour typical of the chicken ethogram. Consequently, it could be possible that genetic selection for productive traits (i.e. muscle growth, breast yield) indirectly produce animals with a reduced kinetic attitude [46].

Table 4. Behavioural macro-categories of birds in the two rearing systems

Housing System	Genotype	Feeding	Locomotor	Social	Resting	Comfort	Stretching
Conventional	Bp	22.56	23.48	4.03	41.65	4.39	4.03
	Rm	17.59	30.80	4.43	38.65	4.28	4.24
	Bp X S	21.63	25.44	3.71	41.90	3.08	4.23
	Rm X S	17.40	20.91	2.97	50.18	5.20	3.35
	Ross	14.99	12.32	0.00	59.21	2.92	11.28
Free-Range	Bp	34.09	41.70	5.98	13.45	3.50	1.52
	Rm	32.20	41.85	3.85	17.37	3.07	1.65
	Bp X S	34.07	41.05	5.44	16.94	1.35	1.14
	Rm X S	40.22	38.75	5.73	11.45	3.23	0.75
	Ross	21.25	13.39	0.18	53.82	5.30	7.45
P Value	Housing	**	**	-	**	**	*
	Genotype	**	**	-	**	**	*
	H X G	*	**	-	*	-	-

* = $P < 0.05$. ** = $P < 0.005$. H = housing system. G = genotype. H x G interaction.

Comfort behaviours were mainly constituted by self-preening and sand bath and were mostly performed in conventional system by LBs and their crossbred, while in free range housing system by Ross birds. We could hypothesize that more active birds like LBs and their crossbred, when reared in confined environment, increase the time spent in performing allowed behaviours, this hypothesis was confirmed by the findings of the same genotypes in free-range lowered the frequencies of comfort behaviour, being free to perform their natural ethogram.

Ross when compared to LBs and crossbred performed more stretching. Leg and wing stretching intended as extending these parts from the body could be performed in order to relieve pain or discomfort due to the posture in meat type birds [47]. On the contrary, other studies have also reported that a lack of stretching could be a sign of compromised bird health [48]. Accordingly, LBs and crossbred performed a lower frequency of stretching, this could be explained by the fact that they were engaged performing other behaviors of their ethogram. Previous studies conducted in laying hens have showed that stretching could be mainly linked to the productive factors of farming, such as the type of soil confirming that HPS are less tolerant, if environmental conditions are not ideal [49, 50].

4 Conclusion

Our results clearly demonstrate that Ross birds increase their mortality in free-range system, whereas in the same system, LBs and their crossbred increased their productive performance with the lowest mortality and then better welfare.

These findings clearly show how HPS such as Ross 308 can give unparalleled meat yields in conventional system, with high quality feed, controlled environment, and short life cycle. If these conditions are not guaranteed, such as in free-range, Ross genotype birds did not take advantage of space and pasture allowance and did not modify their behavioural patterns as opposed to LBs and their crossbred.

We can then state that, LBs and their crossbred can better adapt to alternative farming systems, showing resilience of local poultry production to environmental changes. Moreover, the crossbreeding of LBs with a more productive breed could have a great effect in improving productive performance without impairing animal behaviour and welfare, increasing the economic income for local farmers.

This is useful knowledge both for less developed and developed countries where is important to develop low input farming system to sustain local communities and economies, and to meet consumers demands of animal product with higher welfare and ethical value being.

We can then conclude that less productive chicken breeds can represent a feasible alternative in the appropriate rearing system.

Further experimental trial regarding this argument could be performed as very little previous bibliography is present on this topic, it could be particularly interesting to analyse in more details the increase in performance in relation to strength and resilience of population of LBs crossed with higher productive strains.

This study was supported by the Italian Ministry of Research and University funding project PRIN: "USE OF LOCAL CHICKEN BREEDS IN ALTERNATIVE PRODUCTION CHAIN: WELFARE, QUALITY AND SUSTAINABILITY" (Prot. 2017S229WC).

References

1. FAO, *The Future of Food and Agriculture: Trends and Challenges*, **4** (2017)
2. M. De Vries, I. J. M. De Boer, Livest. Sci., **128**, 1–11 (2010)
3. M. J. Zuidhof, B. L. Schneider, V. L. Carney, D. R. Korver, F. E. Robinson, Poult. Sci. **93**, 2970–2982 (2014)
4. R. Relić, E. Sossidou, A. Dedousi, L. Perić, I. Božičković, M. Đukić-Stojčić, Ankara Univ. Vet. Fak. Derg. **66**, 423–428 (2019)
5. M. Petracci, F. Soglia, C. Berri, pp. 51–75, Woodhead Publishing (2017)
6. A. F. Soleimani, I. Zulkifli, A. R. Omar, A. R. Raha, Poult. Sci., **90**, 1435–1440 (2011)
7. W. M. Muir, H.-W. Cheng, C. Croney, Front. Genet. **5**, 407 (2014)
8. P. Parrot, K. Walley, in *Poultry Quality Evaluation*, pp. 313–334 (2017)
9. A. C. Mancinelli, M. Guarino Amato, D. Meo Zilio, A. Dal Bosco, S. Mattioli, C. Castellini, J. Dairy Vet. Sci. **4**, 555644 (2017)
10. A. Franzoni, M. Gariglio, A. Castillo, D. Soglia, S. Sartore, A. Buccioni, F. Mannelli, M. Cassandro, F. Cendron, C. Castellini, A.C. Mancinelli, S. Cerolini, A. Sayed, N. Iaffaldano, M. Di Iorio, M. Marzoni, S. Salvucci, A. Schiavone, Animals **11**, 629, (2021)
11. H. Nurcahya, S. Darwati, I. J. Tambunan, J. Trop. Biodiv. **1**, 63–73 (2020)
12. I. Hoffmann, Worlds. Poult. Sci. J. **61**, 57–70 (2005)
13. L. Baldinger and R. Bussemas, Org. Agric. **11**, 489–498 (2021)
14. P. Rosa, B. Ávila, I. Angelo, R. Chesini, T. Fernandes, J. Camacho, M. Bugoni, V. Roll, M. Gularte, Br. Poult. Sci. **62**, 387–395 (2021)
15. F. Kaygisiz, B. A. Bolat, and D. Bulut, Rev. Bras. Cienc. Avic. **21**, eRBCA-2019-1060 (2019)

16. FAO, "The Second Reports of the State of the World's Animal Genetic Resources for Food and Agriculture" (2015)
17. A. Castillo, M. Gariglio, A. Franzoni, D. Soglia, S. Sartore, A. Buccioni, F. Mannelli, M. Cassandro, F. Cendron, C. Castellini, Alice Cartoni Mancinelli, N. Iaffaldano, M. D. Iorio, M. Marzoni, S. Salvucci, S. Cerolini, L. Zaniboni, A. Schiavone, Animals **11**, 490 (2021)
18. E. Commission, https://europa.eu/european-union/about-eu/funding-grants_en.
19. Veneto Agricoltura, https://www.venetoagricoltura.org.
20. D. Soglia, S. Sartore, S. Maione, A. Schiavone, S. Dabbou, J. Nery, L. Zaniboni, S. Marelli, P. Sacchi, R. Rasero, Animals **10**, 969 (2020)
21. https://www.venetoagricoltura.org/
22. D. Soglia S. Sartore, S. Maione, A. Schiavone, S. Dabbou, J. Nery , L. Zaniboni, S. Marelli, P. Sacchi, R. Rasero, Poult. Sci. **96**, 2552–2561 (2017)
23. V. Ferrante, C. Mugnai, L. Ferrari, S. P. Marelli, E. Spagnoli, S. Lolli, Ital. J. Anim. Sci. **15**, 303–309 (2016)
24. M. G. Strillacci, M. C. Cozzi, E. Gorla, F. Mosca, F. Schiavini, S. I. Román-Ponce, F. J. Ruiz López, A. Schiavone, M. Marzoni, S. Cerolini, A. Bagnato, Animal **11**, 737–745 (2017)
25. http://www.gallinabianca.it
26. Aviagen, "Ross-Broiler Management Handbook 2018," p. 5 (2018)
27. European Union, "Council Directive 834/2007 on organic production and labelling of organic products" (2007)
28. Feedipedia (2021)
29. Arlington, "AOAC. International Official methods of analysis of AOAC,"
30. THE Commission, OF The, and E. Communities, "L 257/14," *Communities*, no. January, pp. 14–28 (1999)
31. M. Bateson, P. Martin, Behav. Res. Ther. **32**, 898 (1994)
32. StataCorp, "Stata Statistical Software, Release 9.0. College Station, TX 2005"
33. S. Mueller, M. Kreuzer, M. Siegrist, K. Mannale, R. E. Messikommer, I. D. M. Gangnat, Poult. Sci. **97**, 3325–3336 (2018)
34. K. Wang, S. Shi, T. Dou, H. Sun, Poult. Sci. **88**, 2219-2223 (2009)
35. A. Dal Bosco, S. Mattioli, A. Cartoni Mancinelli, E. Cotozzolo, and C. Castellini, Animals, **11**, 1281 (2021)
36. M. Azizian, A. A. Saki, J. Hell. Vet. Med. Soc. **71**, 2087 (2020)
37. K. M. Hartcher, H. K. Lum, Worlds. Poult. Sci. J. **76**, 154–167 (2020)
38. F. S. Guni, S. H. Mbaga, A. M. Katule, and E. H. Goromela, Trop. Anim. Health Prod. **53**, 53–59 (2021)
39. Y. D. Sanka, S. H. Mbaga, S. K. Mutayoba, A. M. Katule, and S. H. Goromela, Trop. Anim. Health Prod. **52**, 3315–3322 (2020)
40. C. Castellini, C. Mugnai, L. Moscati, S. Mattioli, M. Guarino Amato, A. Mancinelli, A. Dal Bosco, Ital. J. Anim. Sci. **15**, 37–46 (2016)
41. https://www.hendrix-genetics.com/en/
42. C. Castellini, A. Dal Bosco, C. Mugnai, and M. Bernardini, Ital. J. Anim. Sci. **1**, 291–300 (2002)
43. S. Bergmann, A. Schwarzer, K. Wilutzky, H. Louton, J. Bachmeier, P. Schmidt, M. Erhard, E. Rauch, J. Vet. Behav. Clin. Appl. Res. **19**, 90–101 (2017)
44. C. W. Tallentire, I. Leinonen, and I. Kyriazakis, Sci. Rep. **8**, 1168 (2018)
45. E. H. van der Waaij, J. Anim. Sci. **82**, 973–981 (2004)
46. A. Cartoni Mancinelli, S. Mattioli, A. Dal Bosco, A. Aliberti, M.G. Amato, C. Castellini, Animals, **10**, 550 (2020)
47. W. Bessei, Worlds. Poult. Sci. J. **62**, 455–466 (2006)

48. G. Li, Y. Zhao, Z. Porter, J. L. Purswell, Animal **15**, 36–38 (2021)
49. J. Carol Petherick, E. Seawright, D. Waddington, I. J. H. Duncan, L. B. Murphy, Anim. Behav. **49**, 1521–1530 (1995)
50. B. L. Nielsen, Br. Poult. Sci., **45**, 306–315 (2004)

A new trend on sustainable animal production

Liang Chou Hsia*

National Pingtung University of Science and Technology, Taiwan, ROC

Abstract. A sustainable animal production can improve the efficiency of animal production and recycle of animal waste, and reduce greenhouse gas emissions. It can be achieved by firstly, properly operation of animal production, such as improvement by animal breeding, synthesis of amino acids, improvement by animal nutrition, good housing design, phase feeding system, liquid feeding system, using left over of one day bakery and biscuit factory, left over of vegetable and fruit market, new methods to plant Leucaena or mulberry to feed goats, grass or fodder plant in hydroponic house for sustainable farming. Secondly, good waste management, such as transferring waste to no contaminant material and even to useful material, reduction of greenhouse gas emissions, liquid waste management, and solid waste treatment. All these subjects also need to consider the interaction between knowledge, e.g. breeding and nutrition need to consider about environmental temperature factors; waste management needs to consider about nutrient requirement by vegetable, fruit, etc. A sustainable animal production is a kind of science integrated different knowledge together, then you can achieve some good results.

1 Introduction

The operation of sustainable animal production tries to: (1) improve the efficiency of animal production and recycle of animal waste, consequently reduce waste produced from animal farms; (2) reduce waste material which cause and increase greenhouse effect of the world [1]. The improvement of efficiency on animal production not only reduces cost of animal production, but also reduces waste to soil and water. It is very important to reduce greenhouse gas which may destroy the world by the increasing of world environmental temperature.

2 Basic principles

The basic principles are, firstly, any method which can improve efficiency and reduce waste to soil, water, and air; secondly, any method which can treat waste efficiently and transfer them to become no contaminant to soil, water or air, or transfer them to become useful feed and fertilizer, etc.; thirdly, it is very important to reduce the production of greenhouse gas [1, 2].

* Corresponding author: lchsia@mail.npust.edu.tw

3 How to achieve the above basic principles

3.1 Animal production part

Feed efficiency = Feed intake / weight gain. If we can reduce the proportion of feed efficiency, then we can reduce waste in soil, water and air. Some of the most exciting examples can be seen in the animal production industry.

3.1.1 Improvement by animal breeding

Broiler is one of the best examples. During last 30-50 years, we can find the growth rate, feed intake and feed efficiency had fast improvement. The results cause better feed efficiency and increase weight gain. This means they waste less feed; that is, they achieve the goal of sustainable animal production.

3.1.2 Synthesis of amino acids

This is really a very important product in sustainable animal production because we may not easily make feed formulation. If we try to balance amino acids in diet, then it is not only more efficient to produce animal products but also reduce waste.

3.1.3 Improvement by animal nutrition

The following two ways are the best methods to improve feed efficiency and reduce waste of nutrient in animal production.

3.1.3.1 Nutrient balance

Nutrient balance contributes great improvement on growth rate and feed efficiency. This is due to balance nutrient can efficiently produce meat, eggs, and milk. They also can reduce waste. The most obvious example is the ideal protein concept. The concept is based on meat, eggs, and milk are constructed by specific amount of amino acids. If we can contribute specific amount of amino acids then it can synthesize specific amount of milk, meat or eggs. The result also can reduce the waste of nutrient, e.g. protein, energy and other nutrients [3].

3.1.3.2 Growth promoter

Antibiotics have been used as growth promoters for animals about 4-5 decades. Since people realize the resistant of bacteria to antibiotics, then the EU become the first area to ban the use of them. Now many Asia countries also ban on antibiotics. We have to thank to antibiotics during last years because they not only improve the efficiency of animal production, but also reduce the mortality during this period of time. This is one of great contribution by antibiotics. Today, we do not use antibiotics as growth promoters and replace by many new products, such as short chain fatty acids, medium chain fatty acids, essential oils, herbal medicine, etc. Some of these products not only can inhibit bacteria growth, but also can improve a great amount of growth performance. Some of them not only can kill bacteria, but also can inhibit virus, e.g. medium chain fatty acids and essential oils or herbal medicine. The good news is we do not use all these new growth promoters in human medicine, so there is no resistant problem.

3.1.4 Good housing design

A good house design can improve the efficiency of animal production, e.g. good insulation of house can reduce heat and cold stress of animals. It also can reduce the waste of feeds with properly feeder design and enough feeding space.

3.1.5 Phase feeding system

Different stages of animals are provided different animal feeds which can reduce waste of feeds, especially the nitrogen loss of feeds. This system called phase feeding program (Fig. 1) [4].

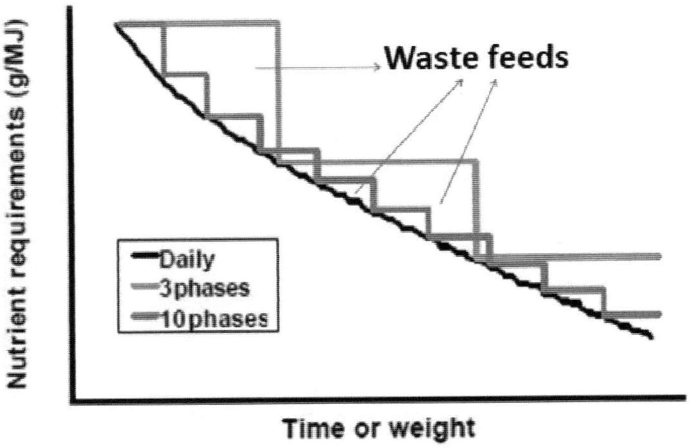

Fig. 1. Phase feeding

3.1.6 Liquid feeding system

This liquid feeding system is not only feeding animal by water + feed. The feeding system is combined feed with by-products of alcohol, cheese, juice, etc. industries.

3.1.7 Left over of one day bakery and biscuit factory

This part can be used for monogastric animals as an energy source. If we can help farmers to make a formula, then the farmers can use this cocked flour (bread and biscuit) to feed animals.

3.1.8 Left over of vegetable and fruit market

Most of the fresh left over can be fed to ruminant animals. It is good for them and they also enjoy to eat.

3.1.9 New methods to plant Leucaena or mulberry to feed goats

The goat is a kind of browsing animal. If we can plant mulberry or Leucaena in systematic way for them, then we can provide more nutritious feed for goats.

3.1.10 Grass or fodder plant in hydroponic house for sustainable farming

This method is an advanced and sustainable way to plant grass or fodder for ruminants. It can save land, water, grass harvesting machine.

3.2 Waste management

3.2.1 Sustainable animal production transfers waste to no contaminant material and even to useful material

No contaminant by animal production is one way to keep clean of the world. To produce useful material is to make animal production even better from sustainable production point of view. There are two ways to make the decision which methods of animal waste treatment could select. The first method is to use water to wash animal waste and clean the pen. The wastewater is stored under the house or outside the house. The second method is to use scraper to clean waste and transfer to outside, then use it to make compost as a fertilizer. Figure 2 shows the flow chart of animal waste treatment and greenhouse gas production. The red characters (CO_2, CH_4 and NH_3) in the figure are greenhouse gas and ammonia.

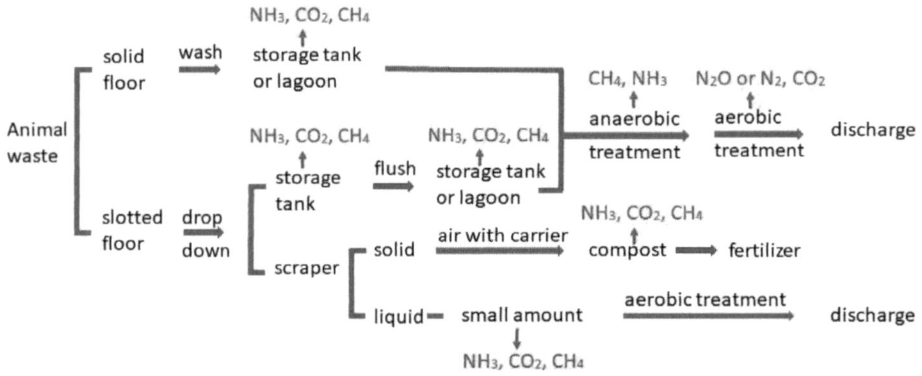

Fig. 2. Flow chart of animal waste treatment and greenhouse gas production

3.2.2 Reduction of greenhouse gas emissions

3.2.2.1 The important greenhouse gas in animal production

When farmers deal with animal waste, they should not only avoid smell but also need to avoid to produce greenhouse gas (CO_2, CH_4, N_2O) emissions in animal production area.

3.2.2.2 The most important waste treatment related to greenhouse gas and ammonia production in the treatment

Figure 3 shows the relationship between treatment methods and the situation which gas production.

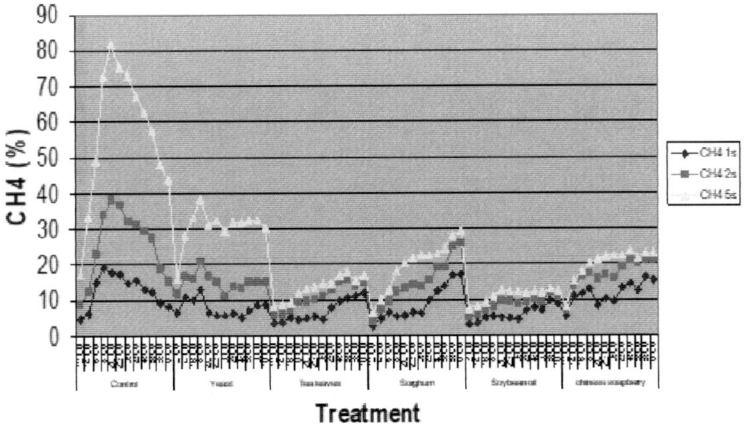

Fig. 3. Interaction between treatments and time on CH_4 production [5, 6]

3.2.3 Liquid waste treatment

The most significant point is to cover the anaerobic treatment and collect methane gas, let them not emit to air. This method also can avoid odor emission to air, then people can use methane as an energy source. It is important to use the heat produced by burning of methane gas to produce electricity. The heat can be used to heat the water which is used to produce algae. Because the algae water has large quantity, a little bit high temperature can let algae reproduce faster. The CO_2 produced by methane burning also can introduce into pond. This can stimulate algae to reproduce even faster, due to photosynthesis in algae [7] needs CO_2. Of course, the CO_2 which produced by burning CH_4 cannot easily consume by algae only. The algae are used to produce methane again, then the air can go into greenhouse and provide for vegetable growth. The purpose of all the procedure is to reduce greenhouse gas but also have positive effect to promote growth of vegetable.

3.2.4 Solid waste treatment

The second way to deal with animal waste is making compost which can be used for plant as a fertilizer. No matter which kinds of livestock or poultry, they need defecation and urination; that is, they cannot digest their food 100% and they have metabolism too. Both defecation and urination will discharge at least 20-40% of nutrient. The nutrient is also the nutrient required by plant, in consequence this is the good example of sustainable animal production. This method is also used by all ancestors since a thousand years ago.

4 Conclusion

A sustainable animal production is not only a simple subject but should be considered as a subject which related to many kinds of subjects on animal production, e.g. animal breeding, nutrition, management, housing, and waste management. All these subjects also need to consider the interaction between knowledge, e.g. breeding and nutrition need to consider about environmental temperature factors; waste management needs to consider about nutrient requirement by vegetable, fruit, etc. A sustainable animal production is a kind of science integrated different knowledge together, then you can achieve some good results.

References

1. L. C. Hsia., in Proceedings of the 2nd International Conference on Greenhouse Gases and Animal Agriculture GGAA2005-Working Papers, Institute of Animal Science, ETH-Zurich, Switzerland, Publication Series, **27**, 286-289 (2005)
2. L. C. Hsia, in Proceedings of The 18th Asian-Australasian Animal Production Congress, Sarawak, Malaysia, p. 15 (2018)
3. R. A. Ballantyne, I. L. Hung, S. P. Su, Y. W. Chen, J. W. Lee, L. C. Hsia, in Proceedings of The 18th Asian-Australasian Animal Production Congress, Sarawak, Malaysia, p. 537 (2018)
4. L. A. den Hartog, G. B. C. Backus, J. A. M. Voermans, in Proceedings of Future of Pig Industry Conference, Pig Research Institute, Taiwan, pp. 51-61 (1991)
5. M. Moncada-Laínez, L. C. Hsia, in Proceedings of the 13th Animal Science Congress of the Asian-Australasian Association of Animal Production Societies, Hanoi, Vietnam, p. 94 (2008)
6. M. Moncada-Laínez, L. C. Hsia, Int. J. Anim. Veter. Adv. **8**, 29-34 (2016)
7. N. Y. Chang, W. C. Kuo, L. C. Hsia, J. Chin. Soc. Anim. Sci. **38** (suppl.), 86 (2009)

A bivalent vaccine of *Eimeria media* and *Eimeria intestinalis* yields protective immunity against coccidiosis during rabbit production

Penny Humaidah Hamid[1*], *Widagdo Sri* Nugroho[2], *Sigit* Prastowo[1], and *Rini* Widayanti[2]

[1]Department of Animal Sciences, Universitas Sebelas Maret, Surakarta 57126, Indonesia
[2]Faculty of Veterinary Medicine, Universitas Gadjah Mada, Yogyakarta 55281, Indonesia

Abstract. Coccidiosis is the costliest disease in the rabbit industry. The kits do not obtain maternal immunity against the diseases and therefore suckling to weaning period being the riskiest time of infection. To date, control of coccidiosis is relied on the use of chemical coccidiostat as a drug or as the substance in feed. Whilst, continuous exposure to antibiotics develops resistance and contamination in the carcass. There is no commercially provided vaccine to prevent rabbit coccidiosis cases. In this report, we processed precocious lines from the strain of Yogyakarta origin by using selection pressure to attenuate *Eimeria* spp. The vaccine candidate composed of 50% of each *Eimeria media* and *Eimeria intestinalis* were given orally to initiate protective immunity against rabbit coccidiosis. The vaccinated groups with 5 x 10^2 oocyst dosage exhibited 95% less total oocyst excretion compared to the non-vaccinated group. Our experiment showed no mortality and without any significantly detrimental response of vaccinated rabbits during the observation time ($P<0.0001$). Since the oocysts in the vaccinated rabbits are excreted in dropped faecal samples, it provides recirculation and provokes a group immunity within the flock. A vaccine candidate is a promising tool as a more environmentally friendly strategy for sustainable rabbit production.

1 Introduction

Rabbit production in the world has been a part of efforts in fulfilling for meat consumption. Rabbits' husbandry is not only for the small family need in rural areas, but becoming a large industrial business [1]. Awareness about health and nature properties of consumed products are increasing since 2011 [2]. Some of the concern is related to development of antibiotic resistance. Bacteria affecting i. e. *Staphylococcus aureus* is evidently resistance to tetracycline, macrolides and multidrug, in both rabbits and farm worker [3]. Nevertheless, the use of antibiotic is unavoidably for health management in an animal production system. However, uncontrolled usage of the drugs is not only associated with meat quality and environmental issue afterwards but also high cost for farm efficiency.

* Corresponding author: pennyhumaidahhamid@staff.uns.ac.id

Among the most costliest diseases in rabbit is coccidiosis [1]. The impact of the disease is not only due to the mortality of young kits but merely on subclinical forms. The surviving animals is not efficiently gaining the optimum body weight with the normal diet for fattening rabbits. The coccidiosis cases are worldwide ranging from small scale husbandry to large industry. Coccidiosis is caused by *Eimeria* spp. *Eimeria intestinalis* and *Eimeria media* are among the species found in prevalent. *E. intestinalis* is reportedly as the medium to pathogen species, whereas *E. intestinalis* is included in the most pathogen in rabbit [2]. These species are also found in our previous report [3] which is circulated among rabbit colonies in central part of Java, Indonesia. Although it is known that coccidiosis is threatening, the prevention of coccidiosis by using vaccine is not yet applied in the field. The farmers mainly use the sulfa-based antibiotic which is basically for avian coccidiosis.

2 Material and methods

Part of the experimental design regarding precocious lines was submitted and approved by ethical committee of LPPT Universitas Gadjah Mada, Indonesia. For all of the experiment we use New Zealand rabbit breed and raised in Veterinary Medicine, Universitas Gadjah Mada. The parents' rabbits and the kits were tested for coccidia free and treated with diclazuril [4]. The faecal samples were periodically observed under microscope for Eimerias. Pellets and water were provided *ad libitum* without chemical coccidiostat during experiment related to infection.

E. media and *E. intestinalis* were isolated from our work and previously published [3]. Oocyst isolation with selection pressure were generated by method previously described [5]. To analyse the reduction of pathogenic effect on reduced prepatent time, infection was performed to 3 kits with 5×10^5 oocysts. The kits were necropsied and the ileum of small intestine were analysed histologically.

After reduction of prepatent period obtained, each species isolated was then mixed in the same amount, for serial immunization with dosage: 5×10^2, 1×10^3, 2×10^3, 4×10^3 and 8×10^3 oocysts orally. Challenge infection dosage was 5×10^5 wild type oocysts on day 14 after immunization of precocious lines. Collection of faeces started from day 3 to 12 post challenge infection and presented as the total oocyst production from experimental groups. Clinical manifestations were evaluated including lethargy, reduced appetite and fecal sample consistency. All of the data report for significance and graphical presentation were analysed by GraphPad Prism.

3 Results and discussion

Coccidiosis led to many consequences on rabbit husbandry. From the price due to excessive medication to long-term subclinical intestinal disorder causing lower efficiency on food conversion ratio [4]. Because the coccidia infection normally occurred in younger rabbit after around 4 weeks old, the effect is more likely due to the mortality of the younger rabbit. Instead, kits survive from the diseases frequently incapable to pass the transition from suckling to weaning. Differ from chicken coccidiosis with a well-developed vaccination for prevention [7], commercially available vaccines for rabbit coccidiosis have not available yet. Since the excessive price of medication, vaccination becomes very strategically critical to reduce infection rate and therefore maintain the most capacity of rabbit husbandry to be worthwhile.

Eimeria oocysts decreased the prepatent period *E. media* and *E. intestinalis*. Decreased pathogenicity was observed in Fig. 1. The precocious lines were able to complete their life cycle within the host as observed during *in vivo* infection but less compared to normal

Eimeria. The fecundity also obtained with gammons which led to possibility of producing more oocysts for other rabbits. However, all the complete cycle within the host do not cause severe detrimental effect within intestine and therefore rabbits showed no clinical manifestation during the pathogenicity test of precocious lines.

The oocysts of wild-type and processed for precociousness selection were our isolate previously isolated [6]. Protective immunity was observed significantly from oocyst excretions between divalent (BV)-vaccinated groups compared to control (C) when challenged with 5 x 10^5 WT (Fig. 2). Total oocysts excreted were significantly lower ($P \leq 0.001$) from the vaccinated group with 4 x 10^2, 8 x 10^2 and 1.6 x 10^3 BV dosages. Whilst, vaccination with lower dosages i.e., 1 x 10^2 and 4 x 10^2 were not significantly reduce oocysts shedding with $P \geq 0.05$. Rabbit infected with 5 x 10^5 WT oocysts without vaccination (C group) showed mild diarrhoea, reduced appetite, and lethargy during 5 to 8 days post-challenge infection. Whilst, no clinical manifestations were observed in the 95% of vaccinated groups after challenge infection. The BV in our experiment also showed fecundity from the dosage of 4 x 10^2 infection. BV oocysts shedding in the environment showed enrichment ability and indicating that the attenuate-oocyst may recirculate again to infect the other rabbit further induce a flock immunity. Coccidiosis in general affects young rabbits simply after weaning (5- to 6-week-old) which are not protected by the immunity from their mother [8]. Therefore, the vaccination in around thirty days old is quite critical to prevent clinical coccidiosis and hence keeping off utilization of continuous coccidiostat in the feed and water.

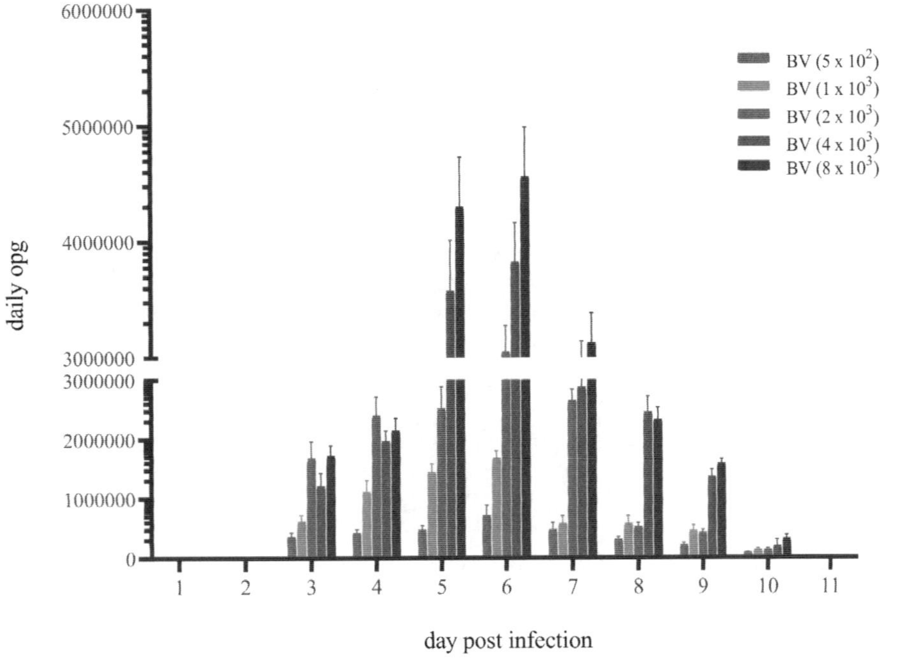

Fig. 1. Vaccination of live attenuated *E. media* and *E. intestinalis* secreted less oocysts in their faeces compared to wild-type.

The immunity against *Eimeria* is reportedly species specific [8]. Immunization strategy with more than one species will give benefit to obtain more complete protection in the field. This strategy was applied in the commercially available poultry coccidian [7]. Vaccination strategy with live rabbit *E. media* and *E. intestinalis* in this paper showed protective immunity against clinical coccidiosis from homolog infection. The immunity against *Eimeria* is reportedly species specific [8]. Immunization strategy with more than one species will give benefit to obtain more complete protection in the field. This strategy was applied in the commercially available poultry coccidian [7]. Vaccination strategy with live rabbit *E. media* and *E. intestinalis* in this paper showed protective immunity against clinical coccidiosis from homolog infection.

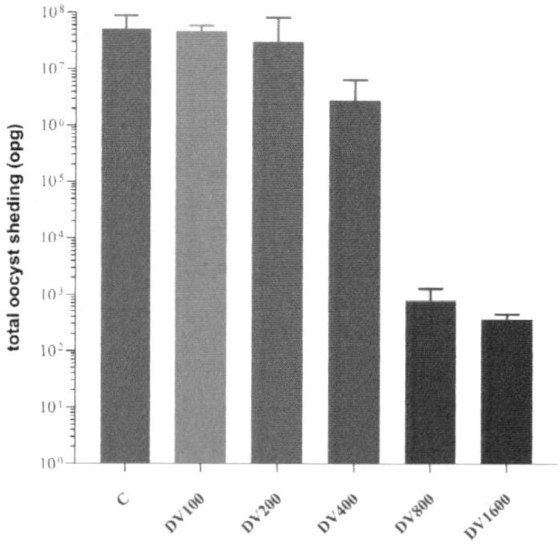

Fig. 2. Vaccination of live attenuated *E. media* and *E. intestinalis* with dosage of 4 x 10^2, 8 x 10^2 and 1.6 x 10^3 were able to reduce total oocyst excretion significantly ($P \leq 0.0001$) from the wild-type infection.

The excreted oocysts in the vaccinated rabbits may also provoke recirculation for a herd immunity if ingested by other. Protection to naturally occurred infection is difficult to be achieved during the research timeframe. Therefore, we evaluated the protection of challenge infection under laboratory condition before to ensure total numbers of oocyst administrated. Furthermore, prolonged observation time and bigger size of population test are valuable to be performed. We conclude in this report that the vaccine candidate was without detrimental effect to rabbit under field condition during the onset of coccidiosis i.e., until 14 days of infection.

4 Conclusion

Live-attenuated of *E. media* and *E. intestinalis* oocysts yields immune protection against coccidiosis infection. Immunity is provided by a single dosage of vaccination starting from 8 x 10^2 live attenuated of mixed oocysts. *E. media* and *E. intestinalis* may act as vaccine candidate for prevention of rabbit coccidiosis. Further studies on field trials and combination of more than 2 species can provide more information on the wide application of vaccine, its prospect and wider protective immunity.

We thank to LPDP who support and finance our work by a grant for university and industry research platform to PHH number PRJ-105/LPDP/2019.

References

1. A. Trocino, E. Cotozzolo, C. Zomeño, M. Petracci, G. Xiccato, C. Castellini, Ital. J. Anim. Sci. **18**, 1361-1371 (2019)
2. R. Schleenbecker, U. Hamm, Appetite **71**, 420-429 (2013)

3. A. R. Attili, A. Bellato, P. Robino, L. Galosi, C. Papeschi, G. Rossi, E. Fileni, M. Linardi, V. Cuteri, F. Chiesa, P. Nebbia, Antibiotics **9**, 673 (2020)
4. T. K. Bhat, K. P. Jithendran, N. P. Kurade, World Rabbit Sci. **4**, 37-41 (1996)
5. D. W. Duszynski, L. Couch, *The Biology and Identification of the Coccidia (Apicomplexa) of Rabbits of the World* (Elsevier Inc., London, UK, 2013)
6. P. H. Hamid, P., S. Prastowo, Y. P. Kristianingrum, Vet. World **12**, 1256-1260 (2019)
7. T.A. Ahmad, B.A. El-Sayed, L.A. El-Sayed, Trials Vaccinol. **5**, 38-47 (2016)
8. M. Pakandl, Folia Parasitol. **56**, 153-166 (2009)

Analysis of vitamin C in probiotic yoghurt with the addition of podang urang mango (*Mangifera indica* L.)

Dyah Nurul Afiyah[1,*], *Irma Isnafia* Arief[2], *Tuti* Suryati[2], and *Riska Nurtantyo* Sarbini[3]

[1]Department of Animal Science, Faculty of Agriculture, Kadiri Islamic University, Kediri 64128, Indonesia
[2]Department of Animal Production and Technology, Faculty of Animal Science, IPB University, Bogor 16680, Indonesia
[3]Department of Computer Engineering, Faculty of Engineering, Kadiri Islamic University, Kediri 64128, Indonesia

Abstract. Yoghurt is a functional drink that is beneficial to health. However, nutritional losses, e.g. vitamin C, occurs during the thermal process. Therefore, vitamin C supplementation is needed to replace the lost amount during processing by adding the extract of podang urang mango (*Mangifera indica* L.). The yoghurt was made by adding 3% (v/v) of starter culture (*Lactobacillus delbrueckii subsp bulgaricus* RRAM-01 and *Streptococcus salivarus subsp thermophillus* RRAM-01) and mango extract at 0%, 2%, 4%, 6%, 8% and 10% (v/v) in four replications. The determination of vitamin C content was performed using titration method. The results showed that heating process at 85-90 °C in 35 minutes lowered the amount of vitamin C in the final product. However, the addition of podang urang mango extract at 6%, 8% and 10% could increase the amount of vitamin C.

1 Introduction

Functional food become a very important issue and people awareness increase in consuming a good functional food which is supporting for their health [1]. Yoghurt drink is considered as one of functional food and can be used to improve and maximize digestive system of human body. Yoghurt is the most popular fermented dairy product all over the world [2]. Yoghurt obtained from fermented milk using *Lactobacillus bulgaricus* and *Streptococcus thermophilus* bacteria or other lactic acid bacteria as appropriate, with or without the addition of other foodstuffs and permitted food additives [3].

Some researchers have been conducted on yoghurt drink related to its processing or making technique [4]. Long processing of yoghurt will result in reduced physicochemical content of the yoghurt [5]. Sterilization causes loss of a number of vitamins, folic acid and causes denaturation of serum proteins. Long and unprotected yoghurt storage will result in

* Corresponding author: dyahnurula@gmail.com

reduced. Therefore, there needs to be supplementation in yoghurt to increase the content of vitamins lost during the manufacturing process.

Supplementation or innovation of yoghurt ingredient is important with the purpose for providing a variety nutritional and flavored value in yoghurt drink to make it more attractive to consumers, both children or adult to consume yoghurt. Today, consumers demanding food products with natural preservatives of high quality and good shelf-life stability. Natural preservatives may be more effective and efficient in protecting food products from deterioration by physical, chemicals and spoilage by microorganisms compared to synthetic additives [6]. Innovation of yoghurt drink can be obtained by utilizing ingredient that has high nutrition and not yet widely used in the surrounding environment, one of which is podang urang mango.

Podang urang mango is one of the local fruits that are abundant in Kediri. According to BPS-Statistics of Kediri Municipality [7], the mango production in Kediri was 647.30 tons in 2020. The highest fruit production occurs in the harvest season, which is in September until December [8]. Podang urang mango has a reddish yellow fruit skin color. Fruit with length of 12.5 cm and a width of 7 cm, the thickness of the flesh of the fruit is 2-2.5 cm. The color of the fruit meat is orange, soft textured, and the taste of the fruit is sweet [9]. Experts believe mangoes are a great source of carotenoids called beta cryptoxanthin, which is a good cancer-suppression ingredient. Mangoes are also rich of vitamins such as vitamins C and E and antioxidants [10].

High levels of vitamin C in podang urang mango is expected to be able to benefits for humans. Recently vitamin C is required for the prevention of scurvy and maintenance of healthy skin, gums, blood vessels, etc. [11]. Because of benefits, it is necessary to do research on the addition of extracts, podang urang extract on probiotic yoghurt to see the vitamin C value.

2 Materials and methods

2.1 Preparation of samples and extraction of podang urang mango [12]

Podang urang mango fruit obtained from the district of Banyakan Kediri. The making of podang urang mango extract is done by maceration using mixture of hexane:acetone:ethanol (2:1:1 v/v). Filtrate (macerate) is then added with 10.0 ml aquadest and extracted using a split funnel until 2 layers are formed. The polar layer is extracted again until a clear filtrate is obtained. The obtained non-polar layer is collected and evaporated until dry extract is obtained.

2.2 Preparation starter cultures [13]

Starter cultures that used in this research were *Lactobacillus delbrueckii subsp bulgaricus RRAM-01* dan *Streptococcus salivarus subsp thermophillus RRAM-01*. Starter rejuvenation is done by inoculating yoghurt culture as much as 10% into sterilized milk first on the autoclave with a temperature of 115°C for three minutes. Furthermore, it is incubated at a temperature of 37 °C for 18 hours until coagulation is formed so that a work culture is obtained.

2.3 Making probiotic yoghurt with the addition of podang urang mango extract [14]

Full milk is heated at 85-90 °C for 35 minutes, then cooled until the temperature reaches 40-45 °C. *Lactobacillus delbrueckii subsp bulgaricus* RRAM-01 and *Streptococcus salivarus subsp thermophillus* RRAM-01 are added to milk. Population of bacteria used more than 107 cfu mL^{-1}, incubated at 37 °C for 18-24 hours until produce coagulation or thickened (plain yoghurt). After that added podang urang mango extract as much as 0% (P1), 2% (P2), 4% (P3), 6% (P4), 8% (P5) and 10% (P6). Yoghurt is stored at room temperature (29-30 °C).

2.5 Analysis of vitamin C [15]

Provide 10 gram of sample (mango, milk or yoghurt) of each treatment on 100 mL of water. Add 5 drops amylum indicator, then titration with iodine (I2) 0,01 N to blue. Vitamin C content can be calculated by formula:

$$Vitamin\ C\ (\frac{mg}{100g}) = \frac{Vol\ I2\ x\ 0.88\ x\ Fp)\ x\ 100}{W\ sample\ (g)} \quad (1)$$

Vol I2 : Volume Iodine (mL)
0.88 : 0.88 mg Vit C
Fp : Dilution Factors
Ws : Sample Weight (g)

2.6 Statistical analysis

The experimental design used in addition of podang urang mango extract in yoghurt is completely randomized design (CRD) and the data was analyzed using Duncan`s post hoc test [16]. The vitamin C data were presented as mean with standard deviation from at least six independent experiments.

3 Results and discussion

Mango, especially podang urang mango plays an important role in balancing and improving the diet of human being by providing such of nutrition like energy, carbohydrates, sugar, dietary fiber, protein, vitamin A, vitamin C, calcium, magnesium, phosphorus, potassium [17]. Vitamins are group of organic complex compounds that the everybody needs in small quantities. Vitamins must be supplies from outside because the human body cannot synthesize them. One of important vitamin is Vitamin C [18].

Vitamin C plays a vital role in our lives because many reasons. First reasons, it contributes to the synthesis of the collagen tissue around bones, teeth, cartilage, skin, and damaged tissue. Second reasons, this vitamin is needed to activate various enzymes related to the nervous system, hormones, and detoxification of medicine and poison in the liver. Third reasons, its role as an antioxidant is well-known in society; its solubility enables it to work as antioxidant within our bodily fluids. Fourth reasons, Vitamin C increases the rate of absorption of iron, calcium, and folic acid. Fifth reasons, it reduces allergic reactions, boosts the immune system, stimulates the formation of bile in the gallbladder, and facilitates the excretion of various steroids [19].

Vitamin C is highly soluble in water and alcohol, and very easily to oxidized. The oxidation of Vitamin C occurs quickly in a base environment at especially high temperatures. Light and heat damage Vitamin C in fruits and vegetables [20]. Qualitative proof of vitamin

C can be done using the Iodometry method. Because this solution is able to reduce vitamin C. Vitamin C is a powerful and powerful reducing agent. Simple can be titrated with a raw solution of iodine as an oxidizer [21]. Thus, Then the content of vitamin C in figs and passion fruit can be determined by the method of Iodometri.

Vitamin C in podang urang mango extract 28.32 mg/100 g. Titratable acid in podang urang mango extract was 0.12 %. Its lower than those reported by Amiarsi and Mulyawanti [22] who found that vitamin C in mango was 33.62-58.59 mg/100 g and titratable acid was 0.11%. This acidity was caused by organic acid such as acetic acid, ascorbate, citrate, butyrate, and propionate [23].

Table 1. Vitamin C in milk

Treatment	Vitamin C (mg/100 g)
Raw Milk	0.77 ± 0.005
Pasteurized Milk	0.14 ± 0.006

From the Table 1 it was observed that Vitamin C in milk is 0.77 mg/100 gram. It slightly lower than National Dairy Council that describe vitamin C in milk was 0.94 mg/100 g. There was a decrease in vitamin C from raw milk to pasteurized milk. [24] reported that heat or sterilization damage the composition of vitamin C. The higher the temperature used for the sterilization process, the more vitamin C getting down. Saleh [25] also concluded that the most vitamin loss/ decrease in the process of processing milk with heat is vitamin C. milk processing with sterilization in polyethylene causing a decrease in vitamin C 50% greater than vitamin B12 loss and folic acid. Vitamin C also decreases if food is left warm or hot for too long. Vitamin C amount is significantly reduced if stored at room temperature [26].

Table 2. Vitamin C in yoghurt with the addition of podang urang mango extract

Treatment	Vitamin C (mg/100 g)
P1	0.22 ± 0.05^{a}
P2	0.25 ± 0.06^{a}
P3	0.25 ± 0.06^{a}
P4	0.35 ± 0.06^{b}
P5	0.35 ± 0.06^{b}
P6	0.38 ± 0.05^{b}

^{ab}Different letters in a column indicate statistically significant differences ($P<0.05$)

Table 2 showed that there is a decrease of vitamin C content in the yoghurt than in form of milk. Decreased vitamin C results in yoghurt influenced by several things, among others, in the process of making yoghurt, fermentation, and storage. In the process of making yoghurt occurs the process heating at so that it can damage vitamin C, while the way storage at less stable room temperatures affects damage Vitamin C [27].

Based on the measurement of vitamin C content, show that the increase of podang urang mango extract led to an increasing vitamin C content of this yoghurt. The addition podang urang mango extract at 6%, 8% and 10% significantly increasing vitamin content than 0%, 2% and 4%. This increase is due to an increase in the amount of podang urang mango extract that is added to yoghurt. The increase in the addition of podang urang mango also aims to improve the quality of organoleptic. Suitable with Suriasih [28] that increasing vitamin C levels in cow's milk yoghurt can also act as a flavor enhancer and can also be a special attraction for society (for consumers). Effendi [29] suggests that the main component in yoghurt is the acidic nature of lactic acid and its aroma substance lactic acid bacteria produced. However, with the addition of podang urang mango extract, aroma will also change.

Based on the data showed in Table 2, P2 (2%) dan P3 (4%) give not significantly different with P0 (yoghurt plain, without the addition of podang urang mango extract). This proves that vitamins are easily damaged components. After harvesting fruit then in the storage of a number of vitamins can disappear, depending on temperature, exposure to the air and sun, and the length of storage. The higher the temperature, the longer exposure to air and sun, the longer it is stored, the more missing vitamins [30].

There are not adequate amount of vitamins A and C in low fat dairy products, especially yoghurt drink which are not fortified. Usually, low fat milk and other dairy products (such as yoghurt drink) are enriched or fortified with vitamin A but not with vitamin C. Fortification of dairy products especially yoghurt drink with vitamins A and C leads to improvement in their nutrition quality and also increases their organoleptic quality [31]. Vitamin C stimulates the oxidation process and also stimulate the production of collagen which is essential for adrenal function. Vitamin C helps the body absorb micronutrients such as iron and calcium. This vitamin also helps in regeneration of vitamin E [32]. Vitamin C deficiency can lead to fatigue, immune body decrease, impaired osteoblast activation, reduced collagen synthesis in the body and triggering hair breakage.

4 Conclusion

The results showed that heating process at 85-90 °C in 35 minutes lowered the amount of vitamin C in the final product. However, the addition of podang urang mango extract at 6%, 8% and 10% could increase the amount of vitamin C.

This research was supported by the 2021 Penelitian Kerjasama Perguruan Tinggi (Joint Research) from Indonesian Ministry of Education, Culture, Research, and Technology.

References

1. K. Menrad, J. Food Eng. **56**, 181-188 (2003)
2. A. N. Madhu, N. Amrutha, S. G. Prapulla, Probiotics Antimicrob. Proteins **4**, 90-97 (2012)
3. National Standardization Agency of Indonesia, *Indonesian National Standard 2981: 2009 Yoghurt* (National Standardization Agency of Indonesia, Jakarta, Indonesia, 2009)
4. M. Kumasaki, T. Shoji, Chem. Eng. Trans. **31**, 259-264 (2013)
5. J. Rorong, H. F. Aritonang, F. P. Ranti, Chem. Prog. **1**, 9-18 (2008)
6. M. Giner, S. Vegara, L. Funes, N. Martí, D. Saura, V. Micol, M. Valero, J. Sci. Food Agric. **92**, 1917-1923 (2012)
7. BPS-Statistics of Kediri Municipality, *Kediri Municipality in Figures 2021* (BPS-Statistics of Kediri Municipality, Kediri, Indonesia, 2021)
8. Baswarsiati, Yuniarti, Buletin Plasma Nutfah **13**, 62-69 (2007)

9. Decree of Minister of Agriculture of the Republic of Indonesia Number 336/Kpts/TP.240/2/2003 regarding Release of Podang Urang Mango as a Superior Variety (2003)
10. A. A. Kabiru, A. A. Joshua, A. O. Raji, Int. J. Res. Rev. Appl. Sci. **15**, 41-50 (2013)
11. G. V. Kumar, K. K. Ajay, P. G. R. Raghu, S. Manjappa, Int. J. Pharm. Life Sci. **4**, 2489-2491 (2013)
12. S. Wijana, F. M. Arie, A. P. Adelia, Studi Proses Pengolahan Bubuk Mangga Podang (Kajian Jenis dan Konsentrasi Filler), Undergraduate Thesis, Universitas Brawijaya Malang, Indonesia (2017)
13. E. L. S. Suharto, I. I. Arief, E. Taufik, Media Peternakan **39**, 82-89 (2016)
14. D. E. Sihombing, I. I. Arief, S. Budiarti, Adv. J. Food Sci. Technol. **8,** 251-255 (2015)
15. A. El-Ishaq, S. Obirinakem, Int. J. Chem. Biomol. Sci. **1**, 17–21 (2015)
16. R. G. D. Steel, J. H. Torrie, *Principles and Procedures of Statistica Biomedical Approach 3rd Edition* (McGraw Hill Inc., Singapore, 1995)
17. S. D. Rajmane, S. L.Korekar, Int. J. Sci. Res. **8**, 225-228 (2016)
18. R. Mansour, M. Aljoubbeh, Int. J. ChemTech Res. **6**, 1124-1130 (2014)
19. N. A. Destianti, S. N. Fatimah, S. Dewi, Althea Med. J. **4**, 173-177 (2017)
20. C. F. Song, Z. W. Cui, G. Y. Jin, A. S. Mujumdar, J. F. Yu, Dry. Technol. **33**, 583-590 (2015)
21. E. Mulyani, Pharmauho: Jurnal Farmasi, Sains, dan Kesehatan **3**, 14-17 (2018)
22. D. Amiarsi, I. Mulyawanti, J. Hort. **23**, 255-262 (2013)
23. I. M. S. Utama, N. L. Yulianti, I. G. N. A. Aviantara, G. Arda, in Prosiding Seminar Nasional PERHORTI dan PERAGI, 14 November 2016, Makassar, Indonesia (2016)
24. N. F. Dewi, Effect of Waiting Time and Sterilization Temperature on the Quality of Commercially Sterilized Flavored Liquid Milk, Master Thesis, IPB University, Bogor, Indonesia (2018)
25. E. Saleh, *Basic Milk Processing and Livestock Products* (USU Digital Library, Medan, Indonesia, 2004)
26. A. B. Tawali, Zainal, Jurnal Sains & Teknologi **4**, 72-82 (2004)
27. A. Sepyani, Sugar Test and Vitamin C Levels in Cow's Milk Yoghurt Boyolali with the Addition of Young Coconut Water (*Cocos nucifera*) and Soursop Fruit Extract (*Annona muricata*), Undergraduate Thesis, Universitas Muhammadiyah Surakarta, Sukoharjo, Indonesia (2013)
28. K. Suriasih, Agrotekno **11**, 3088 (2005)
29. M. H. Effendi, Media Kedokteran Hewan **17**, 144-147 (2001)
30. S. Almatsier, *Basic Principle of Nutrition Science* (Gramedia Pustaka Utama, Jakarta, Indonesia, 2001)
31. R. I. Dave, N. P. Shah., Int. Dairy J. 7, 435-443 (1997)
32. B. Willershausen, A. Ross, M. Försch, I. Willershausen, P. Mohaupt, A. Callaway, Eur. J. Med. Res. **16**, 514-518 (2011)

Effect of using fresh palm oil and used palm oil as a fatliquoring agent to the physical quality of tilapia fish skin in vegetable tannery

Ambar Pertiwiningrum[1*], *Kurniawan Eko* Nugroho[1], *Muhammad Aziz Nur* Roufi[1], *Viagian* Pastawan[1], *Ragil* Yuliatmo[2], *Mohammad Zainal* Abidin[1], and *Yuny* Erwanto[1]

[1]Faculty of Animal Science, Universitas Gadjah Mada, Yogyakarta 55281, Indonesia
[2]Department of Leather Processing Technology, Politeknik ATK Yogyakarta, Yogyakarta 55185, Indonesia

Abstract. Fish skin is one of the alternative materials to replace the common animal skin that used in the leather tannery. This research was carried out by vegetable tanning to the tilapia fish skin using selected concentration of fresh palm oil and used palm oil. The study aims to find out the physical quality of leather through the using of fresh and used palm oil as fatliquoring agent in the vegetable tannery process. The same concentration of fresh palm oil (5%) and used palm oil (5%) were used in this study. The parameter of tensile strength, elongation, and shrinkage temperature as physical quality was observed in triplicates. Based on the results, tensile strength and leather elongation were significantly increased, while the shrinkage temperature was not significantly affected. In conclusion, the addition of palm oil caused the increasing tilapia fish skin quality that can be used for commercial products, and it became the alternative material that could be use in vegetable tannery processing.

1 Introduction

Tilapia is a type of freshwater fish consumption and is sold as a fresh fish. There are several companies that process tilapia into fillets, and by-product of processing tilapia into fillets is skin. Tilapia skin can have high economic value in the form of tanned tilapia skin [1]. One of the factors that play an important role in the tanning process is the using of tanning agents. The tanning material has advantages and disadvantages, and the properties of tanning agent affect the physical quality. The quality of skin to be a leather depends on the character of the fresh skin itself, especially for fish skin, tilapia fish skin has a good and unique character to be processed in vegetable tannery compared to common others animal fresh skin (Table 1).

Fat liquoring process is part of the tannery which aims to put the oil molecules in space that exists between the fibres of the skin and can serve as a lubricant. Oil is an important component in the skin that functions as a lubricant for skin tissue in the tanning process to make the skin soft [2]. Oils or fats can change the important properties of the skin, including the skin becomes softer, more supple, stretchy, soft, and has a smoother tattoo surface. Plant

* Corresponding author: artiwi@mail.ugm.ac.id

oils are widely used as ingredients for skin oiling because they are affordable and easy to obtain. Some examples of vegetable oils are rubber seed oil, soybean oil, sunflower seed oil and palm oil. Sivakumar *et al.* [2] have used vegetable oil as a fatliquor for leather application. Wang *et al.* [3] and Kasmudjiastuti *et al.* [4] also produced and characterized fatliquoring agent from the vegetable oil such as palm oil. Vegetable oils, animal oils, and fish oils contain triglycerides. Palm oil consists of triglycerides which are esters of glycerol with three fatty acid molecules. The use of palm oil in the tanning process can function as an emulsion stabilizer and a good lubricant to produce a limp leather [5-7].

Table 1. Simply character of tilapia fish skin as a tannery material [8-9]

	Fresh Skin	**Tilapia Fish Skin**
Form	Cleavage	Fish scales
Color	As a common animal skin	Various color
Appearance	Fresh	Fresh
Character	Easily rotten	Easily rotten
Section arrangement	Epidermic, hypodermic, cutis	Epidermic, hypodermic, cutis

In this study, we investigated the using of fresh palm oil and used palm oil as a fat liquoring agent in the tannery of tilapia skin. The study aims to find out the physical quality of leather through the using of fresh and used palm oil as fatliquoring agent in the vegetable tannery process. The physical quality of tilapia leather was observed and evaluated by examining parameters of tensile strength, elongation, and shrinkage temperature.

2 Materials and methods

2.1 Tannery process of tilapia fish skin

The study was used 100 pieces of tilapia fish skin in tannery process that was conducted in two treatments of fat liquoring process. The materials used in this study included fresh tilapia skin, water, crystal salt, antiseptic, teepol, lime ($Ca(OH)_2$), feliderm bate PB, pancreobat enzymes, degreasing agent, formic acid, ZA, FE, sulfuric acid (H_2SO_4), mimosa, syntan, napo solution, fresh and used palm oil, wetting agent, ammonia, leveling agent, dyestuf, sulphonated oil, formic acid (HCOOH), and antifungal.

The fat liquoring process is to make the skin softer, has high elongation, and to defend the skin from being damaged by the affection of water, because the skin, that has processed by the fat liquoring, has very good absorption and repulsion to water molecules. The process of tannery was used the same concentration of fresh palm oil (5%) and used pam oil (5%) as fat liquoring agent. Although the common oil material used in fat liquoring proses is lipoderm or paradol as much as 6%. This research replaces the common material using those palm oil as an altenative material that provides the same leather quality. The tanning process was carried out in three stages consisting of pre-tanning, tanning, and finishing. The tilapia tannery process is using vegetable tannery materials those are syntan and mimosa as shown in Figure 1.

Preparing raw materials (preserve using 600% water, 30% salt, and 05% antiseptic) and Sorting (selecting all of fresh skin)

↓

Fleshing (removing the remains of meat on the skin) and Washing (cleaning the skin using the running water)

↓

Weighing (determining the fresh weight of raw material)

↓

Liming (tilapia fish skin soaked in 600% water, 1% teepol, and 5% lime)

↓

Reliming (tilapia fish skin soaked in 600% water, 1% teepol, and 5% lime)

↓

Deliming (tilapia fish skin soaked and washed in 200% water, and 1% ZA)

↓

Bating (tilapia fish skin soaked in 4% pancreobat enzyme, and 1% teepol)

↓

Degreasing (tilapia fish skin was added by 1% degreasing agent)

↓

Pickling (tilapia fish skin soaked in 150% water, 15% salt, 1% FA, and 0.5% H_2SO_4)

↓

Tanning (tilapia fish skin soaked in 100% water, 4% mimosa, and 8% syntan)

↓

Repickling (tilapia fish skin was added by 15% salt, 1% FA, and 0.5% H_2SO_4)

↓

Retanning (tilapia fish skin soaked in 100% water, 4% mimosa, and 8% syntan)

↓

Neutralization (tilapia fish skin was neutralized in 0.5% nafo solution)

↓

Fatliquoring (tilapia fish skin was added by 10% each of fresh palm oil or used palm oil, and 1% FA)

↓

Wetting back (tilapiah fish skin soaked in 400% water, 1% wetting agent, 0.5% amonia)

↓

Dyeing (tilapia fish skin soaked in 100% water 60°C, 1% lavelling agent, 1% dye, 3% sulphonation oil, 1.5% HCOOH, and 0.02% anti-fungi)

↓

Drying and Finishing (Tanned tilapia leather was dried over night)

Fig. 1. Vegetable tannery process of tilapia fish skin using palm oil as fat liquoring agent

2.2 Tensile strength

Tensile strength is the maximum force required to pull the skin apart. The unit of tensile strength is expressed in N/cm2. Tensile strength measurement is carried out using a tool of tensil strength meter. Tensile strength is used as a parameter to determine the magnitude of the bond between the collagen fibers and the tanning material. The tensile strength value can be calculated by the following calculation:

$$Tensile\ strength = \frac{G}{txl}\ N/cm^2 \quad (1)$$

G = maximum tensile load on the sample (kg)
t = sample thickness (cm)
l = sample width (cm)
Converted into N/cm^2 (1 kg = 9.8066 N)

2.3 Elongation

Elongation is the increase in the length of the leather when the leather is pulled until it breaks divided by the original length. The unit of elongation is expressed in percent (%). Elongation measurement is carried out using the same tool of tensil strength measurement. Elongation is related to the slackness of the obtained skin. Elongation is one of the physical parameters used in the manufacture of commercial products. The elongation value can be calculated by the following calculation:

$$Elongation = \frac{Li-Lo}{Lo}\ x\ 100\% \quad (2)$$

Li = final length of sample at break (cm)
Lo = initial sample length (cm)

2.4 Shrinkage temperature

Shrinkage temperature is the temperature at which the leather contracts in hot water. The leather sample was cut to a size of 1 x 4 cm. The measurement of shrinkage temperature is using the tool of shrinkage meter. The sample is attached to the second end of the shrinkage meter clamp. The glycerine solution was put into a glass beaker, then the solution is heated using a stove until it reaches a temperature of 50°C. The function of glycerine is to increase the boiling point, then the temperature of the solution will increase gradually as it is heated. The thread in the sample that shifts indicates that the sample is shrinking. The shrinkage that occurs is recorded as the shrinkage temperature.

2.5 Data analysis

Data on the results of tilapia fish skin tannery using same concentrations of fresh palm oil and used palm oil were analyzed using a T-Test.

3 Results and discussion

3.1 Tensile strength

We observed the tensile strength of the tilapia leather with the addition of 5% fresh palm oil and used palm oil. The results of tensile strength measurement with fresh and used palm oil were 3612.9 N/cm^2 and 2709.4 N/cm^2, respectively. The addition of fresh palm oil has a higher result compared to the used palm oil, but the using of those palm oils was not significantly different (Figure 2.a). The standard tensile strength of tilapia leather based on the Indonesian National Standard (SNI) 06-6121-1999 for the upper leather of footwear products is minimum of 1600 N/cm^2, this indicated that the addition of either fresh palm oil or used palm oil with the concentration of 5% is able to provide tensile strength in accordance to SNI for leather upper footwear products.

The more used of palm oil in the fatliquoring process, the more the surface of the leather fibre is lubricated with oil, so that the leather becomes supple and stretch easily. This situation causes the leather fibre bonds to loosen, then the skin's ability to withstand tensile loads decreases [10-11]. Fatliquoring was occurred in two phases, first the oil is taken by the skin mechanically, then followed by an oxidation process. An important binding process is the presence of at least two double bonds in the molecule. In the oxidation process, the double bond takes two oxygen atoms and forms peroxides. Some of the peroxides can react with the amino groups of collagens [12].

The higher concentration of tannery material used in tanning process makes the higher tear strength of the leather. This can occur due to the binding of tannins in mimosa into protein molecules arrange the tilapia skin collagen, that is the amino acids glycine, alanine, and glutamate which results in the formation of cross-links between Gambir tanners and polypeptide chains in the collagen fibres of tilapia skin, which determines the level of strength the physical appearance of the tilapia leather itself. The increasing concentration palm oil as a treatment affected the increase and decrease in the value of tear strength [13]. Concentrations of 5% palm oil decreased the value of tensile strength due to reduced levels of tannins that entered the skin.

The difference between fresh and used palm oil is that used palm oil has undergone hydrolysis, oxidation, and polymerization. Habib and Alshammari [14] have differentiated the fresh and used (by deep frying) cooking oil. They found increased peroxidation after deep frying with oil. The increasing peroxide value could decrease the emulsifying stability, that can decrease the penetration to the skin. Table 1 shows the lower physical properties (tensile strength, elongation, and shrinkage temperature) of the skin fatliquored with used palm oil compared to the other treatments. It can be caused by low penetration of fatliquoring agent.

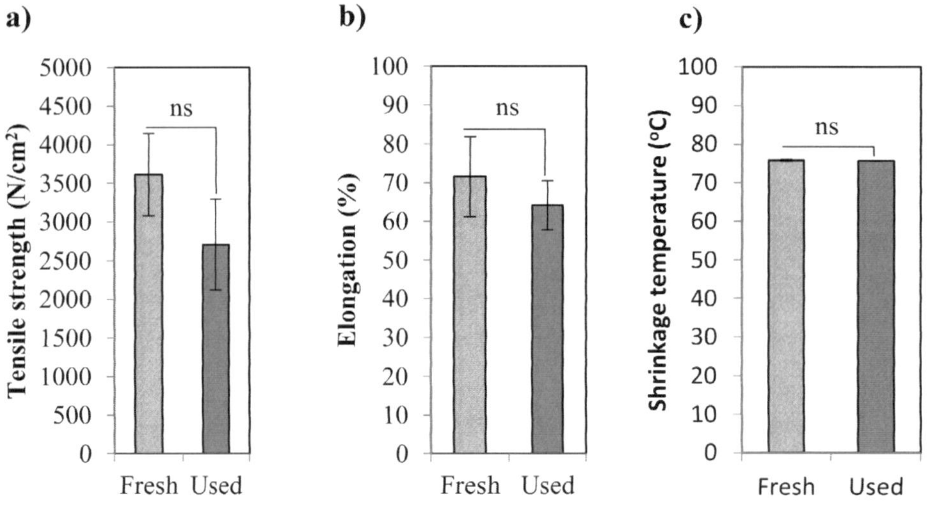

Fig. 2. The physical quality of tilapia leather: a) Tensile strength, b) Elongation, and c) Shrinkage temperature, using 5% of fresh and used palm oil as fat liquoring agent (Error bars represent standard deviation of mean). *ns: non-significant.

3.2 Elongation

Next, we observed the elongation of the tilapia leather with the addition of 5% fresh palm oil and used palm oil. The elongation test for tilapia leather with the addition of fresh palm oil and used palm oil was 71.5% and 64.2%, respectively. The addition of fresh palm oil has a higher elongation value on tilapia leather compared to the used palm oil, but the using of those palm oils was not significantly different (Figure 2.b). The elongation test results from all treatments exceeded the SNI for tanned tilapia skin, which is a minimum of 50%. The number of skin's fibers coated with oil emulsion in low concentration will result in a low leather elongation value. In addition, the elongation of a leather is mainly determined by the amount of oil absorbed by the skin during the fatliquoring stage and also the amount of cross-linking formed between tannery material and collagen. The elongation of the leather will increase when the oil is more absorbed [15-16].

The elongation value with the use of palm oil on the milkfish leather was ranged from 57.23% to 70.03%, which was smaller than that of synthetic oil (70.11%). The high elongation value of the skin is due to the large amount of oil that lubricates the surface of the skin fibers so that the skin becomes more flexible and easily indented [8]. Leather elongation is related to the elasticity, this occurs because there is a reduction of elastin in the calcification process and skin protein erosion. Elastin is a protein that forms very elastic fibers because it has amino acid chains that form angles. The angles become straight when the stress is applied and will return to normal when the stress is released. Loss of elastin in the skin proteins can reduce skin elasticity. The elongation of the skin is also affected by the high protein composition of the fiber. The degree of elongation and slackness is also affected by the finishing process such as toggling, milling, and buffing. The elasticity or looseness of leather is due to the structure of the skin tissue, the process of softening, re-tanning, and fatliquoring [17]. The elongation parameter affects the product quality of the leather goods, such as shoes when they are processed with a machine. If the elongation rate is low, then the leather will be crack, otherwise if the elongation rate is high, then the skin will change shape and increase in size [18]. From the results above, it shows that the tilapia leather that was tanned with syntan and mimosa with the addition of fresh and used palm oil cannot be used to make products that require hard leather materials, because the leather that was produced tends to be limp and stretchy.

3.3 Shrinkage temperature

Finally, we observed the shrinkage temperature of the tilapia leather with the addition of 5% fresh palm oil and used palm oil. The shrinkage temperature measurement for tilapia leather with the addition of fresh palm oil and used palm oil was 75.8% and 75.6%, respectively. The addition of fresh palm oil has a similar shrinkage temperature value on tilapia leather to the used palm oil, and the using of those palm oils was not significantly different (Figure 2.c). The standard of shrinkage temperature for tanned stingray skin based on the Indonesian National Standard (SNI) 06-6121-1999 for finished goods is at a minimum of 70°C. The shrinkage temperature test results from all treatments were in accordance with SNI for tilapia leather. This indicates that the concentration of palm oil either fresh or used palm oil showed the enough value to the standard for product.

Leather products with a shrinkage temperature greater than 70°C will be more durable because they have higher hydrothermal resistance [18]. This means that tilapia skin tanned with syntan and mimosa with the addition of fresh and used palm oil is the leather that is resistant to heat temperature. In addition, during the tanning process using synthetic tanning agents and mimosa, it will form a hydrogen bonds and covalent bonds are formed between the tannin acids from mimosa and the amino acids that presents in the collagen of the tilapia leather. The covalent bonds between collagen and tannins play the most important role in the increasing of shrinkage temperature of the leather. These tannins will form a cross-links with collagen by hydrogen and covalent bonds which produce a leather with a high shrinkage temperature [19].

4 Conclusion

It concluded that the addition of palm oil caused the increasing tilapia fish skin quality that can be used for commercial products, and it became the alternative material that could be use in vegetable tannery processing.

This research was funded by The UGM Strategic Research Program, called Final Project Recognition (RTA).

References

1. A. J. Farid, P. H. Riyadi, U. Amalia, Indones. J. Fish. Sci. Technol. **10**, 80-83 (2015)
2. V. Sivakumar, R. P. Prakash, P. G. Rao, B. V. Ramabrahmam, G. Swaminathan, J. Clean. Prod. **16,** 549-553 (2008)
3. C. Wang, T. Li, S. Feng, Asian J. Chem. **24**, 63-67 (2012)
4. E. Kasmudjiastuti, G. Griyanitasari, D. Rahmawati, Sugihartono, Majalah Kulit, Karet, dan Plastik **34**, 19-26 (2018)
5. L. Strijbos, R. Saumweber, M. Hess, C. Gabagnou, J. Fennen, World Leather **20**, 19-21 (2012)
6. W. Kassahun, Preparation of Leather Fatliquor Cum Filler from Fleshing Waste for Retanning Process in Leather Manufacture, Master Thesis, Addis Ababa University, Addis Ababa, Ethiopia (2014)
7. A. D. Covington, *Tanning Chemistry: The Science of Leather* (The Royal Society of Chemistry, Cambridge, UK, 2009)
8. N. Haq, Yunizal, Suherman, *Teknologi Pengawetan dan Penyamakan Kulit Ikan* (Pusat Penelitian dan Pengembangan Eksplorasi Laut dan Perikanan, Jakarta, Indonesia, 2000)
9. E. Purnomo, *Penyamakan Kulit Reptil* (Kanisius, Yogyakarta, Indonesia, 2001)
10. P. A. Sitorus, P. H. Riyadi, E. Susanto, Jurnal Ilmu dan Teknologi Perikanan **2**, 57-64 (2020)
11. M. Gutterres, L. M. D. Santos, J. Soc. Leather Technol. Chem. **93**, 171-175 (2009)
12. R. M. Judoamidjojo, *Dasar Teknologi dan Kimia Kulit* (Institut Pertanian Bogor, Bogor, Indonesia, 1981)
13. S. Mustakim, A. S. Widati, A. P. Kurniawan, Jurnal Ternak Tropika **11**, 38-50 (2010)
14. M. A. Habib, A. G. Alshammari, Indian J. Chem. Technol. **24**, 198-205 (2017)
15. D. Nurdiansyah, eJurnal Mahasiswa Universitas Padjadjaran **1**, 1599 (2012)
16. R. D. Herhady, R. Sukarsono, in Risalah Seminar Ilmiah Aplikasi Isotop dan Radiasi, 12 December 2006, Jakarta, Indonesia, pp. 67-73 (2006)
17. R. A. Budiyanto, Pengaruh Kadar (Cr_2O_3) terhadap Mutu Kulit Ikan Kakap (*Lutjonus* Sp) Tersamak, Undergraduate Thesis, Institut Pertanian Bogor, Bogor, Indonesia (2010)
18. L. Sahubawa, A. Pertiwiningrum, A. T. Pamungkas, Majalah Balai Besar Kulit, Karet, dan Plastik Yogyakarta, **27**, 38-45 (2011)
19. O. Suparno, A. D. Covington, C. S. Evans, *Teknologi Baru Penyamakan Kulit Ramah Lingkungan: Penyamakan Kombinasi Menggunakan Penyamak Nabati, Naftol dan Oksazolidin* (Institut Pertanian Bogor, Bogor, Indonesia, 2012)

Diversity of milk traits in 25 goat breeds

Ferdy Saputra[1,*] and *Anneke* Anggraeni[1]

[1]Indonesian Research Institute for Animal Production, Bogor 16720, Indonesia

Abstract. Goats are livestock that is mostly raised by small farmers in Indonesia because they are easier to raise. Apart from having the potential to become meat, several breeds of which are kept as milk-purpose. Milk traits of each breed differ from one another. Therefore, this study tried to observe genetic differences of 25 goat breeds with statistical approach. Information about milk traits from 25 goat breeds is obtained from published journal. Multidimensional preference analysis and average linkage cluster analysis were performed using SAS 9.4 to determine the differences in goat breeds from three traits, namely milk yield, fat content, and protein content. Multidimensional preference analysis was able to see the advantages of breeds from the three observed traits. Goat breeds with superior milk yields are Saanen, Camosciata delle Alpi, and Charmoisée. Sarda Primitiva, Sarda, Etawah Grade have high fat content in milk. In addition, Arsi-Bale and Somali have high protein content. Average linkage cluster analysis is able to observe the genetic relationship of goat breeds based on three traits. According to average linkage cluster analysis, we found four clusters for goat breeds in this study. With existing statistical approaches, we can evaluate genetic diversity in milk traits.

1 Introduction

Goats are livestock that are widely used as a producer of meat, milk, and skin in the world. Goat milk is a food ingredient that is used for human health. Goat's milk has a niche market because the price is more expensive than cow's milk. Moreover, goat's milk has more benefit than cow's milk in some important nutrients: vitamin A, niacin, choline, and inositol [1]. The quality of milk is highly dependent on the type of feed used in dairy farming [2]. The quality of milk can be determined from value of protein and fat. Furthermore, goat milk has been an important food of human nutrition, because of the greater similarity of goat milk to human milk, softer curd formation, higher proportion of small milk fat globules, and different allergenic properties compared with cow milk [3].

In the world many goat breeds are developed for high milk production or high fat content depending on the use in making dairy products. Crosses with superior breeds are mostly done to increase the productivity of local goats that are more adaptive to the environment. In the world, Saanen is a popular breed known as a goat breed with high productivity, but other breeds need to be explored, especially for the milk trait. Saanen had higher milk yield compared to Alpine and Nubian for 305 days of lactation [4]. Among the Mediterranean goats, Murciano-Granadina goats had the highest milk yield, fat, and

* Corresponding author: ferdy44saputra@gmail.com

protein contents compared to Maltese, Sarda, and Sarda Primitiva [5]. In addition, there are many dairy goat breeds that have not been explored properly to create crossbreeding programs with local goats. Therefore, this study was conducted to explore the diversity of milk traits, especially milk production, fat content, and protein content.

2 Materials and methods

This research used the data of 25 breeds from nine published journal (Table 1). The observed variables were milk production, fat content, and protein content. All unit measures are equated in this study.

Table 1. Dataset of milk traits in this study

No	Breed	Milk Yield (kg/d)	Fat Content (%)	Protein Content (%)	Reference
1	Alpine	2.08	3.47	3.08	[6]
2	Anglo-Nubian	0.90	3.71	3.29	[7]
3	Arsi-Bale	1.13	5.15	4.80	[8]
4	Barbari	0.59	4.66	4.04	[9]
5	Boer	1.41	4.70	4.05	[8]
6	Camosciata delle Alpi	2.57	4.07	3.50	[5]
7	Canaria (Canary)	0.79	3.96	3.72	[7]
8	Charmoisée	2.55	3.40	2.84	[7]
9	Cross (Toggenburg x Arsi-Bale)	0.93	3.65	4.08	[8]
10	Damascus	1.88	4.46	3.82	[7]
11	Etawah Grade	0.46	5.57	4.19	[10]
12	Girgentana	0.70	3.93	3.48	[7]
13	Jamunapari	0.87	4.81	3.75	[9]
14	Malagueña	1.47	5.49	3.40	[7]
15	Maltese	1.83	4.05	3.35	[5]
16	Murciano-Granadina	2.18	4.48	3.46	[5]
17	Nordic	1.92	4.28	2.87	[7]
18	Polish White	1.38	3.38	3.12	[8]
19	Saanen-Croatia	2.63	3.25	3.01	[6]
20	Saanen-Italy	2.79	3.8	3.40	[5]
21	Sapera	1.55	6.00	2.87	[11]
22	Sarda	0.95	5.32	3.98	[5]
23	Sarda Primitiva	0.87	5.83	4.09	[5]
24	Somali	0.85	4.90	4.34	[8]
25	Toggenburg	1.82	3.37	2.96	[7]

Multidimensional preference analysis (MDPREF) and average linkage cluster analysis were performed using SAS 9.4 with PROC PRINQUAL and PROC CLUSTER, respectively. Multidimensional preference analysis is a principal component analysis which aims to project the high-dimensional ranking data into biplot. average linkage cluster analysis is a group analysis method that construct a hierarchical group of data by calculating the distance between two cluster which is referred to as the average distance where the distance is calculated on each cluster. Both of these analyzes to compare the diversity of the milk traits.

3 Results and discussion

Based on multidimensional preference analysis, milk yield, protein content, and fat content have no correlation shown in the opposite direction (Fig. 1). Saanen from Italy, La Mancha, Saanen from Croatia, Camosciata delle Alpi, Charmoisée, Murciano-Granadina, and Alpine are dairy goat breeds that have high milk production. In Indonesia, crosses of Saanen and Etawah Grade (Sapera) are carried out to obtain a breed with high milk production with tropical climate resistance [12]. In Brazil, Saanen and Anglo-Nubian breeds were crossed in an effort to strengthen the rusticity [13]. Sapera has a higher production than Etawah Grade [10, 11] and has a higher fat content than Saanen [6, 10, 12]. Sapera, Sarda Primitiva, and Etawah Grade have a higher fat content among other breeds.

According to average linkage cluster analysis, we found four cluster 1 (Sapera and Malagueña), cluster 2 (Sarda, Sarda Primitiva, Etawah Grade, Boer, Somali, Jamunapari, Barbari, Arsi-Bale), cluster 3 (Cross, Girgentana, Canaria, Anglo-Nubian), and cluster 4 (Nordic, Maltese, Murciano-Granadina, Damascus, Saanen-Croatia, Charmoisée, Saanen-Italy, Camosciata delle Alpi, Polish White, Toggenburg, Alpine) (Fig. 2). Cluster 4 is breeds with high milk yield. On the other hand, cluster 2 is breeds with high protein content. Ruminant milk contained 95 % of the proteins that are synthesized from six structural genes that encode well characterized proteins, the whey proteins, and the caseins [14]. Cluster 1 has high fat content. On the other hand, cluster 3 has moderate fat and protein content. This findings found high diversity of milk traits in 25 breeds.

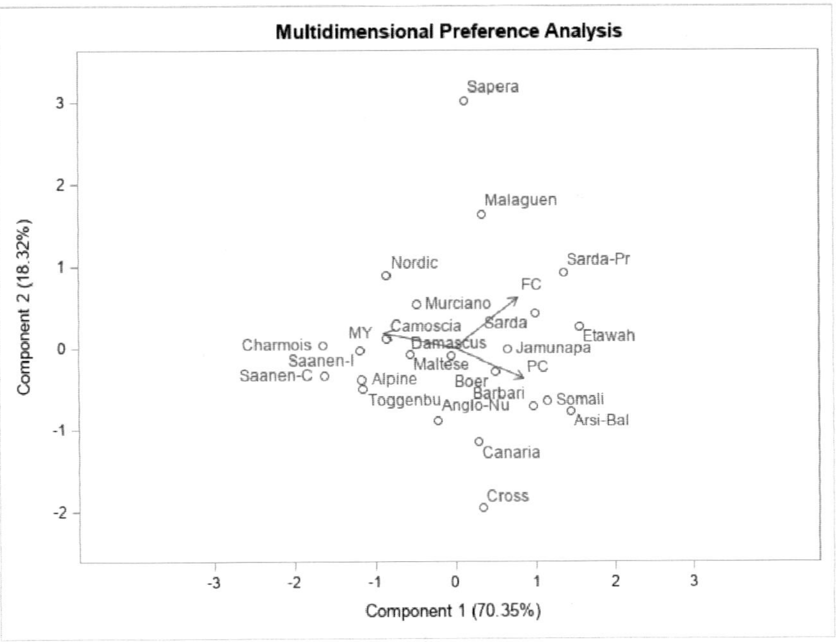

Fig. 1. Multidimentional preference analysis of 25 goat breeds based on milk yield (MY), fat content (FC), and protein content (PC).

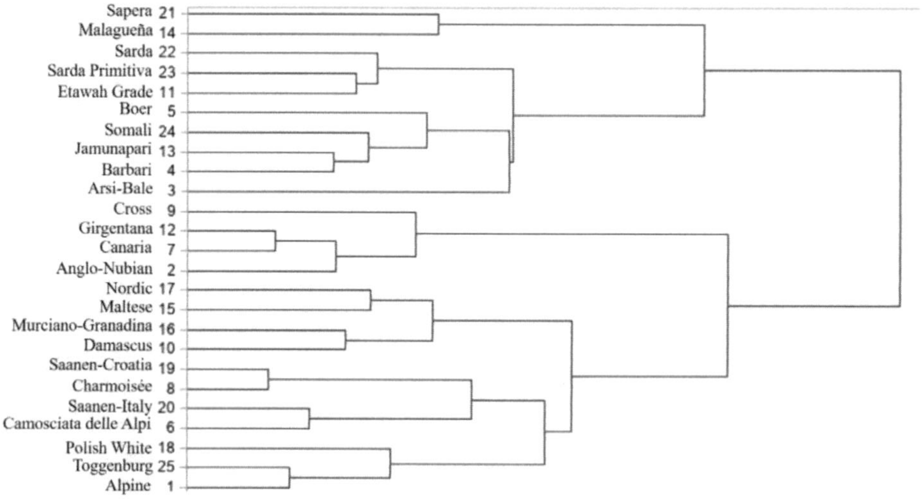

Fig. 2. Average linkage cluster analysis of 25 goat breeds.

4 Conclusion

The dairy goats in this study have various milk traits. The milk traits do not have a positive correlation with other properties. Based on this research, the dairy goat crossing program can increase productivity such as Sapera.

References

1. L. Cyrilla, B. P. Purwanto, A. Atabany, D. A. Astuti, A. Sukmawati, Media Peternakan **28**, 204-211 (2015)
2. Arief, Rusdimansyah, S. Sowmen, R. Pazla, Rizqan, Biodiversitas **21**, 4004-4009 (2020)
3. S. Clark, M. B. M. García, J. Dairy. Sci. **100**, 10026-10044 (2017)
4. A. S. Shuvarikov, O. N. Pastukh, E. V. Zhukova, O. A. Zheltova, IOP Conf. Ser. Earth Environ. Sci. **640**, 032031 (2021)
5. G. M. Vacca, G. Stocco, M. L. Dettori, E. Pira, G. Bittante, M. Pazzola, J. Dairy. Sci. **101**, 7236-7247 (2018)
6. B. Mioč, Z. Prpić, I. Vnučec, Z. Barać, V. Sušić, D. Samaržija, V. Pavić, Mljekarstvo, **58**, 305-313 (2008)
7. M. M. Ferro, L. O. Tedeschi, A. S. Atzori, Transl. Anim. Sci. **1**, 498-506 (2017)
8. T. A. Mestawet, A. Girma, T. Ådnøy, T. G. Devold, J. A. Narvhus, G. E. Vegarud, Small Rumin. Res. **105**, 176–181 (2012)
9. S. N. Kala, B. Prakash, Small Rumin. Res. **3**, 475-484 (1990)
10. R. Rosartio, Y. Suranindyah, S. Bintara, Ismaya, Buletin Peternakan **39**, 180-188 (2015)
11. Supriyati, R. Krisnan, I. G. M. Budiarsana, L. Praharani, Jurnal Ilmu Peternakan dan Veteriner **21**, 88-95 (2016)

12. A. Anggraeni, F. Saputra, A. Hafid, A. B. L. Ishak, Jurnal Ilmu Peternakan dan Veteriner **25**, 48-59 (2020)
13. G. M. Silvia, R. Rossetto, M. G. F. Salles, M. F. S. Sá, A. C. J. S. Rosa e Silva, A. P. R. Rodrigues, C. C Rodrigues, C. C. Campello, J. R. Figueiredo, Anim. Reprod. **11**, 56-60 (2014)
14. M. Selvaggi, V. Laudadio, C. Dario, V. Tufarelli, Mol. Biol. Rep. **41**, 1035-1048 (2014)

Monomorphic allele rs109231213 in 3'UTR PLAG1 gene in purebred of Bali cattle (*Bos javanicus*)

Sutikno Sutikno[1,*], *Jean Pierre* Munyaneza[2], *Santiananda Arta* Asmarasari[3], and *Jakaria* Jakaria[4]

[1]Research Center for Biology, National Research and Innovation Agency (BRIN), Bogor 16911, Indonesia
[2]Division of Animal and Dairy Science, Department of Animal Science, College of Agriculture and Life Sciences, Chungnam National University, Daejeon 34134, Korea
[3]Indonesian Research Institute for Animal Production, Bogor 16720, Indonesia
[4]Faculty of Animal Science, Bogor Agricultural University, Bogor 16680, Indonesia

Abstract. The mutation rs109231213 that is located in 3'UTR of PLAG1 gene is associated with the growth and body weight in several *Bos taurus* and *Bos indicus* breeds. This study aimed to identify SNP rs109231213 in Bali cattle (*Bos javanicus*). The study used 41 samples of Bali cattle. The PLAG1 gene polymorphism was analyzed using PCR and direct sequencing methods. PCR pimers were 5'- TTGCACAGAATCAGTGTGTC-3' and 5'-AGCCTAACGTGGATCTATGG-3'. The results showed that primers successfully amplified the 331 bp fragment at annealing 60^0C that contained rs109231213. SNP was monomorphic in Bali cattle with one allele (G). This study concludes that rs109231213 in 3'UTR of PLAG1 gene can be used as specific marker in purebred of Bali cattle that have never been crossed with *Bos taurus* and *Bos indicus*.

1 Introduction

Bali cattle (*Bos javanicus*) are indigenous beef cattle in Indonesia that were domesticated from Banteng and are most preferred among small farmers [1-2]. Bali cattle are classified as beef cattle [3] with higher carcass percentages (42.1-51.1%) [4], well adapted to tropical climates, good adaptations to intensive and extensive maintenance, and good reproductive ability [5]. The genetic diversity of Bali cattle is one of the important information in evaluating the genetic potential of livestock for the sake of conservation, sustainable use, and be developed into premium beef cattle.

Pleomorphic adenoma gene 1 (PLAG1) is associated with birth weight in Nellore and PO cattle [6-8], also with body size in Jersey, Friesian holstein, Simmental, Red Angus, Angus, Hereford, Limousin, Gelbvieh, Brangus, Maine Anjou cattle, Shorthorn, Charolais cattle [9-11] futhermore reproductive traits in *Bos indicus* and *Bos taurus* cattle [12-14]. Single nucleotide polymorphisms (SNPs) rs109231213 (chr 14:g.23.326.588, C>G forward strand

[*] Corresponding author: sutiknoabdulkadir@gmail.com

based on ensembl) located on the 3'UTR of PLAG1 gene associated with height [15] and body weight in cattle [16]. Thus, the aim of this research was to identify rs109231213 polymorphism in Bali cattle.

2 Materials and methods

2.1 DNA samples

Total of DNA samples were used in this research are forty one. The sample of Bali cattle from the Superior Livestock Breeding and Livestock Forage Center of Denpasar (BPTU-HPT Denpasar). DNA was obtained using the geneaid DNA extraction kit.

2.2 PCR amplification and genotyping

Primers were designed to amplify 331 base pairs fragment (Table 1). The SNP was accessed through ensembl (http://asia.ensembl.org/, SNP rs109231213, chromosome 14 : 23.326.588 C>G and gene sequence PLAG1 (access no: : ENSBTAG00000004022). This set of primers was made with software MEGA7 then evaluated by primer stats (www.bioinformatics.org) to test their suitability.

Table 1. A pair of primer to amplify 3'UTR of PLAG1 gene

Primer	Sequence (5'- 3')	Length	Ta (°C)
Forward	F: TTGCACAGAATCAGTGTGTC	331 bp	60
Reverse	R: AGCCTAACGTGGATCTATGG		

GenBank access No. NC_037341 fragment position PCR : 23.326.380-23.375.751 (331 bp) SNP position rs109231213 (chr 14: 23.326.588, C>G) ensembl rs109231213

The PCR mix volume was 25 μL, containing of 1 μL DNA template (50 ng), 0.3 μL primer forward and reverse (5 pmol), 12.5 μl GoTaq® Green Master Mix (1x Promega), and 10.9 μL nuclease-free water. PCR mix run using Applied Biosystem thermal cycler, with condition pre-denaturation was at 95° C for 5 min, and amplifiction for 35 cycles (denaturation at 95° C for 10 s, annealing at 60° C for 20 s, extension at 72° C for 30 s, and final elongation at 72° C for 5 min. The PCR products visualized by electrophoresis using 2% agarose gel and sequenced by First BASE Laboratories Selangor, Malaysia.

2.3 Data analysis

The frequencies genotype and allele were calculated by software Popgene 3.2 [17].

3 Results and discussion

PLAG1 gene located on 5'UTR of Bali cattle was 100% successfully amplified at 60°C of annealing temperature (Fig. 1). The length of PCR product was 331 bp. Sequencing of PLAG1 gene located on 5'UTR was monomorphic in Bali cattle, only one allele C was found (Fig. 2).

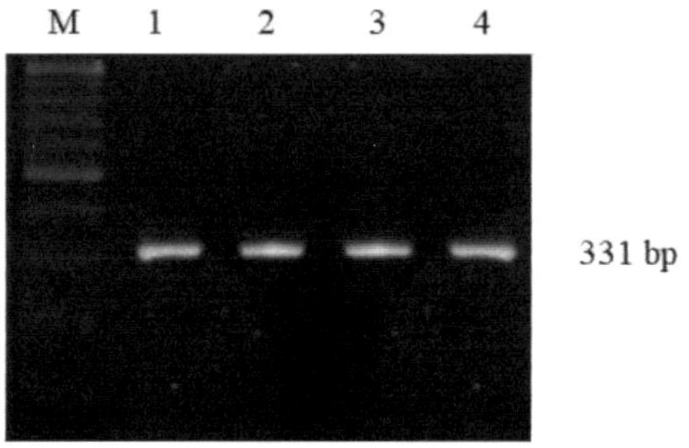

Fig. 1. Visualization of PCR product, M was marker 100 bp, line 1,2,3, 4 were analyzed samples.

```
                                10        20        30        40        50        60        70
NC_037341.1:2332638   TTGCACAGAATCAGTGTGTCTTTTCATAAGGTTTACAATGGTTTATTTTTTTTTACAAGGTTTACGTGTC
Sapi Bali             ......................................................................
Sapi Bali 2           ......................................................................
Sapi Bali 3           ......................................................................

                                80        90       100       110       120       130       140
NC_037341.1:2332638   TCAAAACACACTGTCTTCCCAGTATGTAAGTTAAAATACAAAAAAAAATATTGACCCAAGTTGCTTCCAG
Sapi Bali             ....................................................C.................
Sapi Bali 2           ....................................................C.................
Sapi Bali 3           ....................................................C.................

                               150       160       170       180       190       200       210
NC_037341.1:2332638   CCTACTGAGATCCAGGCGACATTGGAGGAGATCTTTTAGACGTCATGTATTCTTCATTAGCAATGGCAGA
Sapi Bali             ......................................................................
Sapi Bali 2           ......................................................................
Sapi Bali 3           ......................................................................

                               220       230       240       250       260       270       280
NC_037341.1:2332638   CTGTTACCTCTGAGAAGTGAGTGCCTACGTTGAGCCCTTGTCTTATTTTTCTCACGCTGTGTCAGCAGCC
Sapi Bali             ......................................................................
Sapi Bali 2           ......................................................................
Sapi Bali 3           ......................................................................

                               290       300       310       320       330
NC_037341.1:2332638   ACATCAATCACACAGATGGTCTGTGGTGATCCCATAGATCCACGTTAGGCT
Sapi Bali             ...................................................
Sapi Bali 2           ...................................................
Sapi Bali 3           ...................................................
```

Fig. 2. Visualisazion of the 3'UTR sequence alignment of the PLAG1 gene in Bali cattle

The rs109231213 was polymorphic in *Bos taurus*, *Bos indicus* and Tropical Composite cattle with frequency C allele was 0.96, 0.52, 0.68 and G allele 0.04, 0.48, 0,32, respectively. The C allele increased hip height, weight, net food intake, age at puberty, fat depth and decreased IGF-I concentration in blood [9-11,18]. The results show that Bali cattle have genetic potential to be developed into premium beef cattle based on this SNP. Futhermore, PLAG1 became as a key regulator of mammalian growth [6-8].

Previously study of rs109231213 reported that G is ancestral allele and C is derived allele. The C allele was introgressed from *Bos taurus* to Brahman cattle during the grading up of Brahman cattle [19]. Introgression is the transfer of genetic information from *Bos taurus* to Brahman cattle as a result of hybridization between them and repeated backcrossing. In this study Bali cattle did not have the G allele. The results show that Bali cattle have never been crossed with *Bos taurus* (i.e. Hereford, Angus, Shorthorn) and *Bos indicus* (i.e. Brahman, PO).

4 Conclusion

SNP rs109231213 is monomorphic in Bali cattle (C allele). Bali cattle have genetic potential to be developed into premium beef cattle based on this SNP. Furthermore, providing information that Bali cattle have never been crossed with Bos indicus, Bos taurus, and composite cattle.

References

1. P. Gardner, S. Hedges, S. Pudyatmoko, T. N. E. Gray, R. J. Timmins, The IUCN Red List of Threatened Species **2016**: e.T2888A46362970 (2016)
2. H. Martojo, Reprod. Dom. Anim. **47**, 10-14 (2012)
3. J. Jakaria, H. Khasanah, R. Priyanto, M. Baihaqi, M. F. Ulum, J. Indones. Trop. Anim. Agric. 42, 59-65 (2017)
4. H. Hafid, Nuraini, Inderawati, W. Kurniawan, IOP Conf. Ser. Earth Environ. Sci. **119**, 012043 (2018)
5. Sutarno, A. D. Setyawan, Biodiversitas 16, 327-354 (2015)
6. H. Hartati, Y. T. Utsunomiya, T. S. Sonstegard, J. F. Garcia, J. Jakaria, M. Muladno, BMC Genet. **16**, 75 (2015)
7. Y. T. Utsunomiya, A. S. do Carmo, R. Carvalheiro, H. H. R. Neves, M. C. Matos, L. B. Zavarez, A. M. P. O'Brien, J. Sölkner, J. C. Mc Ewan, J. B. Cole, C. P. Van Tassell, F. S. Schenkel, M. V. G. B. da Silva, L. R. P. Neto, T. S. Sonstegard, J. F. Garcia, BMC Genet. **14**, 52 (2013)
8. M. D. Littlejohn, T. Grala, K. Sanders, C. Walker, G. Waghorn, K. MacDonald, W. Coppieters, M. Georges, R. Spelman, E. Hillerton, S. Davis, R. Snell, Anim. Genet. **43**, 591-594 (2012)
9. L. Karim, H. Takeda, L. Lin, T. Druet, J. A. C. Arias, D. Baurain, N. Cambisano, S. R. Davis, F. Farnir, B. Grisart, B. L. Harris, M. D. Keehan, M. D. Littlejohn, R. J. Spelman, M. Georges, W. Coppieters, Nat. Genet. **43**, 405-413 (2011)
10. M. Saatchi, R. D. Schnabel, J. F. Taylor, D. J. Garrick, BMC Genom. **15**, 442 (2014)
11. M. R. S. Fortes, K. Kemper, S. Sasazaki, A. Reverter, J. E. Pryce, W. Barendse, R. Bunch, R. McCulloch, B. Harrison, S. Bolormaa, Y. D. Zhang, R. J. Hawken, M. E. Goddard, S. A. Lehnert, Anim. Genet. **44**, 636-647 (2013)
12. Y. T. Utsunomiya, A. S. Carmo, H. H. R. Neves, R. Carvalheiro, M. C. Matos, L. B. Zavarez, P. K. R. K. Ito, A. M. P. O'Brien, J. S. lkner, L. R. Porto-Neto, F. S. Schenkel, J. McEwan, J. B. Cole, M. V. G. B. da Silva, C. P. V. Tassell, T. S. Sonstegard, J. F. Garcia, Plos One **9**, e88561 (2014)
13. M. R. S. Fortes, A. Reverter, M. Kelly, R. McCulloch, S. A. Lehnert, Andrology **1**, 644-650 (2013)
14. M. R. S. Fortes, S. A. Lehnert, S. Bolormaa, C. Reich, G. Fordyce, N. J. Corbet, V. Whan, R. J. Hawken, A. Reverter, Anim. Prod. Sci. **52**, 143-150 (2012)
15. J. L. Zhong, J. W. Xua, J. Wanga, Y. Wena, H. Niub, L. Zheng, H. Hea, K. Penga, P. Hea, S. Shia, Y. Huang, C. Leia, R. Danga, X. Lana, X. Qid, H. Chena, Y. Huang, Gene **689**, 166-171 (2019)

16. Y. T. Utsunomiya, M. Milanesi, A. T. H. Utsunomiya, R. B. P. Torrecilha, E. S. Kim, M. S. Costa, T. S. Aguiar, S. Schroeder, A. S. do Carmo, R. Carvalheiro, H. H. R. Neves, R. C. M. Padula, T. S. Sussai, L. B. Zavarez, R. S. Cipriano, M. M. T. Caminhas, G. Hambrecht, L. Colli, E. Eufemi, P. Ajmone-Marsan, D. Cesana, M. Sannazaro, M. Buora, M. Morgante, G. Liu, D. Bickhart, C. P. Van Tassell, J. Sölkner, T. S. Sonstegard, J. F. Garcia, Sci. Rep. **7**, 17140 (2017)
17. F. C. Yeh, R. C. Yang, T. Boyle, *Popgene Version 1.31: Microsoft Window-Based Freeware for Population Genetics Analysis* (University of Alberta, Edmonton, Canada, 1999)
18. M. R. S. Fortes, K. Kemper, S. Sasazaki, A. Reverter, J. E. Pryce, W. Barendse, R. Bunch, R. McCulloch, B. Harrison, S. Bolormaa, Y. D. Zhang, R. J. Hawken, M. E. Goddard, S. A. Lehnert, Anim. Genet. **44**, 636-647 (2013)
19. L. Karim, H. Takeda, L. Lin, T. Druet, J. A. C. Arias, D. Baurain, N. Cambisano, S. R. Davis, F. Farnir, B. Grisart, B. L. Harris, M. D. Keehan, M. D. Littejohn, R. J. Spelman, M. Georges, W. Coppieters, Nat. Genet. **43**, 405-413 (2011)

Identification of 19-bp indel of the Pleomorphic Adenoma Gene 1 in Bali cattle population

Sukaryo Sukaryo[1], *Rizka* Augustin[1], *Yuli* Yanti[1], *Joko* Riyanto[1], *Slamet Diah* Volkandari[2], *Pita* Sudrajad[3], and *Muhammad* Cahyadi[1*]

[1]Department of Animal Science, Sebelas Maret University, Surakarta 57126, Indonesia
[2]Research Center for Biotechnology, Research Organization for Life Sciences, National Research and Innovation Agency (BRIN), Cibinong 16911, Indonesia
[3]Assessment Institute of Agricultural Technology-Central Java, Semarang 50552, Indonesia

Abstract. Pleomorphic adenoma gene 1 (PLAG1) is a zinc finger transcription factor gene located on bovine chromosome 14 (BTA14) affecting body size and reproduction traits in cattle. The objective of this study was to identify 19-bp indel of the PLAG1 gene in Bali cattle population. A total of 96 blood samples of Bali Cattle were collected from Balai Pembibitan Ternak Unggul dan Hijauan Pakan Ternak (BPTU-HPT) Denpasar. Genomic DNA was extracted from blood samples and used to detect 19-bp indel of the PLAG1 gene using following primer pair 5'-TCCGAACAACAGGTGAGGGAGAAAT-3' and 5'-CCACTTCAGG-GGTGCTCTAGGTTTG-3'. The polymerase chain reaction (PCR) products using DNA pool samples were sequenced to validate the PCR product and to find out novel polymorphism in Bali cattle population. The result showed that there was no variation found in Bali cattle population based on 19-bp indel of the PLAG1 gene, which is indicated by 123 bp DNA band. However, sequence analysis of the PLAG1 gene resulted in a novel single nucleotide polymorphism (SNP) at nucleotide number 32235 of the PLAG1 gene that changed guanine (G) to adenine (A). This novel SNP could be furthermore genotyped and it might be a potential candidate marker for body size and reproduction traits in Bali cattle.

1 Introduction

Cattle is one of livestock commodity producing meat who are considerable contributing to national beef supply. In 2019, beef production reaches 504,802.29 tons which contributes around 10.32% of total meat production in Indonesia. Beef is the second highest meat commodity after broiler in fulfilling meat national needs [1]. Cattle plays an important role in national meat supply. Therefore, it is very important to improve productivity and population of Indonesian local cattle, especially whose genetically superior such as Bali cattle *(Bos javanicus)*.

Bali cattle is Indonesian native cattle breed which is hypothesized to be originated from domestication of wild Banteng (*Bibos banteng syn Bos sondaicus*) long time ago. Some

* Corresponding author: mcahyadi@staff.uns.ac.id

advantages of Bali cattle are more adaptable with Indonesian tropical environment, high efficiency of production and reproduction such as high fertility (80-85%), calving interval around 12 to 14 months, and high carcass percentage. Moreover, sexual maturity of Bali cattle is started at the age of 18 months, the estrus cycle in heifer ranges from 16 to 23 days, its estrus time is 36 to 48 hours with 18 to 27 hours of fertile period, and Bali cattle's fertility is much higher than *Bos taurus* cattle [2]. In addition, body weight and measurement traits are the easiest growth traits to be observed consisting of birth weight, actual body weight, withers height, body length, chest girth, waist height, and chest depth [3]. Those traits are well-known as quantitative traits which are economically and biologically important traits and able to be used as indicators in selecting superior Bali cattle [4]. The development of recent molecular technique allows human being to improve native cattle production genetically through marker assisted selection.

A candidate gene well-known involving in cattle growth is Pleomorphic adenoma gene 1 abbreviated PLAG1. It is also known as a gene controlling growth and body size of cattle due to directly affecting *Insulin-like growth factor* 1 (IGF1) serum [5-6]. Previous study reported that a 19-bp indel namely rs523753416 located in intron 1 of the PLAG1 gene is associated with growth and body measurement traits of Chinese cattle population [7]. It is also associated with birth weight of Indonesian Peranakan Ongole cattle [8]. In addition, PLAG1 gene affects carcass traits of Nellore cattle [9]. The PLAG1 gene is not well-studied in Indonesian cattle. Therefore, the objective of this study was to identify 19-bp indel of the PLAG1 gene in Bali cattle population.

2 Materials and methods

2.1 Bali cattle population and DNA extraction

A total of 96 Bali cattle were used in this study. They were 2 to 3 years old which are originated from BPTU-HPT Denpasar, the Province of Bali. The blood samples were collected from *vena jugularis* by using 18G vacutainer needle and 3 ml tube containing EDTA. Those blood samples were then used to extract genomic DNA based on high concentrated salt method described by Montgomery and Sise [10]. Moreover, extracted genomic DNA was quantified to evaluate DNA concentration and purity using NanoPhotometer (P-Class®, Implen, Munchen, Jerman). The genomic DNA was stored at -20°C until used.

2.2 Amplification of the PLAG1 gene and sequencing analysis

Amplification of the PLAG1 gene was conducted using primer pairs previously reported by Xu et al. [7] as follows: 5'-TCCGAACAACAGGTGAGGGAGAAAT-3' as forward primer and 5'- CCACTTCAGGGGTGCTCTAGGTTTG-3' as reverse primer. A polymerase chain reaction (PCR) was carried out in total volume of 25 μL containing 12.5μL Go Taq® Green Master Mix (Promega, Madison, USA), 9.5 μL nuclease free water, 1 μL each primer, and 1 μL genomic DNA template. The PCR was initiated by pre-denaturation at 95°C for 5 minutes and followed by 35 cycles of denaturation at 95°C for 30 seconds, annealing at 60°C for 30 seconds, and extension at 72°C for 30 seconds. Final extension was applied at 72°C for 10 minutes to complete the reaction. Furthermore, the PCR product was checked on 2% agarose gels stained by ethidium bromide using submarine electrophoresis system (Mufid ex, Advance, Japan) at 110 volt for 35 minutes. Then, it was visualized on the Glite UV Gel Doc System® (Pacific Image Electronic Co., Ltd., New Taipei City, Taiwan). In addition, three

DNA pool samples containing three different individual genomic DNA was sequenced using 1st BASE sequencing service (APICAL SCIENTIFIC Laboratory, Selangor, Malaysia).

2.3 Data analysis

The PLAG1 gene sequence obtained in this study was analyzed using BioEdit Sequence Aligment Editor Version 7.2.5 to extract and to get clear the PLAG1 gene sequences. To identify 19-bp indel and other polymorphism within the gene, all sequences were aligned using Clustal Omega software which is accessible online (https://www.ebi.ac.uk/Tools/-msa/clustalo/).

3 Results and discussion

3.1 Amplification of 19-bp indel of the PLAG1 gene

Amplification of the PLAG1 gene fragment in Bali cattle population resulted 123 bp PCR product (Figure 1). This result was in line with those predicted for detection 19-bp indel of the PLAG1 gene. It showed that the DNA isolated from Bali cattle blood was adequate to be used for PCR. The success of DNA amplification depends on the accuracy of the primers used to anneal with the right site of the DNA template [11]. In the PCR, the function of oligonucleotide primer is to hybridize with the DNA template, to define the specific fragment of DNA to be amplified, and also at the same time to provide a hydroxy group (-OH) at the 3' end which is needed for the DNA extension [12].

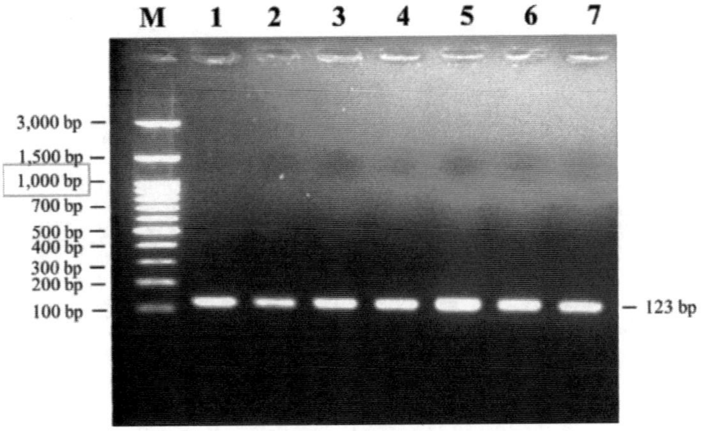

Fig. 1. The PCR product of 19-bp indel of the PLAG1 gene. M is 100 bp marker ladder; 1-7 are individual Bali cattle sample with DD genotype.

The PLAG 1 gene is located on chromosome 14 in cattle and has 4 exons [13]. Identification of 19-bp indel of the PLAG 1 gene in Bali cattle found only one genotype (DD) indicated by 123 bp DNA band (Figure 1). In identification of 19-bp indel in PLAG1 gene, the expected PCR products are 142 bp DNA band for homozygous wild type (WW), 142 and 123 bp for heterozygous (WD), and 123 bp for homozygous deletion (DD) [7]. Bali cattle genotype which tends to be less varied (monomorphic) is due to the phenomenon of phenotypic plasticity as a result of natural selection [14]. In addition, the ability of an individual to display more than one morphology is a response to environmental changes which is known as phenotypic plasticity [15].

3.2 Analysis of the PLAG1 gene sequence

Sequence alignment analysis of the PLAG1 gene fragment identified an intronic single nucleotide polymorphism (SNP) located at nucleotide number 32235 of the PLAG1 gene. This SNP was a kind of substitution mutation that changed Guanine (G) to Adenine (A) (Figure 2A and 2B). The 19-bp indel did not vary since only DD variant found which were proved by identical PCR products size for all samples used in this study (Figure 1) and identical nucleotide sequences based on alignment analysis (Figure 2C). Previous study reported that there are three variants for 19-bp indel location [7]. Moreover, polymorphisms in the PLAG1 gene are associated with body size and growth traits of beef cattle [7,16-18]. Therefore, the novel SNP located at nucleotide number 32235 of the PLAG1 gene could be promising marker to be genotyped in Bali cattle since PLAG1 gene is well-studied gene affecting cattle growth. Unfortunately, there was no restriction enzyme site for this novel SNP, direct sequencing could be the right approach to genotype this SNP.

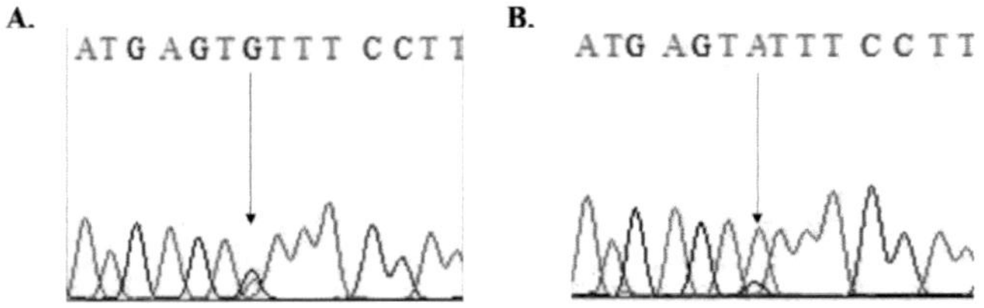

C.

```
AC_000171.1   TCCGAACAACAGGTGAGGGAGAAATCCCGACAGATGAGTGTTTCCTTGTGGGCTGTGTGT  60
B12           TCCGAACAACAGGTGAGGGAGAAATCCCGACAGATGAGTATTT-----------------  43
B22           TCCGAACAACAGGTGAGGGAGAAATCCCGACAGATGAGTGTTT-----------------  43
B32           TCCGAACAACAGGTGAGGGAGAAATCCCGACAGATGAGTGTTT-----------------  43
              ***************************************  ***

AC_000171.1   TCCCTTGTGGGCTGTGTGTTCTCGGGGAGGGGAGGTGTGTGGGTTGAGGACGGCGTGCAA  120
B12           --CCTTGTGGGCTGTGTGTTCTCGGGGAGGGGAGGTGTGTGGGTTGAGGACGGCGTGCAA  101
B22           --CCTTGTGGGCTGTGTGTTCTCGGGGAGGGGAGGTGTGTGGGTTGAGGACGGCGTGCAA  101
B32           --CCTTGTGGGCTGTGTGTTCTCGGGGAGGGGAGGTGTGTGGGTTGAGGACGGCGTGCAA  101
                **********************************************************

AC_000171.1   ACCTAGAGCACCCCTGAAGTGG      142
B12           ACCTAGAGCACCCCTGAAGTGG      123
B22           ACCTAGAGCACCCCTGAAGTGG      123
B32           ACCTAGAGCACCCCTGAAGTGG      123
              **********************
```

Fig. 2. A novel SNP of the PLAG1 gene fragment identified in this study. A. G variant at nucleotide number 32235 of the PLAG1 gene, B. A variant at nucleotide number 32235 of the PLAG1 gene, C. Alignment of PLAG1 sequences.

4 Conclusion

The 19-bp indel of the PLAG1 gene did not vary where only DD genotype identified in Bali cattle population. However, a novel SNP located at nucleotide number 32235 of the PLAG1 could be promising genetic marker to be genotyped for association study.

This study was fully funded by Ministry of Research Technology and Higher Education through National Competitive Basic Research scheme with contract No. 11/E1/KP.PTNBH/ 2021.

References

1. Direktorat Jendral Peternakan, Direktorat Jenderal Peternakan dan Kesehatan Hewan Kementrian Pertanian, (Direktorat Jenderal Peternakan dan Kesehatan Hewan, Jakarta, 2020)
2. N. M. A. G. R. Astiti, *Sapi Bali dan Pemasarannya* (Warmadewa University Press, Bali, Indonesia, 2018)
3. S. D. Volkandari, P. Sudrajad, D. Prasetyo, Subiharta, A. Prasetyo, J. Pujianto, M. Cahyadi, in Prosiding Seminar Nasional Kesiapan Sumber Daya Pertanian dan Inovasi Spesifik Lokasi Memasuki Era Industri 4.0, 9 October 2019, Semarang, Indonesia, pp. 547-551 (2019)
4. E. Handiwirawan, Subandriyo, Wartazoa, **14**, 107-115 (2004)
5. L. Karim, H. Takeda, L. Lin, T. Druet, J. A. C. Arias, D. Baurain, N. Cambisano, S. R. Davis, F. Farnir, B. Grisart, Nat. Genet. **43**, 405-413 (2011)
6. M. R. S. Fortes, K. Kemper, S. Sasazaki, A. Reverter, J. E. Pryce, W. Barendse, R. Bunch, R. McCulloch, B. Harrison, S. Bolormaa, Y. D. Zhang, R. J. Hawken, M. E. Goddard, S. A. Lehnert, Anim. Genet. **44**, 636-647 (2013)
7. W. Xu, H. He, L. Zheng, J. W. Xu, C. Z. Lei, G. M. Zhang, R. H. Dang, H. Niu, X. L. Qi, Gene **675**, 144-149 (2018)
8. H. Hartati, Y. T. Utsunomiya, T. Sonstegard, J. F. Garcia, BMC Genet. **16**, 75 (2015)
9. A. G. T. Pereira, Y. T. Utsunomiya, M. Milanesi, R. B. P. Torrecilha, A. S. Carmo, H. R. Neves, R. Carvalheiro, M. A. Paolo, T. S. Sonstegard, J. Solkner, C. J. C. Castilo, J. F. Garcia, PLoS One **11**, e0158165 (2016)
10. G. W. Montgomery, J. A. Sise, New Zealand J. Agric. Res. **33**, 437-441 (1990)
11. T. Diss, in *Molecular Biology in Cellular Pathology*, edited by J. Crocker, P. G. Murray (John Willey and Sons, West Sussex, United Kingdom, 2003) pp. 193-212
12. D. Handoyo, A. Rudiretna, Unitas **9**, 17-29 (2001)
13. F. Van Dyck, J. Declereq, C. V. Braem, W. J. M. Van de Ven, Int. J. Oncol. **30**, 765-774 (2007)
14. Zulkharnaim, Jakaria, R. R. Noor, Media Peternakan **33**, 81-87 (2010)
15. R. R. Noor, *Genetika Ekologi Laboratorium Pemuliaan dan Genetika Ternak* (IPB University, Bogor, Indonesia, 2002)
16. F. Zhao, S. Mcparland, F. Kearney, L. Du, D. P Berry, Genet. Sel. Evol. **47**, 49 (2015)
17. Y. T. Utsunomiya, M. Milanesi, A. T. H. Utsunomiya, R. B. P. Torrecilha, E. S. Kim, M. S. Costa, T. S. Aguiar, S. Schroeder, A. S. do Carmo, R. Carvalheiro, H. H. R. Neves, R. C. M. Padula, T. S. Sussai, L. B. Zavarez, R. S. Cipriano, M. M. T. Caminhas, G. Hambrecht, L. Colli, E. Eufemi, P. Ajmone-Marsan, D. Cesana, M. Sannazaro, M. Buora, M. Morgante, G. Liu, D. Bickhart, C. P. Van Tassell, J. Sölkner, T. S. Sonstegard, J. F. Garcia, Sci. Rep. **7**, 17140 (2017)
18. J. L. Zhong, J. W. Xua, J. Wanga, Y. Wena, H. Niub, L. Zheng, H. Hea, K. Penga, P. Hea, S. Shia, Y. Huang, C. Leia, R. Danga, X. Lana, X. Qid, H. Chena, Y. Huang, Gene **689**, 166-171 (2019)

Evaluation of cassava leaf meal (*Manihot esculenta* Crantz) in feed with enzymes supplementation on broiler performances

Ririn Angriani[1], *Widya* Hermana[2], and *Muhammad* Ridla[2,*]

[1]Graduate Study Program of Nutrition and Feed Science, IPB University, Bogor 16680, Indonesia
[2]Department of Nutrition and Feed Technology, Faculty of Animal Science, IPB University, Bogor 16680, Indonesia

Abstract. The research aimed to evaluate the effect of cassava leaf meal with enzymes supplementation in the feed on the performance of broiler. The research used factorial completly randomized design 2x3 on 240 Cobb-strain broilers. There were 6 treatments with 2 factors is used cassava leaf meal (CLM) and enzymes (non-starch polysaccharides and protease) with each dose 250 g/ton feed, i.e. P_1: basal ration, P_2: basal ration with enzymes, P_3: ration with 1.5% CLM, P_4: ration with 1.5% CLM and enzymes, P_5: ration with 3.0% CLM, P_6: ration with 3.0% CLM and enzymes. The variables analyzed were body weight, body weight gain, feed intake, feed conversion ratio, and mortality. The data obtained were analyzed by analysis of variance (ANOVA) and if the results indicated significantly different, then the post-hoc test conducted. The results showed that there was an interaction between CLM and enzymes on the feed conversion ratio (P < 0.05). The addition enzymes can decrease feed conversion ratio (from 1.91 to 1.74) siginificantly (P < 0.05). Addition CLM to a level of 3.0% did not interfere with performance. Enzymes can offset the detrimental effect caused by the use cassava leaf meal in the feed.

1 Introduction

Protein is one of the nutrients needed for broiler for growth and feed efficiency. The main source of protein in poultry feed is soybean meal, it has high protein and digestibility. Total import soybean was 2.48 million tons in 2020 [1]. One of the alternatives to overcome the dependence on imports was the utilization of local feedstuff protein source. Cassava is a local plant Indonesia that has high production. Cassava production in Indonesia was 19.3 million tons in 2018 [2]. The part of cassava is the tuber, skin of tuber, stem, and leaf. 10-40% of cassava plants consist of leaves. Cassava leaf contain high crude protein, β-carotene, and minerals [3, 4]. Cassava leaves have active compounds of flavonoids, phenolics, and chlorophyll which is a natural antioxidant [3]. Crude protein content in cassava leaf meal reachs 23.78% [5]. The nutritional content of cassava leaf varies in each age [4]. The older the leaf, the protein and amino acid levels will decrease, while the crude fiber content will

* Corresponding author: hmridla@apps.ipb.ac.id

increase. Cassava leaf meal levels 1.5% and 3.0% can increase weight gain with a lower feed conversion ratio [6].

Cassava leaves have a constrain. Crude fiber content in cassava leaf meal reaches 17.69% [5], while chickens have limitations in breakdown crude fiber due to the simple anatomy of the digestive tract of chickens. Cassava leaf have antinutrients such as cyanide acid (HCN). Several studies have been done to reduced the HCN levels (58 - 99 %) through feed processing such as sun-drying [7]; oven-drying [8]; fermentation [9, 10], but the most effective method was the combination of chopping and sun-drying [11]. Cassava leaves have low digestibility due to the presence of antinutrients and high crude fiber. Low digestibility can be overcome by giving enzymes. The addition of non-starch polysaccharides (NSP) enzymes can optimize the breakdown of crude fiber and protease enzymes in feed can also optimize the use of protein. According to Ding *et al.* [12], the addition of protease enzymes can increase crude protein digestibility, trypsin activity, and intestinal morphology. NSP enzymes can overcome the problem of high fiber broiler feed [13]. This research aimed to evaluate the effect of addition of cassava leaf meal in feed with enzyme supplementation on the performance of broiler. Finally, it can be use as the reference of the utilization of cassava leaf meal in feed.

2 Materials and methods

2.1 Animals and Feed

This research used 1 day old-240 Cobb-strain broilers that kept in cage sized 100 cm × 100 cm. Ration formulation adapted to nutrient requirements of broiler at pre-starter phase (age 1-7 days) based on SNI 8173.1:2015 [14], starter (age 7-21 days) based on SNI 8173.2:2015 [15], and finisher (age 21-35 days) based on SNI 8173.3:2015 [16]. Formulation and nutrients value of the experimental ration presented in Table 1. The ration given to the pre-starter phase-chickens was the basal ration. Furthermore, the treatment ration was given starting from the starter phase until finisher. Feed was given twice a day, at 07.00 and 17.00 WIB (West Indonesian Time) and water was provided ad libitum.

The manufactured of ration began with the manufactured of cassava leaf meal (CLM). The cassava leaves that used were manggu cassava varietas with the red leaf-stem cassava, finger-shaped leaves, and aged 4-5 months old. Cassava leaves were taken from 15-20 leaves on the top of leaf plants. Cassava leaves were separated between the leaves and the stems. Furthermore, the fresh leaves were dried in the shade for 2-3 days, then put into the oven 24 hours with temperature of 60 °C. The dried leaves were mashed by the grinder. The addition of enzymes was carried out with micro-feedstuffs mixtures such as amino acids and minerals during the mixing process. The enzyme used was NSP (Non-Startch Polysaccharide) with Superzyme-CS brand and protease with Concentrase-P brand. The enzyme each level used was 250 g/ton feed. Level of enzyme activity used can be seen in Table 2. The nutrient content that analyzed were dry matter and crude fiber based on method of SNI 01-2891-1992 [17]; crude protein, crude fat, and calcium (Ca) based on method of AOAC 2005 [18]; phosphor (P) based on gravimetry method; and gross energy by bombcalorimeter.

Table 1. Formulation and nutrients value of the experimental ration

Feedstuffs	Pre-starter (%)	Starter (%)			Finisher (%)		
	Basal	K_0	K_1	K_2	K_0	K_1	K_2
Maize	54.00	56.00	56.00	56.00	59.00	59.00	59.00
Rice bran	4.30	3.10	2.10	1.10	4.30	3.30	2.30
Soybean meal	26.00	26.50	25.50	24.50	21.00	20.00	19.00
Cassava leaf meal	0.00	0.00	1.50	3.00	0.00	1.50	3.00
Corn gluten meal	8.00	7.00	7.00	7.00	8.00	8.00	8.00
Crude palm oil	2.50	2.30	2.30	2.30	2.50	2.50	2.50
Fish meal	2.00	2.00	2.50	3.00	2.00	2.50	3.00
Calcium carbonate	1.00	1.00	1.00	1.00	1.10	1.10	1.10
Dicalcium phosphate	1.10	1.30	1.30	1.30	1.30	1.30	1.30
Premix	0.70	0.70	0.70	0.70	0.70	0.70	0.70
L-Lysine	0.10	0.05	0.05	0.05	0.05	0.05	0.05
DL-Methionine	0.10	0.05	0.05	0.05	0.05	0.05	0.05
Total	100	100	100	100	100	100	100
Nutrients value							
Dry matter (%)	85.91	85.77	85.88	85.88	85.70	85.82	85.93
Crude protein (%)	22.33	22.87	22.86	22.86	21.22	21.21	21.20
Crude fat (%)	5.22	5.02	5.08	5.08	5.39	5.45	5.52
Crude fiber (%)	2.86	2.75	2.91	2.91	2.83	2.90	3.14
Gross energy (cal/g)	3070.8	3074.2	3070.9	3070.9	3108.3	3104.9	3101.6
Calcium (%)	0.88	0.88	0.93	0.93	0.91	1.15	1.02
Phospor (%)	0.51	0.51	0.53	0.53	0.50	0.52	0.53

K0: ration without CLM; K1: ration with 1.5% CLM; K2: ration with 3.0% cassava leaf meal

Table 2. Level of enzyme activity

Enzyme	Enzyme activity in product (unit/g)	Enzyme activity in feed (unit/kg)
Protease	25000	6250
Non-Startch Polysaccharide (NSP)		
Xylanase	2400	600
Glucanase	300	75
Invertase	1400	350
Protease	2400	600
Cellulase	1000	250
Amylase	24000	6000
Mannanase	120	30
Pectinase	1700	425

Source : Canadian Bio-System

3 Results and discussion

The results research cassava leaf meal and enzymes on the body weight, body weight gain, feed intake, feed conversion ratio, and mortality as the performance variables for 35 days presented in Table 3.

Table 3. Effect of cassava leaf meal and enzymes levels on broiler performance for 35 days

Parameter	Feed	Enzymes		Average
		Without	With	
Body Weight (g)	K0	792.44 ± 0.77	817.33 ± 15.01	804.89 ± 16.62
	K1	804.00 ± 16.47	816.45 ± 11.76	810.23 ± 14.83
	K2	800.80 ± 15.36	809.32 ± 13.95	805.06 ± 14.33
Average		799.08 ± 13.27^{b}	814.37 ± 12.63^{a}	
Body Weight Gain (g)	K0	748.71 ± 0.12	773.67 ± 15.32	761.19 ± 16.76
	K1	760.48 ± 16.67	773.00 ± 11.75	766.74 ± 14.94
	K2	757.58 ± 15.45	765.64 ± 14.36	761.61 ± 14.46
Average		754.59 ± 13.41^{b}	770.77 ± 12.85^{a}	
Feed Intake (g)	K0	1389.55 ± 4.74	1374.04 ± 37.23	1381.80 ± 23.04
	K1	1423.12 ± 46.56	1383.90 ± 38.29	1403.51 ± 39.32
	K2	1432.67 ± 33.40	1398.49 ± 34.31	1415.58 ± 30.18
Average		1415.11 ± 32.18^{a}	1385.47 ± 28.95^{b}	
Feed Conversion Ratio	K0	1.86 ± 0.01^{b}	1.74 ± 0.01^{e}	1.80 ± 0.07
	K1	1.87 ± 0.02^{b}	1.79 ± 0.02^{d}	1.83 ± 0.05
	K2	1.91 ± 0.02^{a}	1.83 ± 0.00^{c}	1.87 ± 0.05
Average		1,88 ± 0,03	1,78 ± 0,04	
Mortality (%)	K0	0.14 ± 0.24	0.10 ± 0.21	0.12 ± 0.20
	K1	0.10 ± 0.21	0.00 ± 0.00	0.05 ± 0.16
	K2	0.10 ± 0.21	0.10 ± 0.21	0.10 ± 0.19
Average		0.12 ± 0.19	0.07 ± 0.17	

Numbers followed with different superscript in same column-subgroup showed the significantly different ($P < 0.05$).
K0: ration without CLM; K1: ration with 1.5% CLM; K2: ration with 3.0% CLM

There was no interaction between cassava leaf meal and enzymes on body weight and body weight gain, but addition of enzymes increase body weight and body weight gain significantly ($P < 0.05$). Feed with enzymes had higher final body weight and body weight gain, 814.37 g and 770.77 g, respectively. In line with this finding, previous reports also showed that the addition of protease and NSP enzymes can increased body weight gain [13, 20, 21]. Supplementation of protease enzyme in low protein diets resulted in increased body weight due to the ability protease enzyme to increase absorption of nutrients [22]. This could be because the protease enzyme was able to increase protein digestibility so that protein from cassava leaf meal was utilized optimally. The used of NSP enzyme can also increase the digestibility of feed ingredients that have high fiber resulted in body weight gain [23, 24].

The final body weight of this study is below the Cobb standard, the body weight of 35 days old Cobb chickens can reachs 2 kg. Addition cassava leaf meal up to level 3.0% did not interfere with performance. Ration without casssava leaf meal also have body weight below the Cobb standard. Environmental factors can cause body weight to not reach the standard. The temperature of the cage during the study ranged from 22.8 - 36.2 °C with a humidity of 48 - 89%. The ideal temperature for broiler rearing is 10 – 22 °C for achieving optimum body weight and 15 – 27 °C for ration efficiency [25]. According to Fatmaningsih *et al.* [26], chickens that feel uncomfortable with their environment can affect the physiological changes of chickens. The high environmental temperature can result in the accumulation of heat in the body, which will cause the livestock to experience heat stress. Stress reduction efforts have been carried out in this study, namely opening the curtains and turning on the fan at a high temperature and at night using the lamp as a heater. In addition, the provision of vitamins in drinking water is also carried out.

Addition of enzymes decrease feed intake significantly ($P < 0.05$). Feed with enzymes had a lower average of feed intake, but there was no interaction between cassava leaf meal and enzymes on feed intake. The research in line with Flores [27] and Kalmendal and Tauson [28] who report that protease and NSP enzymes can reduced feed intake. This was due to the adequacy of chicken in meeting the protein needs of the body. According to Risnajati [29], chickens have the ability to regulate feed consumption. The requirement of energy and protein affect the amount of feed consumption [6]. Protease enzyme can hydrolyze complex proteins into simpler compounds so as to optimize absorption [30].

The results showed that there was an interaction between cassava leaf meal and enzymes significant effect on the feed conversion ratio ($P < 0.05$). The results showed that cassava leaf meal increase FCR significantly ($P < 0.05$). The higher level of cassava leaf meal, the increased in feed conversion ratio. The research in line with Salu and Paembonan [6] shown that cassava leaf meal with levels 1.5% and 3.0% could increase weight gain with a lower feed conversion ratio, and more effective than 4.5% and 6.0% levels. This was due to the presence of cyanide acid in cassava leaf meal, so the more cassava leaf meal will make the cyanide acid (HCN) level higher. Based on the test results of cyanide acid content in cassava leaf meal, it was found that the addition of 1.5% CLM resulted in 1.19 mg/kg levels of HCN, while 3.0% CLM produced 2.38 mg/kg levels of HCN.

The addition of cassava leaf meal with enzymes has a low feed conversion ratio. The addition enzymes can decrease feed conversion ratio (from 1.91 to 1.74) significantly ($P < 0.05$). Feed conversion ratio is a parameter in seeing feed efficiency. It can be seen that the decrease in feed intake was in line with the decrease in feed conversion ratio and increased in body weight. Chickens without enzymes on their feed had high average conversion ratio than with enzymes. Previously, it was also showed that the use of protease and NSP enzymes can reduce feed conversion ratio [13, 21, 27, 28]. This was because protease enzymes were able to increase protein and amino acid digestibility so feed can be utilized properly [22]. Protease enzyme can also reduce the negative effects of anti-nutrients [12]. The used of NSP enzymes was able to reduce the detrimental effects of cassava leaf meal which has high fiber. Hydrolysis by NSP enzymes can release nutrients to cell walls and breakdown starch into simple sugars thereby increased nutrient absorption and feed digestibility efficiency [21]. The combination of NSP and protease can also optimize feeds containing high fiber, so increase the digestibility [31, 32].

There was no interaction between cassava leaf meal and enzymes on mortality. Based on the data the percentage of mortality was low about 0.00 – 0.14%. Addition of cassava leaf meal and enzymes did not affect mortality. Most of the deaths in this study occurred during brooding. The brooding period is a critical period in the maintenance of DOC (Day Old Chick) from the age of 1-14 days. The ration given to the pre-starter phase-chickens was the basal ration, no addition cassava leaf meal and enzyme. Cause of the death occurs because management and disease [33]. Factors that affect health and performance in chickens are daily growth, type of light adaptation, and type of drinking place [34]. Yerpes *et al.* [35]

divided the causes of death of broiler into two factors, they were internal factors (breeder age, sex, and type of DOC) and external factors (type of cage, season, and drinking water).

4 Conclusion

The ration with cassava leaf meal and enzymes supplementation has the best performance with the lowest value feed conversion ratio and does not adversely affect the broiler performance.

References

1. BPS – Statistics Indonesia, *Imports of Soybean by Major Countries of Origin*, 2010-2020 (2021) https://www.bps.go.id/statictable/2019/02/14/2015/impor-kedelai-menurut-negara-asal-utama-2010-2019.html
2. Ministry of Agriculture of the Republic of Indonesia, *Cassava Production by Province, 2014-2018* (2018) https://www.pertanian.go.id/home/?show=page&act=view&id=61
3. S. Hasim, Falah, L. K. Dewi, Curr. Biochem. **3**, 116-127 (2016)
4. C. Wobeto, A. D. Correa, C. M. P. Abreu, C. D. Santos, J. R. Abreu, Ciênc. Tecnol. Aliment. **26**, 865-869 (2006)
5. N. K. Morgan, M. Choct, Anim. Nutr. **2**, 253-261 (2016)
6. B. Salu, B. H. Paembonan, AgroSaint UKI Toraja, **1**, 36-42 (2010)
7. N. Madalla, N.W. Agbo, K. Jauncey, Tanzan. J. Agric. Sci. **15**, 1-12 (2016)
8. E. N. Junior, R. C. Chistea, R. S. Pena, Food Res. Int. **119**, 517-523 (2019)
9. Hermanto, Fitriani, Jurnal Riset Teknologi Industri **12**, 169-180 (2018)
10. B. R. S. Santos, E. F. R. Silva, L. A. C. Minho, G. C. Brandao, A. M. P. Santos, W. P. C. Santos, W. V. L. Silva, W. N. L. Santos, Microchem. J. **152**, 104271 (2019)
11. A. O. Fasuyi, Pak. J. Nut. **4**, 37-42 (2005)
12. X. M. Ding, D. D. Li, R. Li, J. P. Wang, Q. F. Zeng, S. P. Bai, Z. W. Su, K. Y. Zhang, Livest. Sci, **193**, 26–31 (2016)
13. Z. Makhdum, H. Rehman, J. M. Larik, P. Bux, A. Hameed, J. Appl. Anim. Res. **41**, 218-222 (2013)
14. National Standardization Agency of Indonesia, *SNI 8173.1-2015 Pakan Ayam Ras Pedaging (broiler) - Bagian 1: Sebelum Masa Awal (Pre Starter)* (National Standardization Agency of Indonesia, Jakarta, Indonesia, 2015)
15. National Standardization Agency of Indonesia, *SNI 8173.2-2015 Pakan Ayam Ras Pedaging (Broiler) - Bagian 2: Masa Awal (Starter)* (National Standardization Agency of Indonesia, Jakarta, Indonesia, 2015)
16. National Standardization Agency of Indonesia, *SNI 8173.3-2015 Pakan Ayam Ras Pedaging (Broiler) - Bagian 3: Masa Akhir (Finisher)* (National Standardization Agency of Indonesia, Jakarta, Indonesia, 2015)
17. National Standardization Agency of Indonesia, *SNI 01-2891-1992 Cara Uji Makanan dan Minuman* (National Standardization Agency of Indonesia, Jakarta, Indonesia, 1992)
18. AOAC, *Official Methods of Analysis of the Association of Analytical Chemists International, 18th edition* (Association of Official Analytical Chemists, Maryland, USA, 2005)

19. R. G. D. Steel, J. H. Torrie, D. Dickey, *Principles and procedures of Statistics: A Biometrical Approach, 3rd Edition* (McGraw Hill Inc., New York, USA, 1997)
20. A. K. Singh, J. F. D. Berrocoso, Y. Dersjant-Li, A. Awati, R. Jha, Anim. Feed Sci. Technol. **232**, 16-20 (2017)
21. L. Yuan, M. Wang, X. Zhang, Z. Wang, PLOS ONE **12**, e0173941 (2017)
22. M. Mohammadigheisar, I. H. Kim, J. Appl. Anim. Res. **46**, 1377-1388 (2008)
23. K. Schedle, E. Humer, R. Leitgeb, G. Freudenberger, K. M. Ebner, C. J. Schwarz, Die Bodenkultur : J. Land Manag. Food Environ. **67**, 185-198 (2016)
24. Z. Zduńczyka, J. Jankowskib, D. Mikuliskib, P. Zduńczyka, J. Juśkiewicza, B. A. Slominskic, Anim. Feed Sci. Technol. **263**, 114448 (2020)
25. K. Nova, J. Anim. Sci. **10**, 117-121 (2008)
26. R. Fatmaningsih, Riyanti, K. Nova, Jurnal Ilmiah Peternakan Terpadu **4**, 222-229 (2016)
27. C. Flores, M. Williams, J. Pieniazek, Y. Dersjant-Li, A. Awati, J. T. Lee, J. Appl. Poult. Res. **25**, 328–337 (2016)
28. R. Kalmendal, R. Tauson, Poult. Sci. **91**, 1387–1393 (2012)
29. D. Risnajati, Jurnal Ilmiah Ilmu-Ilmu Peternakan **14**, 62-67 (2011)
30. A. D. Anggraini, F. Poernama, C. Hanim, N. D. Dono, Buletin Peternakan **41**, 243-249 (2017)
31. A. K. Singh, J. F. D Berrocoso, Y. Dersjant-Li, A. Awati, R. Jha. Poult. Sci. **98**, 5571–5581 (2019)
32. M. Toghyani, S. B. Wu, R. A. P´erez-Maldonado, P. A. Iji, R. A. Swick, Poult. Sci. **96**, 3960–3972 (2017)
33. A. Nurmi, M.A. Santi, N. Harahap, M.F. Harahap, Jurnal Ilmiah Peternakan Terpadu **6**, 134-139 (2018)
34. T. Limbergen, S. Sarrazin, I. Chantziaras, J. Dewulf, R. Ducatelle, I. Kyriazakis, P. McMullin, J. Méndez, J. Niemi, S. Papasolomontos, P. Szeleszczuk, J. Van Erum, D. Maes, BMC Vet. Res. **16**, 287 (2020)
35. M. Yerpes, P. Llonch, X. Manteca, Animals **10**, 310 (2020)

Effect of dietary inclusion of black soldier fly larvae (*Hermetia illucens*) on broiler performance : A meta-analysis

Sandi Nayohan[1,*], *Irwan* Susanto[1], *Dessy* Permata[1], *Raiza Tri* Pangesti[1], *Mardiah* Rahmadani[1], and *Anuraga* Jayanegara[1]

[1]Nutrition and Feed Technology Department, IPB University, Bogor 16680, Indonesia

Abstract. Black Soldier Fly (BSF, *Hermetia illucens*) is an insect species that potentially can be used as a protein source for broiler nutrition. This study aimed to evaluate different inclusion level of BSF on broiler performance and blood metabolites by integrating various data from related studies using meta-analysis. Total of 12 studies that comprised of 31 data points were integrated in the database. Parameters observed included Live Weight (LW), Average Daily Gain (ADG), Daily Feed Intake (DFI), Feed Conversion Ratio (FCR), blood metabolites, and carcass proportion. The database was statistically analyzed by using mixed model method. Different studies were considered as random effects, and the level of BSF was treated as fixed effects. The model statistics used were the p-values and the Akaike information criterion. The significance of an effect was stated when its p-value was < 0.05. The results revealed that BSF feeding had significant linearly reduction effect on LW, ADG, and FCR of broiler ($P < 0.05$) and significant linearly increase effect on DFI and carcass proportion. However, BSF did not significantly affect on blood metabolites of broilers. It can be concluded that BSF negatively affect the performance of broiler.

1 Introduction

One of the determining factors for the success of the livestock industry is the availability of quality animal feed. About 50-70% of production costs are used for feed [1]. Protein source is the most expensive component compared to other nutrient sources. Generally, in some developing countries, protein source come from animal and vegetable proteins such as fish, blood, and soybean. However, the price of these feed ingredients is relatively high. It can burden farmers in terms of production costs. Alternative protein sources that have been researched in recent years are insects. Insects contain high unsaturated fatty acids and can be used as an alternative feed that does not compete with humans.

One of the insects that may be used as an opportunity supply of protein is the Black Soldier Fly (BSF). BSF in poultry rations as protein source is more potential because its availability is relatively abundant in nature and can adapt to a tropical climate. The protein

* Corresponding author: nayohansandy5@gmail.com

contained in the BSF may differ depending on the growth medium and the feed given to the magot. According to [2] BSF larvae contain relatively high protein of around 40-50%, with a fat content of about 29-32%, and the digestibility of BSF pupae is 77.7%.

Efforts to utilize alternative protein source feeds are needed to reduce the importation of protein source. BSF larvae mixed in feed is expected to reduce the cost of feed consumption so that farmers get maximum profit. A meta-analysis study to examine the effect of BSF administration on broiler performance has never been done. Therefore, this study aimed to evaluate the effect of different levels of BSF on the performance of broilers and blood metabolites through a meta-analysis study. This study integrates data from various related journals using a mixed model methodology.

2 Materials and methods

2.1 Research method

This study was carried out by looking for articles and published data on the utilization of BSF levels on broilers. The literature obtained comes from Scopus and Science Direct 2020-2021. The keywords used in the literature search were "bsf" and "broiler." Based on these keywords, 46 papers were found. Based on the selection results, there were 12 relevant studies used in this study consist of [3-14]. Table 1 provides a statistical overview of the database. The parameters observed in this study include live weight, average daily gain, daily feed intake, feed conversion ratio, blood metabolites, and carcass proportion.

Table 1. Statistical summary of the database regarding effect of dietary inclusion of black soldier fly larvae (*Hermetia illucens*) on broiler performance

Parameter	Unit	n	Mean	SD	Minimum	Maximum
LW	g	28	2298.1300	498.23793	1490.00	3756.00
ADG	g/d	24	69.7100	34.00925	40.40	207.30
DFI	g	24	114.6538	41.83722	61.20	212.40
FCR		24	1.7096	.27420	1.47	2.34
Metabolite	mg/dl	3	79.0300	8.66720	69.13	85.25
Carcass	%	14	73.6021	10.72389	57.70	93.72

2.2 Data analysis

The statistical analysis methodology used in study used a mixed model [15, 16]. The level of BSF was regarded as a fixed effect in this study, whereas study differences were treated as a random effect. Initially, the quadratic mixed-model method was applied to see the effect of level BSF on the experimental parameters.

3 Results and discussion

3.1 Live weight and average daily gain

Black Soldier Fly (BSF) is a type of fly (Diptera) from the family *Stratiomyidae* originating from tropical, subtropical, and warm temperature zones in America. BSF larvae have good potential to be used in feeds, especially poultry rations. According to [17], the crude protein level of BSF larvae meal is 44.9 percent, the crude fat content is 29.1 percent, the crude fiber content is 16.4 percent, and the ash content is 8.1 percent. It indicates that BSF larvae are capable of being used as an alternative feed source of insect protein. [18] concluded that BSF meal could substitute fish meal for broiler feed mixtures without affecting dry matter digestibility. According to the findings of the investigation, the administration of BSF meal resulted in substantially different ($P < 0.05$) outcomes in the body weight and body weight increase of chickens (Table 2). These results follow [19] which states that livestock fed 13% and 20% BSF larvae meal had a significantly higher difference in body weight than those fed non-BSF larvae meal. The body weight obtained in this study has a bodyweight that tends to be higher than the results of the [20] using the same percentage of larvae meal, which is 15% of 1785 grams of fish meal. Good body weight and weight gain results can be obtained from an excellent essential amino acid profile.

Table 2. Effect of dietary inclusion of black soldier fly larvae (*Hermetia illucens*) on broiler performance

Response Parameter	Unit	n	Intercept	SE Intercept	Slope	SE Slope	P-Value	RMSE	R^2
LW	g	28	2324.43	174.94	-0.3	0.1	< 0.01	93.28	0.98
ADG	g/d	24	60.43	5.73	-0.005	0.02	< 0.01	2.34	0.83
DFI	g	24	113.71	15.75	0.03	0.01	< 0.01	1.69	0.99
FCR		24	1.72	0.09	-0.01	0.001	< 0.01	0.039	0.98
Metabolite	mg/dl	3	95.15	6.88	-2.68	1.06	0.24	4.51	0.86
Carcass	%	14	71.59	4.49	0.02	0.02	< 0.01	1.46	0.99

3.2 Daily feed intake and FCR

In the current study, the different levels of dietary BSF (*Hermetia illucens* L.) significantly affect on variables daily feed intake ($P < 0.01$) (Table 2). The results of the analysis showed that broiler feed consumption increased following the diet BSF inclusion level. Increased feed intake It is also likely that the chitin content of the BSF worked as a prebiotic in the gut of the broilers, resulting in lower DM digestibility and decreased LW and ADG [21, 22]. Therefore, to meet the metabolic needs, broiler chickens increase their feed intake.

The results were analyzed for variables FCR during the study. FCR showed significantly ($P < 0.01$) affected by the level of dietary black soldier fly (Table 2). The reduction in feed efficiency for every increase in the level of dietary BSF in broiler rations has been obtained, indicating that lauric acid in BSF has a significant antibacterial activity as well as growth promoter potential with minimal toxicity [21]. In this study, feed intake was increase and feed efficiency were enhanced. This shows that there is no effect of dietary The effect of BSF on broiler palatability and the prospect of utilizing BSF in broiler chickens as an alternative to standard lipid sources [7].

3.3 Blood metabolites and carcass proportion

This study found that supplementing with BSF meal had no effect on the carcass percentage of broilers (Table 2). The addition of a BSF meal in terms of carcass percentage had no detrimental effect on chicken meat quality. The previous studies have shown no significant differences in carcass traits by BSF. A broiler diet with 15% inclusion of BSF meal showed a negative effect on growth performance and carcass [10]. Other studies on broilers, [9] there was no effect of BSF meal inclusion on carcass weight or carcass percentage in broiler chickens given BSF larvae meal during the growth and finishing stages. [4] for all treatments, the inclusion of BSF meals in the broiler diet had no significant influence on the dressed weight of carcasses (as well as the relative weight of internal organs (liver, heart, spleen, gizzard, and intestines).. In agreement with their result, the present study reported that broiler chicken in the BSF meal group did not affect the carcass cut yield and carcass percentage [3] .

The carcass percentage of broilers trait by BSF meal did not significantly different might be due to the chitin in the exoskeleton of HI larval Chitin was shown to have a detrimental impact on protein digestion. BSF meal has a lot of chitin, which is indigestible and can reduce protein digestibility. Chitin, which cannot be digested, and the microwave drying process of BSF meal might explain these reduced carcass features. This conclusion is consistent with a recent study that found that the carcass weight fell linearly with the inclusion levels (5, 10, and 15%) of BSF meal and significantly with a high inclusion level (15 percent). As a result, including a modest level of BSF meal into the broiler diet may help to preserve carcass features. Overall, a modest inclusion level (up to 10%) of BSF meal looks adequate for broiler diets in terms of chicken meat growth performance, intestinal morphology, and fatty profile of chicken meat [5]. [23] assumed that chitin content in BSF meals was the reason for the crude protein in the BSF meal diet being considerably lower. Chitin in BSF meal influences normal nitrogen-to-protein conversion rates (usually 6.25 for organic matter). [24] determined that a conversion rate of 4.76 is better for predicting crude protein content from measured BSF meal nitrogen levels. Even with a conversion rate of 6.25, the crude protein content of the BSF meal broiler diet is lower than that of the other experimental diets. Nonetheless, there were no negative impacts on carcass percentage or other meat quality factors.

This study found that include BSF meals in the diet had no effect on blood markers ($P >$ 0.05) (Table 2). These findings are consistent with the findings of [9], who found no significant differences in blood parameters between broiler chickens and turkey poults given BSF feed [7]. The blood parameters reveal information about the birds health. [7] proposed that the broiler chickens were under physiological circumstances and confirmed that BSF meal and BSF fat had no effect on the animals' health state. [7] This action might be attributed to the chelating effect of chitin in the BSF larva feed, according to one study. Chitin, a major ingredient of arthropod exoskeletons, is found in BSF larvae. Chitin concentration in BSF larvae ranges between 5.9 and 8.7 percent.

4 Conclusion

The results revealed that BSF feeding had a considerable lowering effect on broiler live weight and average daily growth while increasing daily feed intake and feed conversion ratio. However, BSF did not significantly affect on blood metabolites and carcass proportion of broilers. this show that increasing BSF level on ration negatively affects the performance of broiler.

References

1. F. A. Katayane, B. Bagau. F. R. Wolayan, J. Zootek, **34**, 27–36 (2014)
2. G. Bosch, S. Zhang, D. G. A. B. Oonincx, W. H. Hendriks, J. Nutr. Sci, **3** (2014)
3. T. Popova, E. Petkov, Comptes Rendus L'Academie Bulg. des Sci, **73**, 1617–1626

(2020)
4. M. M. Mutisya, A. K. Mawufe, J. N. Kinyuru, C. M. Tanga, M. Gicheha, G. Hailu, D. Salifu, Z. Khan, S. Niassy, Poult. Sci, **100**, 420–430 (2021)
5. B. Kim, H. R. Kim, S. Lee, Y. C. Baek, J. Y. Jeong, H. T. Bang, S. Y. Ji, S. H. Park, Animals, **11**, 1–12 (2021)
6. A. F. Ipema, W. J. J. Gerrits, E. A. M. Bokkers, B. Kemp, and J. E. Bolhuis, Appl. Anim. Behav. Sci, **230** (2020)
7. S. Dabbou, A. Lauwearts, I. Ferrocino, I. Biasato, F. Sirri, M. Zampiga, S. Bergagna, G. Pagliasso, M. gariglio, E. Colombino, C. G. Narro, F. Gai, M. T. Capuchio, L. Gasco, L. Cocolin, A. Schiavone, Animals, **11** (2021)
8. M. Gariglio, S. Dabbou, M. Crispo, I. Biasato, F. Gai, L. Gasco, F. Piacente, P. Odetti, S. Bergagna, I. Placha, E. Valle, Elena. Colombino, M. T. Capucchio, A. Schiavone, Animals, **9** (2019)
9. A. Schiavone, S. Dabbou, M. De Marco, M. Cullere, I. Biasato, E. Biasibetti, M. T. Cappucchio, S. Bergagna, D. Dezzutto, M. Meneguz, F. Gai, A. Dalle Zotte, L. Gasco, Animal, **12**, 2032–2039 (2018)
10. S. Dabbou, F. Gai, I. Biasato, M. T. Cappucchio, E. Biasibetti, D. Dezzutto, M. Menuguz, I. Placha, L. Gasco, J. Anim. Sci. Biotechnol., **9** (2018)
11. B. Kim, H. T. Bang, K. H. Kim, M. J. Kim, J. Y. Jeong, J. L. Chun, S. Y. Ji, J. Anim. Sci. Technol, **62**, 187–197 (2020)
12. B. Kierończyk, J. Sypniewski, M. Rawski, W. Czekała, S. Swiatkiewicz, D. Józefiak, Ann. Anim. Sci, **20**, 157–177 (2020)
13. A. Schiavone, M. Cullere, M. De Marco, M. Meneguz, I. Biasato, S. Bergagna, D. Dezzutto, F. gai, S. Dabbou, L. Gasco, A. D. Zotte, Ital. J. Anim. Sci, **16**, 93–100 (2017)
14. A. Schiavone, S. Dabbou, M. Petracci, M. Zampiga, F. Sirri, I. Biasato, F. Gai, L. Gasco, Animal, **13**, 2397–2405 (2019)
15. D. Sauvant, P. Schmidely, J. J. Daudin, and N. R. St-Pierre, Animal, **2**, 1203–1214 (2008)
16. N. R. St-Pierre, J. Dairy Sci, **84**, 741–755 (2001)
17. A. Jayanegara, N. Yantina, B. Novandri, E. B. Laconi, Nahrowi, M. Ridla, J. Indones. Trop. Anim. Agric, **42**, 247–254 (2017)
18. V. Rambet, J. F. Umboh, Y. L. R. Tulung, Y. H. S. Kowel, Zootec, **35**, 13 (2015)
19. J. P. Anankware, R. Ayizanga, O. Opoku, D. Obeng-Ofori. Glob. Adv. Res. J. Agric. Sci, **7**, 272–280 (2018)
20. J. Hwangbo, E. C. Hong, A. Jang, H. K. kang, J. S. Oh, B. W. Kim, B. S. park, Journal of Environmental Biology, **30**, 609-614 (2009)
21. J. de Souza Vilela, N. M. Andrenicos, M. Kolakshyapati, M. Hiliar, T. Z. Sibanda, N. R. Andrew, R. A. Swick, S. Wilkinson, S. Ruhnke, Anim. Nutr, **7**, 695–706 (2021)
22. M. E. Abd El-Hack, M. E. Shafi, W. Y. Alghamdi, S. A. Abdelnour, A. M. Shehata, A. E. noreldin, E. A. Ashour, A. A. Swelum, A. A. Al-Sagan, M. Alkhateeb, A. E. Taha, A. M. E. Abdel-Moneim, V. Tufarelli, M. Ragni, Agric, **10**, 1–31 (2020)
23. B. A. Altmann, R. Wigger, M. Ciulu, D. Mörlein, J. Sci. Food Agric, **100**, 4292–4302 (2020)
24. R. H. Janssen, J. P. Vincken, L. A. M. Van Den Broek, V. Fogliano, C. M. M. Lakemond, J. Agric. Food Chem, **65**, 2275–2278 (2017)

Phylogenetic tree analysis for Bali Cattle based on partial sequence 16S rRNA Mitochondrial DNA

R. Misrianti[1], *S. H.* Wijaya[3], *C.* Sumantri[2], *J.* Jakaria[2*]

[1]Graduate Program in Animal Production and Technology, Faculty of Animal Science IPB University, JL Agatis, Dramaga Campus, Bogor 16680 – Indonesia
[2]Department of Animal Production and Technology, Faculty of Animal Science IPB University, JL Agatis, Dramaga Campus, Bogor 16680 – Indonesia
[3]Department of Computer Science, Faculty of Mathematics and Natural Science IPB University, JL Agatis, Dramaga Campus, Bogor 16680 – Indonesia

Abstract. Mitochondria DNA (mtDNA) as a source of genetic information based on the maternal genome, can provide important information for phylogenetic analysis and evolutionary biology. The objective of this study was to analyze the phylogenetic tree of Bali cattle with seven gene bank references (*Bos indicus, Bos taurus, Bos frontalis*, and *Bos grunniens*) based on partial sequence 16S rRNA mitochondria DNA. The Bayesian phylogenetic tree was constructed using BEAST 2.4. and visualization in Figtree 1.4.4 (tree.bio.ed.ac.uk/software/figtree/). The best model of evolution was carried out using jModelTest 2.1.7. The most optimal was the evolutionary models GTR + I + G with p-inv (I) 0,1990 and gamma shape 0.1960. The main result indicated that the Bali cattle were grouped into *Bos javanicus*. Phylogenetic analysis also successfully classifying *Bos javanicus*, *Bos indicus*, *Bos taurus, Bos frontalis* and *Bos grunniens*. These results will complete information about Bali cattle and useful for the preservation and conservation strategies of Indonesian animal genetic resources.

1 Introduction

Bali cattle as Indonesian native cattle have an important role in Indonesia. These cattle are very adaptive in tropical conditions and most suitable for sustainable small farming in Indonesia [1]. It's necessary to have information genetic of these cattle as basic information in a breeding program. Mitochondrial DNA (mtDNA) has been used in genetic studies because have a high mutation rate, absence of introns, and lack of recombinant event [2, 3, 4]. The 16S rRNA gene is a part of mtDNA has been reported as an advanced genetic marker for animal genetic diversity and phylogenetic studies. Yan *et al.* [5] reported 16S rRNA as a suitable marker for cattle phylogenetic analyses of Chinese cattle breeds. No

* Corresponding author: jakaria@apps.ipb.ac.id

reports about 16S rRNA sequenced in Bali cattle. Most phylogenetic studies in Bali cattle using d-loop [6, 7], Cyt-b sequenced [8, 9], COI [10, 11]. Thus, the objective of this research was to identified a phylogenetic tree among Bali cattle and four cattle breeds (*Bos indicus, Bos taurus, Bos frontalis, and Bos grunniens*) based on partial sequenced of 16S rRNA gene.

2 Materials and methods

2.1 DNA materials

The total of DNA samples was used in this study is twenty-two. The sourced of the sample from Nusa Penida Island-Bali. Furthermore, we also employed 16S rRNA sequence from GenBank, including *Bos indicus Sahiwal* (GeneBank Acc. KX575711), *Bos indicus Wayanad* (MF667931), *Bos indicus breed Nellore* (AY126697.1), *Bos Frontalis* (MW763078.1), *Bos grunniens* (MZ313873.1), and *Bos taurus* (GU947013.1).

2.2 Primer, amplification, sequencing

Primer design using primer 3 software based on Bos Indicus mitochondrial complete sequence (*GenBank Accession No. AY126697*) (**http://primer3.ut.ee**). Information about the sequenced and annealing temperature of the primer is shown in Table 1. The final volume of Polymerase Chain reaction (PCR) is 28 ul, containing 2 μL DNA samples, 12,5 μL GoTaq® Green Master Mix (Promega, United States), 0,2 μL of each Primer forward and reverse (25 pmol/μL), and 10.1 μL nuclease-free water.

Table 1. Sequenced of Primer for Amplification Partial Sequence16S rRNA Gene

Primer	Sequenced (5'- 3')*	Length	Ta(°C)
16srRNA-1	F:AAC AAA GCA TCC AGT TTA CAC C	846pb (1038-1883pb)	55
16srRNA-1	R: TTT CCT TAG ATG CAC TCC TGT G		

*GenBank access No. AY126697

PCR mix run using *Applied Biosystems* thermal cycler, with condition pre-denaturation was at 95°C for 5 min, and amplification for 35 cycles (denaturation at 95°C for 10 s, annealing at 57°C for 20 s, extension at 72°C for 30 s), and final elongation at 72°C for 5 min. The PCR products were visualized by electrophoresis using 1.5% agarose gel and sequenced by 1st base-Selangor Malaysia (http://www.base-asia.com/fragment_analysis/).

2.3 Data Analysis

The sequenced of 16srRNA was analyzed using Sequences were analyzed using MEGA X 10.2.6 software. The Bayesian phylogenetic tree was constructed using BEAST 2.4. and visualization in Figtree 1.4.4 (tree.bio.ed.ac.uk/software/figtree/). The best model of evolution was carried out using jModelTest 2.1.7. The most optimal evolutionary models GTR + I + G with p-inv (I) 0,1990 and gamma shape 0.1960.

3. Results and Discussion

The 16S rRNA gene of Bali cattle was successfully amplified with annealing temperature 55⁰C. The primer amplified at position 1038 bp to position 1883 bp (according to GeneBank Acc number No. AY126697) with fragment prediction 846 bp. The frequency of A base was the highest (30.8) among three other bases (T (23.90%); C(23.12%); G(19.18%)). No significant differences in nucleotide composition of the mtDNA 16S rRNA gene were found among the Bali cattle. [5] reported the stable GC content in 16S rRNA sequenced was helpful in phylogenetic analysis marker.

The nucleotide composition of A-T was higher than G-C. This result was conducted well with the previous study in Bali cattle, including d-loop sequenced [6, 7], Cyt-b sequenced [8, 9], COI [10]. This finding is also consistent in another breed of cattle. [5] also identified the rich composition of A+T in the 16S rRNA gene of Chinese cattle.

Table 2. Nucleotide composition of Partial sequences 16S rRNA of Bali Cattle

Sample	%						Total
	T(U)	C	A	G	T+A	C+G	
B1	23.90	23.11	33.82	19.17	57.72	42.28	887
B2	23.90	23.11	33.82	19.17	57.72	42.28	887
B3	23.90	23.11	33.71	19.28	57.61	42.39	887
B4	23.90	23.11	33.82	19.17	57.72	42.28	887
B5	23.90	23.11	33.82	19.17	57.72	42.28	887
B6	23.90	23.11	33.82	19.17	57.72	42.28	887
B7	23.90	23.00	33.93	19.17	57.84	42.16	887
B8	23.90	23.34	33.60	19.17	57.50	42.50	887
B9	23.90	23.11	33.71	19.28	57.61	42.39	887
B10	23.90	23.11	33.71	19.28	57.61	42.39	887
B11	23.90	23.11	33.82	19.17	57.72	42.28	887
B12	23.90	23.11	33.82	19.17	57.72	42.28	887
B13	23.90	23.11	33.82	19.17	57.72	42.28	887
B14	23.90	23.11	33.82	19.17	57.72	42.28	887
B15	23.90	23.11	33.82	19.17	57.72	42.28	887
B16	23.90	23.11	33.82	19.17	57.72	42.28	887
B17	23.90	23.11	33.82	19.17	57.72	42.28	887
B18	23.90	23.11	33.82	19.17	57.72	42.28	887
B19	23.90	23.11	33.82	19.17	57.72	42.28	887
B20	23.90	23.11	33.82	19.17	57.72	42.28	887
B21	23.90	23.11	33.82	19.17	57.72	42.28	887
B22	23.90	23.11	33.82	19.17	57.72	42.28	887
Avg.	23.90	23.12	33.80	19.18	57.77	42.23	887

The phylogenetic tree showed Bali cattle were separated from other cattle breeds (Bos Frontalis, Bos grunniens, Bos indicus, and Bos taurus) based on partially sequenced 16S rRNA gene (Figure 1). Hall (2001) identified the change of nucleotide sequence identified by the length of branch in the phylogenetic tree. [1] reported the Bali cattle are the domesticated direct descendant of the wild Banteng still surviving as an endangered species in three National Wild Reservation Parks (Ujung Kulon, Baluran, and Blambangan) in Java.

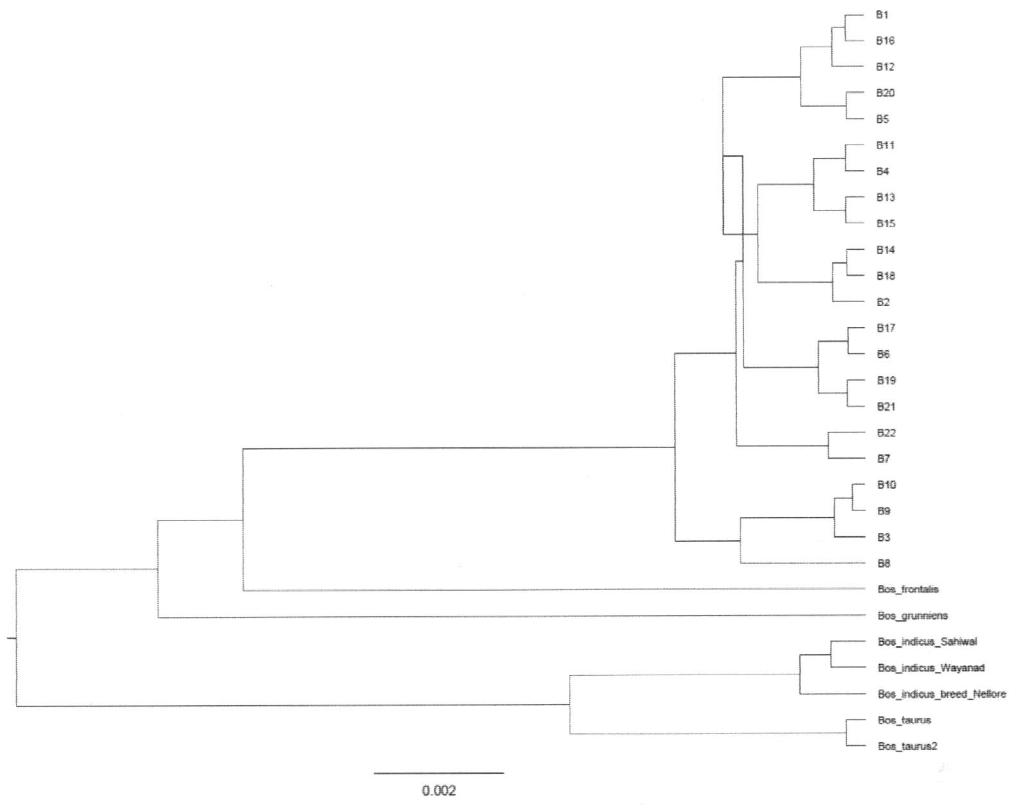

Fig. 1. The Bayesian Phylogenetic Tree of Bali Cattle and Other Cattle Breed Based on Partial Sequenced of 16S rRNA Gene (B: Bali cattle)

This result was consistent with a previous study, Jakaria *et al* (2019) identified Bali cattle as a different cluster with *Bos taurus* and *Bos indicus* based on the d-loop sequence. [12] also identified Bali cattle as a different group with *Bos frontalis, Bos indicus*, and Bison based on cyt-b sequenced.

4 Conclusion

The composition of the A-T base in the partially sequenced 16S rRNA gene was higher than C-G. The phylogenetic analysis using partial sequenced 16S rRNA gene indicated Bali cattle have a different cluster with *Bos indicus, Bos taurus, Bos frontalis, and Bos grunniens*.

This work was financially supported by the Directorate General of Resources for Science, Technology and Higher Education, Ministry of Research, Technology and Higher Education Fiscal Number: 2000/IT3.L1/PN/2021 date 15 March 2021.

References

1. Martojo, Reprod Dom Anim, **47** (2012)
2. P. Di Lorenzo, H. Lancioni, S. Ceccobelli, L. Colli, I. Cardinali, T. Karsli, PLoS ONE, **13** (2018)
3. M. Rusdin, D. D Solihin, A. Gunawan, C. Talib, C. Sumantri, Tropical Animal Science Journal, **43** (2020)
4. C. Ginja, L. T. Gama, O. Cortés, *Sci Rep,* **10** (2020)
5. L. Yan, Y. She, M. A. Elzo, C. Zhang, XF, H. Chen, Arch. Anim. Breed, **62** (2019)
6. J. Jakaria, T. Musyaddad, S. Rahayu, M. Muladno, C. Sumantri, JITAA, **44** (2019)
7. Hikmawaty, Jakaria, A. Gunawan, M. I. A. Dagong, L. Rahim, *IOP Conf. Ser.: Earth Environ. Sci,* **492** (2020)
8. T. Hartatik, D. N. H. Hariyono, Y. Adinata, Biodiversitas, **20** (2019)
9. P.W. Prihandini, A. Primasari, M. Luthfi, J. Efendy, D. Pamungkas, JITV, **25** (2020)
10. N. N. W. Susari, P. Suastika, K. K. Agustina, Biodiversitas, **22** (2021)
11. A. Wulandari, V. M. A. Nurgiartiningsih, Kuswati, T. E. Susilorini, P. A. Partogi, IntlResearch J Advanced Engineering and Science, **4** (2019)
12. E. T. Margawati, S. D. Volkandari, Indriawati, M. Ridwan, JITV, **23**(2018)

The influence of entrepreneurship competence to entrepreneurships orientation on beef cattle farmer in Karanganyar regency

Sutrisno Hadi Purnomo[1*], *Eka* Handayanta[1], *Agni* Hanifa[1], *Ayu Intan* Sari[1], *Shanti* Emawati[1], and *Endang Tri* Rahayu[1]

[1]Dept. of Animal Science, Faculty of Agriculture, Universitas Sebelas Maret, Jl Ir Sutami 36A Surakarta 57126, Indonesia.

Abstract. This research aims to analyze the influence and relationship of entrepreneurial competence to entrepreneurship orientation of beef cattle business. The research was conducted in Tasikmadu, Karanganyar, Central Java Indonesia at February-April 2021. Methods of data collection using survey method with questionnaire as primary data retrieval tool. The method of determining the respondents using census method on all farmers who attended the training with the number of respondents 42 people. Method of determining the location by purposive sampling. Data analysis used was test of validity and reliability, multiple linear regression analysis continued test of determination (R^2), F test, t test and classical assumption test. Validity test results show valid results $r_{hitung} > r_{tabel}$ is 0.304 and reliability value $\alpha >$ 0.6 on all variables. Multiple linear regression analysis obtained value of equation $Y = 0.741 + 0.350X_1 + 0.246X_2 + 0.264X_3 + 0.047X_4 + 0.011X_5 + 0.038X_6$. The value of determination coefficient (R^2) is 0.608, which means that entrepreneurship competence variable can explain the entrepreneurship orientation variable equal to 60.8%. The entrepreneurial competence variable affects together toward the orientation with the value $F_{count} > F_{table}$. It can be concluded that entrepreneurial competence consisting of communication, problem solving, initiative, planning and organization, self-awareness and technology simultaneous influence to entrepreneurial orientation.

1 Introduction

Indonesia as an agricultural country has a lot of potential resources that can be used as a business. One of the businesses that can be developed in Indonesia is beef cattle farming. The beef cattle business in Indonesia is still promising in line with the high demand for beef and the government's program to limit beef imports [1]. Livestock business, especially beef cattle farming in Indonesia is still promising because the market opportunities are quite open [2]. The livestock sector is a business sector that plays a very large role in meeting the needs of animal protein and various industrial needs. Beef cattle farming is one part of the agricultural

* Corresponding author: sutrisnohadi@staff.uns.ac.id

sector that needs to be developed and utilized optimally, so it is necessary to continue to strive for its development in order to make a real contribution to the Indonesian economy [3]. In order to realize the fulfillment of beef needs in Indonesia, it is necessary to increase the entrepreneurial spirit of beef cattle farmers.

Entrepreneurship can be interpreted as a spirit, attitude and behavior as an example in the courage to take calculated risks based on one's own will and ability. According to [4], an entrepreneur is an individual who has the character to achieve higher than most other individuals. An entrepreneur must have entrepreneurial competence because with the entrepreneurial competence in himself entrepreneurs are expected to be able to build businesses and achieve business success [5]. Competence is the attitude, knowledge, skills and abilities needed by farmers to maximize their business performance. Farmers who have high competence will affect their entrepreneurial orientation. Entrepreneurial orientation is the tendency of farmers to innovate, be proactive and dare to take risks in managing their business [6].

Tasikmadu District is one of the areas in Karanganyar Regency which has a fairly high beef cattle population of around 5678 heads [7]. Beef cattle are cattle that are kept with the main purpose of producing meat. Most of the smallholder beef cattle business is a hereditary business with a maintenance pattern according to the ability of the breeder [8]. The development of beef cattle farming which is still in the traditional or traditional farming pattern results in a lack of entrepreneurial competence in the community [6]. Low business competence means lack of orientation so that business performance will not be optimal [9]. Therefore, it is necessary to conduct research to analyze the effect of entrepreneurial competence on the entrepreneurial orientation of beef cattle.

2 Materials and methods

This research was conducted using a survey method in Tasikmadu District, Karanganyar Regency on February – April 2021. The location of the research was determined by purposive sampling, located in Tasikmadu District, Karanganyar Regency. This is because the location is one of the centers of beef cattle business which has a large area of land and the availability of abundant feed so that the location is very supportive when used for data collection of beef cattle entrepreneur research. Sampling method using purposive sampling method, namely the technique of determining the sample with certain considerations [10]. The sample used as a respondent is a farmer who has at least 2 beef cattle in the District of Tasikmadu, Karanganyar Regency, members of the Subur Lestari Livestock Farmer Group who participated in the training with a total of 42 peoples.

The data analysis used in this research is quantitative data analysis using SPSS 22 software for windows. This study uses variables dépendent of entrepreneurial orientation and 6 independent variables, namely communication, problem solving, initiative, planning and organization, self-awareness and technology. The analysis of this research includes validity analysis to measure the validity of a question or questionnaire, if (r-count > r-table) then the question item is valid [11], reliability analysis to measure a questionnaire which is an indicator of the variable, if the value cronbach's alpha > 0.6 data is considered reliable [12].

Multiple linear regression analysis was used to determine the effect of the variables of communication, problem solving, initiative, planning and organization, self-awareness and technology on entrepreneurial orientation. Statistical test with coefficient of determination (R^2) to measure how far the ability of a model to explain the variation of the dependent variable, F test to find out the independent variables jointly affect the dependent variable at a significance level of 5% (0.05), t test to test the variable partially independent effect on the dependent variable, t-count > t-table then Ho is rejected and Ha is accepted which means significant, t-count < t-table then Ho is accepted which means it is not significant. Classical assumption test of residual normality to test the residual value of the regression model normally distributed or not, multicollinearity to see whether there is a correlation between independent variables, heteroscedasticity to determine whether there is a variance inequality from residual to other observations [12].

3 Results and discussion

Tasikmadu is one of the sub-districts in Karanganyar Regency which is located between $7^0 30' - 7^0 45'$ South Latitude and $110^0 - 110^0 45'$ East Longitude, which consists of 13 villages with an area of 5.508.43 Ha. The total population of Tasikmadu Subdistrict in 2020 was recorded at 62.501 people, consisting of 30.588 male residents and 31.913 female residents, with a population density of 1.135 people/km. The number of households is 18.196 and the average per household is 3-4 people.

Tasikmadu District has a total land area of 5.508.43 Ha of which 2.479.83 Ha is agricultural land. Tasikmadu District is a fertile area with abundant water availability so it has the potential to be used as agricultural land. Extensive land in the form of rice fields produces agricultural waste that can be used as animal feed. This makes it easier for farmers to find animal feed and reduces the costs incurred in the beef cattle business. The main livelihood of the people of Tasikmadu District is as farmers and has a side job, namely raising livestock, both ruminants and poultry. The type of ruminant livestock that is mostly kept is beef cattle of 7.552 heads [7].

3.1 Characteristics of respondents

Characteristics of respondents analysed in this study include gender, age, education level, occupation, length of raising and number of cattle ownerships. The characteristics of the respondents of entrepreneurship training participants can be seen in Table 1.

Based on Table 1, the results of the study show that most of the respondents have an age ranging from 25 to 65 years, this means that most of the respondents are of productive age. According to [1] that the productive age group is 15 to 64 years old and the unproductive age group is more than 64 years old. The most recent education taken by respondents was at the high school level as many as 24 people or 57.14% which indicates that awareness of public education is quite high.

The main occupations of the respondents are farmers, as many as 19 respondents or 45.24%, while 9 people or 21.43% have their main job as livestock entrepreneurs. This is because livestock entrepreneurs are mostly only used as side jobs. According to [3] that to deal with business risks such as production failure, farmers do side business as a source of income to meet the needs of their families. The majority of respondents had less than 5 livestock, which was 73.81% or 31 of the total number of respondents. This is in accordance with the opinion of [13], the scale of smallholder beef cattle farming is described by the small number of livestock ownership. This means that only 2-3 cattle owned by a farmer.

3.2 Research instrument test

The validity test criteria if the value of rcount > rtable, then the statement item is said to be valid (Abdulrahman et al., 2011). The validity test of each item shows the value of rcount > 0.304 (r-count > r-table) it can be concluded that the statement item is valid. Cronbach's alpha value of all variables has the lowest value of 0.711 and the highest value of 0.872 so that all variables are said to be reliable. This is in accordance with the opinion of [14] if ($\alpha > 0.6$) then the statement item is reliable.

Table 1. Characteristics of respondents

Characteristics	Number of people)	Percentage (%)
Age		
25-35	9	21.4
36-45	9	21.4
46-55	9	21.4
56-65	13	31
>65	3	4.8
Education		
Elementary school	4	9.53
Junior high school	5	11.9
Senior High School	24	57.14
S1/D3	9	21.43
Long time raising		
<5	21	50
5-10	14	33.3
>10	7	16.7
Number of cattle ownerships		
<5	31	73.81
5-10	8	19.05
>10	3	7.14

Source: Processed Primary Data, 2021

3.3 Analysis of factors affecting entrepreneurial orientation

Factors that influence entrepreneurial orientation can be measured by multiple linear regression tests. The results of multiple linear regression can be seen in Table 2.

Based on Table 2, the following multiple linear regression equation is obtained:

$Y = 0.741 + 0.350X_1 + 0.246X_2 + 0.264X_3 + 0.047X_4 + 0.011X_5 + 0.038X_6 + e$

The results of the multiple linear regression test in this study obtained a constant value (α) of 0.741 with a positive constant value. This means that if the independent variable is constant, the Entrepreneurial Orientation (Y) of fattening beef cattle will increase by 0.741 [9]. The coefficient value for the Communication variable (X1) obtained a value of 0.350 and is positive. This means that communication has a positive influence on the entrepreneurial orientation and if the perception of communication increases by 1 unit, while other variables are considered constant, then there is an effect on increasing the value of the entrepreneurial orientation of beef cattle fattening by 0.350.

Table 2. Multiple linear regression test results

Variable	Regression Coefficient	t-count	t-table	Sig.t (<0.05)
X1 (Communication)	0.350	1.526	2,030	0.136
X2 (Troubleshooting)	0.246	1.071	2,030	0.292
X3 (Initiative)	0.264	1.458	2,030	0.154
X4 (Planning and organization)	0.047	0.195	2,030	0.847
X5 (Self-awareness)	0.011	0.073	2,030	0.942
X6 (Technology)	0.038	0.224	2,030	0.824
Constant	0.741			
F count	9,064			0.000
R square (R^2)	0.608			
Dependent variable = Y (Entrepreneurship Orientation)				

Source: Processed Primary Data, 2021

The coefficient value for the Problem Solving variable (X2) obtained a value of 0.246 and is positive. This means that the Problem Solving variable has a positive influence on the Entrepreneurial Orientation and if the perception of problem solving increases by 1 unit, while other variables are considered constant, then there is an effect on increasing the value of the Entrepreneurial Orientation of beef cattle fattening by 0.246. The coefficient value for the Initiative variable (X3) obtained a value of 0.264 and is positive. This means that the Initiative has a positive influence on the Entrepreneurial Orientation and if the perception of the Initiative increases by 1 unit, while other variables are considered constant, then there is an influence on the increase in the value of the Entrepreneurial Orientation of beef cattle fattening by 0.264. The coefficient value for the Planning and Organization variable (X4) obtained a value of 0.047 and is positive. This means that Planning and Organization have a positive influence on Entrepreneurial Orientation and if the perception of Planning and Organization increases by 1 unit, while other variables are considered constant, then there is an effect on increasing the value of the Entrepreneurial Orientation of beef cattle fattening by 0.047. The coefficient value for the Self-Awareness variable (X5) is obtained a value of 0.011 and is positive. This means that Self-Awareness has a positive influence on Entrepreneurial Orientation and if the perception of Self-Awareness increases by 1 unit, while other variables are considered constant, then there is an effect on the increase in the value of Entrepreneurial Orientation for beef cattle fattening by 0.011. The coefficient value for the Technology variable (X6) is obtained a value of 0.036 and is positive. This means that the use of technology has a positive influence on the Entrepreneurial Orientation and if the perception of the use of technology increases by 1 unit, while other variables are considered constant, then there is an effect on the increase in the value of the Entrepreneurial Orientation of beef cattle fattening by 0.036.

The coefficient of determination (R2) essentially measures how far the ability of a model to explain variations in the dependent variable. The value of determination is between zero and one [12]. Based on Table 2, the determination value is 0.608. This shows that the independent variable contributes to the influence of the dependent variable by 60.8% and the rest is influenced by other variables not examined in this study.

According to [14] if F-count < F-table then Ho is accepted and if F-count > F-table then Ho is rejected. Based on Table 2, it can be seen that the F-count value is 9.064 and the F-table value can be seen based on the F statistic table with a 5% significance level of 2.372. The data shows the value of F-count > F-table which is 8.705 > 2.372, this means that simultaneously the independent variables (communication, problem solving, initiative, planning and organization, self-awareness and technology) have a significant influence on the dependent variable (Entrepreneurship Orientation). In other words, the six independent variables simultaneously influence the entrepreneurial orientation of beef cattle fattening.

Based on Table 2, it can be seen that the t-table of this study is 2,030; if t-count > t-table, it means that the independent variable has a significant effect on the dependent variable [9] The results showed that the t-count of all independent variables (communication, problem solving, initiative, planning and organization, self-awareness and technology) partially had no effect on the entrepreneurial orientation of beef cattle fattening in Tasikmadu District because all tcount < ttable. The majority of respondents in Tasikmadu District only make beef cattle fattening a business or a side job to increase the economic value of their families, because the scale of their beef cattle fattening business is still small, with an average of raising 1-5 cattle. According to respondents, the scale of the business has not been able to support the family's economy, if it is used as a business or main job. This statement is in accordance with [15] that most Indonesian people make beef cattle as a side business to increase family income from their main livelihood as farmers, besides that beef cattle are traditionally kept and used as savings. This causes some of the variables used in this study, do not have a significant effect on the entrepreneurial orientation of the farmer members of the Subur Lestari Farmer's Group.

3.4 Classic assumption test

Classical assumption tests in this study include multicollinearity test, heteroscedasticity test and autocorrelation test. The multicollinearity test can be seen from the tolerance value or the Variance Inflation Factor (VIF) value. The tolerance value limit is not less than 0.1 and the VIF value is not more than 10, so it can be concluded that the model used is free of multicollinearity between the independent variables in the regression [14]. The results show the tolerance value of all independent variables has a tolerance value> 0.1 and the VIF value is less than 10. These results can be concluded that there is no multicollinearity between independent variables in the regression model.

According to [12], to test the autocorrelation can use the Durbin Watson statistical test (dw test) using SPSS. The decision making for the absence of autocorrelation in the Durbin Watson test is du < dw < 4-du, where the value of du is obtained from the Durbin Watson statistical table. The value of dw will be compared with the table value with a significance value of 5%. The number of samples is 42 and the number of variables is 6, the values observed from table dw are dl 1.202 and du 1.845. The test results obtained Durbin Watson has a value of 2,050. The test results show that du < dw < 4-du, where 1.845 < 2.050 < 2.155. It can be concluded that there is no autocorrelation.

The heteroscedasticity test in this study uses the Glejser test, which is regressing absolute residuals with each independent variable. The results of the Glejser test on the significance value will be compared with a value of 0.05, if the significance value is > 0.05 then heteroscedasticity is free [12]. The significance results of all variables have a value greater than 0.05. It can be concluded that there is no heteroscedasticity in the regression model.

4 Conclusion

Based on the research that has been done, it can be concluded that the entrepreneurial competence variables consisting of communication, problem solving, initiative, planning and organization, self-awareness and technology have a simultaneous effect on entrepreneurial orientation, but individual variables have no effect on entrepreneurial orientation.

The authors would like to acknowledge the research grant from the Kementerian Pendidikan dan Kebudayaan Riset dan Teknologi Indonesia, through research grant Penelitian Terapan UNS, 2021.

References

1. H. Y. Prawira, Muhtarudin dan R. Sutrisna. Jurnal Ilmiah Peternakan Terpadu. **3**, 250-255 (2015)
2. S.H. Purnomo, A. I. Sari, S. Emawati and E.T. Rahayu. IOP Conference Series: Earth and Environmental Science, **518**, 012054 (2020)
3. H. Afriani, N. Idris dan Fatati. Jurnal Ilmiah Ilmu-Ilmu Peternakan. **17**, 77-83 (2014)
4. Mudjiarto. *Membangun Kepribadian dan Kewirausahaan* (Graha Ilmu, Yogyakarta, 2006)
5. S.H. Purnomo, E. T Rahayu, A. I Sari and S. Emawati. IOP Conference Series: Earth and Environmental Science, **250**, 012093 (2019)
6. A.D. Ratnasari. eJournal Administrasi Bisnis. **5** (2017)
7. Badan Pusat Statistik (BPS) Kabupaten Karanganyar. Statistik Daerah Kabupaten Karanganyar. Badan Pusat Statistik Karanganyar. Kab. Karanganyar (2020)
8. U. Argoyogi, Umiyasih, D. B. Wijono, dan D. Wahyono. Prosiding Seminar Hasil Penelitian T.A 1998/1999. Balai Penelitian Tanaman Pangan Karangploso. Malang (2000)
9. T. D. Santosa. Jurnal Sainstech Politeknik Indonesia. **1**, 17-27 (2016)
10. Sugiyono. *Statistika Untuk Penelitian* (Alfabeta, Bandung, 2012)
11. M. Abdulrahman, A.M. Sambas, dan S. Anting. *Dasar-Dasar Metode Statistika Untuk Penelitian* (Pusat Setia, Bandung, 2011)
12. I. Ghozali. *Aplikasi Analisis Multivariate Dengan Program SPSS Cetakan Ke-4* (Badan Penerbit Universitas Diponegoro, Semarang, 2011)
13. M. S. Sosroadmidjodjo, dan Soeradji. *Peternakan Umum* (CV Yasaguna, Jakarta, 2000)
14. D. Priyatno. *Statistika Terapan* (Gadjah Mada University Press, Yogyakarta, 2016)
15. Suryana. *Kewirausahaan: Pedoman Praktis, Kiat dan Proses Menuju Sukses* (Salemba Empat, Jakarta, 2011)

Using fuzzy time series with and without markov chain: to forecast of edible bird nest exported from Indonesia

Osfar Sjofjan[1*] and *Danung Nur* Adli[1,2]

[1]Faculty of Animal Science, Universitas Brawijaya, Jl. Veteran, Malang 65145, Indonesia
[2]Animal Feed and Nutrition Modelling Research Group (AFENUE), Department of Animal Nutrition and Feed Technology, Faculty of Animal Science, IPB University, Bogor 16680, Indonesia

Abstract. Edible bird nest (EBN) were traditional medicine consumed by the Tiongkok. This study compared two-algorithm method. Fuzzy time series and Markov chain as forecast method the number of bird nest exported from Indonesia. The secondary data between 2012 and 2018 were from Bureau Central Statistic (BPS). The scope using in this study were bird nest between 2012 until 2018, with a unit of volume kilograms (Kg). Used secondary export data, collected from BPS of Indonesia. Data were analysed using Fuzzy Time Series with and without Markov Chain using R Studio. The result showed that Fuzzy Time Series with and without Markov Chain method performs better in the forecasting ability in short-term period prediction and the values of Mean Absolute Percentage Error (MAPE) and Mean Square Error (MSE) tends to be smaller than the Fuzzy Time Series without Markov Chain. It can be concluded the number of exported can be used Fuzzy time series.

1 Introduction

One of them is influenced by the composition and nutrition of the feed ingredients used in the feed mixture so that it can accelerate of poultry [1-3]. Feed material is any food that can be eaten, preferred, partially or completely digested, absorbable and beneficial for livestock [4, 5]. Edible bird nest (EBN) were traditional medicine consumed by the Tiongkok. Edible bird nest were the most expensive animal products, which products by salivation of swiftlets (*Aerodramus fuciphagus*). Swiflets in Indonesia called *wallet* and edible bird nest were *sarang wallet*. The swiftlets were speeded between entire rest of the world. The exported of this product were increase during recently five years. During five years a rural economic were helps to maintenance balancing of rural economical farmers in urban area [1]. There are inconsistently data such as Indonesia, there is a lot of method that can be used in the livestock data. For example are meta-analysis that uses to coin a single experiment by used various models instance a dataset [6, 7]. But, later we forget to conduct the estimated or rearrange plant of the idea of the research. One of the method is calling fuzzy time series. Time series

* Corresponding author: osfar@ub.ac.id

forecasting is the process of analyzing time series data using statistics and modelling to make predictions and inform strategic decision-making [8, 9].

In accordance to increase a number of imported number there are predict and mathematical models to use it. Russian experts named A. A. Markov in 1906 first developed the Markov Chain (Markov Chain) [10]. Conceptually the Markov chain can be illustrated by assuming $\{Xn, \text{n} = 0, 1, 2 \ldots\}$ as a finite stochastic forecasting value [11].

2 Materials and methods

There is a largely lacking data of the livestock sector that can be used to be exported. We try to reach data more stable, until we selected this sector as the data. The scope using in this study were bird nest between 2012 until 2018, with a unit of volume kilograms (kg). Used secondary export data, were collected from BPS of Indonesia [3]. Data were analysed using Fuzzy Time Series with and without Markov Chain. The steps were divided into 4 steps there were: determine the set of universes, create fuzzy sets, perform fuzzification, create FLR, create FLRG and defuzzify using software. Furthermore, library "AnalyzeTS" "Fuzzynumbers" and "fuzzysTs" are installed. Moreover, our data scientist are using ggplot2 library to conduct also the map origin that exported to origin contries.

Table 1. Edible bird nest of swiflet Export Data 2012- 2018

Time	Period	Netweight (tonnes)
1	2012	405.321
2	2013	536.787
3	2014	636.104
4	2015	761.216
5	2016	992.161
6	2017	1286.648
7	2018	1291.987

Source: [3], various years from 2012 to 2018, after fitted data to probes 0.25; 0, 75; 1.00

2.1 Time series analysis

The actually main function time series analysis were to forecast or predictions of the data. The assumption first derivative as exploration of time. In this experiment were not determine random component but stationary data. In order to forecast future values of exported data [1]. The main objective in analysing time series is to understand, interpret and evaluate changes in the phenomena in the hope of more correctly anticipating the course of future events.

2.2 Fuzzy set and forecasting

The fuzzy sets theory can be defined as a mathematical formalization that enables us to eliminate indefiniteness and deal with incomplete, inaccurate information of both qualitative and quantitative data [3]. The fuzzy set were classical mathematical method introduction from [12].

2.3 Fuzzy time series (FTS)

Fuzzy Time Series (FTS) is a forecasting data method that uses the concept of fuzzy set as the basis of calculations. The forecasting system with this method works by returning patterns from actual data and then used to project future data. The process also does not require a learning system from a complex system, as is the case with genetic algorithms and neural networks [13]. Furthermore, Fuzzy Time Series (FTS) has terms that are published, namely Fuzzy Logic Relations (FLR) and Fuzzy Logic Relations Group (FLRG). Fuzzy Logic Relations (FLR) is a fuzzy logic that has a relationship between the series of associations that have been set in the previous data and continued. Then FLRG is only the same as FLR, but only what is determined is the group making from FLR which is the same as the weighted group.

The algorithm as following step from Devore, *et al.* [14]

$$Ai = f Ai\ (uk)/ uk + f Ai\ (uk)/ uk + \ldots + f Ai\ (uk)/uk \quad (1)$$

Step 1. Finding discourse and the intervals. as:

$$U = [Dmin - D1, Dmax + D2] \quad (1)$$

Notes: D1 and D2 are the positive number that determines the effective long interval is the average based Fuzzy Time Series model [14].

Step 2. Separated between the intervals. Partition universal discourse U into several equal intervals. Let the universal discourse U be partitioned into n equal intervals; the difference between two successive intervals can be defined as ℓ as follows:

$$\ell = [(Dmax + D2) - (Dmin - D1)] / n \quad (1)$$

Step 3. Classify the fuzzy sets.

Step 4. Fuzzy the data. Fuzzy the actual data.

$$\ell = [(Dmax + D2) - (Dmin - D1)] / n \quad (3)$$

$$F(t) = (m1 + m2 + mj)/n \quad (7)$$

3 Results and discussion

3.1 Comparison of fuzzy time series and markov chain

In the beginning the author try to reach data of livestock sector that present exported product but later there is inconsistently data that provided on exported on livestock sector. Generally, fuzzy time series methods are based on three stages [13-15]. These are fuzzification, determination of fuzzy relation, and defuzzification [8]. In the literature [9-11], the fuzzy time series methods are improved by employing various artificial intelligence techniques in these three stages. The genetic algorithms, particle swarm optimization, and fuzzy c-means methods are used in the fuzzification stage. Feed forward neural networks are used determining fuzzy relation and defuzzification stage. Many time series in actual life to defined uncertain observations. This kind of the time series is called fuzzy time series.

Table 2. The forecasted values of Indonesia bird nest swift let

Time	Year	Actual data	Fuzzy Data	Time	Year	Actual data	Fuzzy Data
1	2012	405.321	A1	16	2012	405.321	A2
2	2013	536.787	A1	17	2013	536.787	A3
5	2016	636.104	A1	20	2016	636.104	A4
7	2018	761.216	A2	30	2018	761.216	A1

Forecasting using Fuzzy Time Series (FTS) is done to get the appropriate comparison model. The difference between Fuzzy Time Series-Markov Chain (FTS-MC) and FTS is the forecasting step. Forecast calculations using FTS use equations (Rule1), (Rule2), and (Rule3). For example, for forecasting in 1990 the data seen is the previous year's data, namely 1989 where the state transitioned from $A1$ to $A1$, the forecasting calculation uses equation (7) because of Fuzzy. Our fuzzy time series found that the actual data from 2012 until 2018 following forecasted value were at near to the actual value which is range 2016 -2018 were at 429,919-1429,919. While, the adjusted forecasting value started from 2012 until 2018 are 308.730; 848.41, 0, 0, 0, 0, 4.83 respectively.

Table 3. Forecasting of Indonesian birdnest Export

Year	Actual data	Forecasted Value	Adjusted Value	Adjusted Forecasting value
2012	405.321	395.321	-384.89.41	308.730
2013	536.787	636.787	0	848.41
2014	636.104	845.214	0	0
2015	761.216	341.216	0	0
2016	992.161	1092.161	429,919.83	0
2017	1286.648	1106.648	1429,919.83	0
2018	1291.987	1491.987	429,919.83	4.83

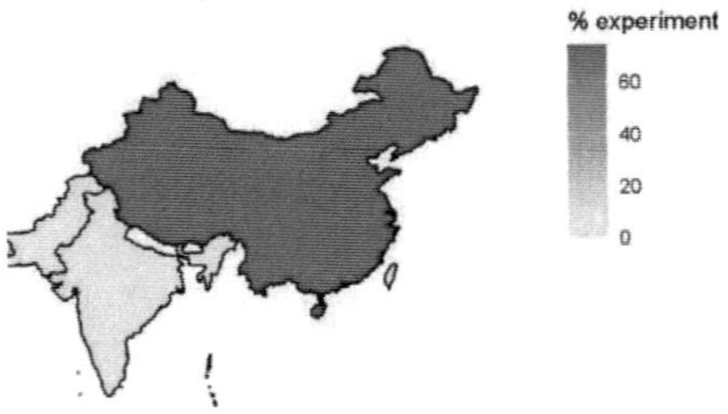

Fig. 1. Actual and estimated of the edible bird nest exported from Indonesia on countries origin target. (Figures are belong to author itself).

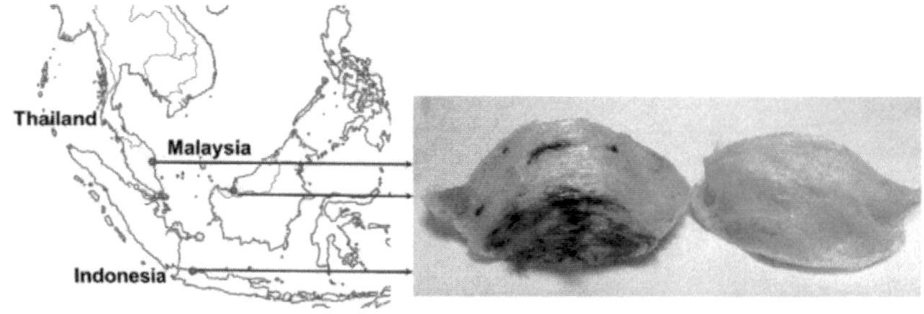

Fig. 2. The location of edible bird nest (EBN) based on the countries origin adapted from [16].

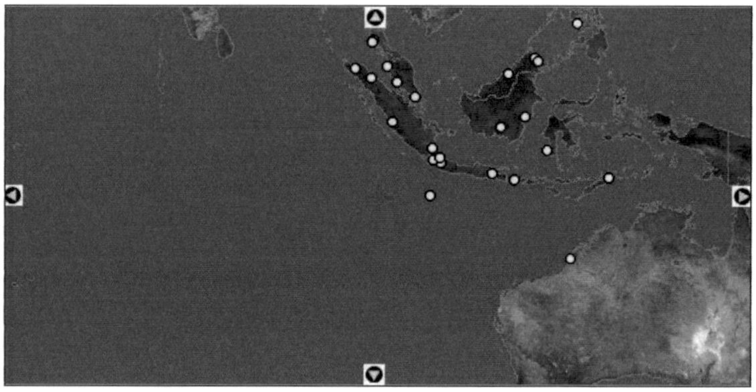

Fig. 3. Pin point view of *Aerodramus fuciphagus*, the dataset are Indonesia countries.

Map center: NAD83 Lat-long 10.5°S 99°E; UTM 47 500000E 8839306N Resolution 0.125 degrees/pixel, Target 10.2°S 99°E. Source [17]

3.2 Challenges future after using fuzzy time series on edible bird nest

Fuzzy Markov Chain in regard to the proposed method is 2.1% and 1.26%. That the forecasting error of Mean Absolute Percentage Error (MAPE) and Mean Square Error (MSE) Fuzzy Time Series proposed method is 1.0% and 0,5 which is better than that of the other methods. Moreover, that according to figure 1. Our edible nest are mostly exported to china mainland, Taiwan, Hong Kong, Japan, and Korea. Following by new emerging market that started from 20-40% exported are in the Middle East. Thus, research in the edible nest are largely unknown. Our hypothesis the total value of the edible bird nest (EBN) are only reached tens millions of total production. Whether, this are only informal sector of Indonesia's economy that is difficult to estimate. But, this sector are underestimated from this country. The unregulated industry are raised through year of year (yoy) that Indonesia swiftlet farmers could producing more than just nest. The bird nest must be utilisation in accordance to fulfil requirement. Another, obstacle are since 2003 the virus H5N1 or later called avian flu are most trouble for responsibilities 146 dead. Later, it was coined as global fears of a pandemic. The number of toll in Indonesia are reached highest in the world (Fig 2). The problem in the future are caused by diseased [16, 20]. Water pump must be acquired maintenance to inhibit mosquitoes [17-19].

4 Conclusion

Because edible bird nest (EBN) are mostly potential of exported end product from livestock sector since the china mainland are using for herbs it should be optimise of it potential both government, private sector, or even scientist. There is a lot fuzzy time series can be used as this model such as Markov chain, arima, Chen, Singh, hybrid, grey, or even novelty of this fuzzy time series. We suggest for doing this research in the future. Last, It can be concluded the number of exported can be used Fuzzy time series.

We would to say thanks to central bureau statistics to provide an open access data for exported edible bird nest, later we say thanks to whose providing library R studio from BPPT Cran.

References

1. O. Sjofjan, D. N. Adli, M. H. Natsir, Y. F. Nuningtyas, I. Bastomi, F. R. Amalia, J. Indones. Trop. Anim. Agric. **46** (2021)
2. O. Sjofjan, D. N. Adli, M.H. Natsir, Y. F. Nuningtyas, T. S. Wardani, I. Sholichatunnisa, S. N. Ulpah, O. Firmansyah, Jurnal Ilmu Ternak dan Veteriner, **26** (2021)
3. Badan Pusat Statistik, Jakarta. Indonesia (2018)
4. O. Sjofjan, M. H. Natsir, D. N. Adli, D. D. Adelina, L. M. Triana, IOP Conf. Ser. Earth Environ. Sci. **465** (2020)
5. D. N. Adli, O. Sjofjan, M. H. Natsir, Y. F. Nuningtyas, N. Sholikah, A. C. Marbun, Livestock Res Rural Dev. **32** (2020)
6. O. Sjofjan, D. N. Adli, M. M. Sholikin, A. Jayanegara, A. Irawan, J. Anim. Feed Sci. **30** (2021)
7. O. Sjofjan, D.N. Adli, IOP Conf. Ser. Earth Environ. Sci, **478** (2020)
8. J. A. Rice, Int. J. Appl. Math, (2007)
9. C. H. Wang, Tourism Management, **25** (2004)
10. X. Zeng, L. Shu, J. Jiang, International Journal of Applied Mathematics, **46** (2016)
11. O. Sjofjan, D. N. Adli, R. P. Harahap, A. Jayanegara, D. T. Utama, A. P. Seruni, F1000Research, **10** (2021)
12. E. Eğrioğlu, Adv. Fuzzy Syst. (2012)
13. S. C. Albright, C. J. Zappe, W. L. Winston, *Data Analysis and Decision Making* (2011)
14. J. L. Devore, K. N. Berk, *Modern Mathematical Statistics with Application* (2007)
15. H. Guney, M. A. Bakir, C. H. Aladag, International Journal of Fuzzy Systems, **20** (2018)
16. E. K. Shim, S. Y. Lee, *Spectroscopy Europe*, **29** (2017)
17. L. Discover, Australia: Thomson Brooks/Cole. (2021)
18. D.N. Adli, Y. Chi, J. W. Lee, O. Sjofjan, nt. j. adv. eng. res. sci. **4** (2019)
19. D. N. Adli, Livestock Res Rural Dev, **33** (2021a)
20. D. N. Adli, Livestock Res Rural Dev, **33** (2021b)

Characteristics of reduced fat mayonnaise using pumpkin flour (*Cucurbita moschata*) as fat replacer

Hemas Azizila Nidhal[1], *Herly* Evanuarini[1*], and *Imam* Thohari[1]

[1] Faculty of Animal Science, Universitas Brawijaya Malang, 65145, Indonesia

Abstract. Reduced fat mayonnaise has the disadvantage of destabilization of emulsion. Emulsion destabilization that occurs is the separation of oil and water. Pumpkin flour is used to increase the viscosity and stability of the emulsion. The objective of this study was to determine the best percentage of the use of pumpkin flour in reduced fat mayonnaise based on pH, emulsion stability, moisture content, and sensory evaluation. The material used in this research was mayonnaise made from egg yolk, canola flower oil, vinegar, pumpkin flour as a fat replacer carbohidrates based, and other optional ingredients. The method used was an experimental laboratory with a completely randomized design with 4 treatments and 4 replications. The treatments consisted of without the use of pumpkin as a control, the use of pumpkin flour 2%, 4%, and 6% of the total oil used. The variables measured were pH, emulsion stability, moisture content, and sensory evaluation. The results showed that the use of pumpkin flour on mayonnaise gave highly significant effect on pH, stability emulsion, moisture content and increased panelists preferences. The conclusion of the study that using 6% pumpkin flour produced the best reduced-fat mayonnaise.

1 Introduction

Mayonnaise is one of the processed products which is an oil in water (o/w) emulsion in semi-solid form. Mayonnaise has a thick and creamy texture. The color of mayonnaise is generally white to yellowish-white with a distinctive sour aroma. The type of mayonnaise generally consists of a traditionally mayonnaise of at least 66% oil, reduced-fat mayonnaise of at least 50% oil, and low-fat mayonnaise of at least 25% oil [1]. Vegetable oil is used full-fat mayonnaise production in high quantities, causing the consumption of mayonnaise in the community to be limited. Foods that contain high fat are harmful to health of consumers. Reduced-fat mayonnaise is prepared by reducing the use of oil in full-fat mayonnaise. Reducing the volume of vegetable oil in reduced-fat mayonnaise can cause destabilization emulsion between the oil phase and the water phase. The emulsion system in mayonnaise is formed in the presence of a mixture of two or more food ingredients, namely emulsifiers and stabilizers. The addition of a stabilizer in reduced-fat mayonnaise will form

* Corresponding author: herlyfptub@ub.ac.id

a stable emulsion system. Stabilizers have the main function of stabilizing emulsions, binding water, improving texture, and thickening for food [2]. One source of natural stabilizers that is abundantly available in Indonesia is pumpkin which contains pectin. Pumpkin contains carotene, vitamins, minerals, fiber, and pectin which are good for health [3]. Pectin contained in pumpkin is hydrocolloid or water binding. Hydrocolloids play an important role in the food industry, one of which is as a stabilizer [4]. The content of pectin in food shows antioxidant activity, besides that it also contains vitamin C, carotenoids, and phenolic compounds. Pumpkin contains 1.27% pectin with a weight of 2.4 kg [5]. Pumpkin contains 92.24% water content, 0.15% fat, 0.98% protein, 0.76% ash, 0.56% crude fiber, and 5.31% carbohydrates [6]. Some natural stabilizers that have been used previously are the use of banana peel waste in the form of flour which is added to reduced-fat mayonnaise [17]. The addition of 6% watermelon rind flour as a natural stabilizer was able to increase the physicochemical quality from reduced-fat mayonnaise [18]. Based on the content and properties possessed by pumpkin, it can be added to reduced-fat mayonnaise in the form of flour. The use of pumpkin flour as a stabilizer can be used as a fat replacer from carbohydrates and has hydrocolloid properties. The purpose of this study was to determine the appropriate percentage of addition of pumpkin flour to reduced-fat mayonnaise on pH values, moisture content, emulsion stability, and sensory evaluation.

2 Materials and methods

2.1 Materials

The ingredients for making mayonnaise are canola flower oil (70% for control and 50% for reduced-fat mayonnaise), 5% vinegar, 20% egg yolk, 3% sugar, 1% salt, 0.5% white pepper powder, and 0.5% mustard. Addition of pumpkin flour according to the treatment, namely 2%, 4%, and 6%. The research materials for variable testing were distilled water, pH 4 buffer, and pH 7 buffer. The tools used for variable testing were analytical balance, pH meter, beaker glass, petri dish, oven, desiccator, test tube, centrifuge and questionnaire sheet. While the tools used to make mayonnaise are mixers, spatulas, and analytical scales.

2.2 Methods

The method used in this study is a laboratory experiment method with a completely randomized design with four treatments and four replications presented in Table 1.

Table 1. Experimental design and research treatment

Treatments	Descriptions
Control	FFM control with 70% oil
T1	RFM with 50% oil and 2% pumpkin flour
T2	RFM with 50% oil and 4% pumpkin flour
T3	RFM with 50% oil and 6% pumpkin flour

2.3 Pumpkin flour preparation

The pumpkin used for the study was purchased from the traditional market in Pudjon District, Batu City, East Java Province. The pumpkin is first peeled and seeds removed. Pumpkin was cut into 3 x 3 cm sizes and put in an oven at 60°C for 48 hours until the weight was constant. The dried pumpkin was mashed using a dry mill, sieved using an 0,177 millimeters sieve, and the pumpkin flour was ready to use.

2.4 Mayonnaise preparation

Mayonnaise production begins by mixing the ingredients into a plastic basin. The ingredients mixed are salt, sugar, mustard, and powdered white pepper using a mixer at 1500 rpm for 1 minute. Added egg yolks into a bowl, then added oil interspersed with vinegar. The ingredients are mixed to form an emulsion. Add the apple peel flour and pumpkin to the mayonnaise and mix for 10 seconds.

2.5 Data analysis

Data were analyzed statistically using analysis of variance (ANOVA). If there are results that show a difference in influence between treatments, it is followed by Duncan's Multiple Range Test.

3 Results and Discussion

3.1. pH value

The results of ANOVA show that the addition of pumpkin flour in different rations had a highly significant effect on the pH value of mayonnaise ($P < 0.01$). Table 2 shows the average pH of mayonnaise with pumpkin flour added.

Table 2. The average pH value of reduced-fat mayonnaise

Treatments	pH value
Control	4.91 ± 0.08^{b}
T1	4.52 ± 0.09^{a}
T2	4.66 ± 0.07^{a}
T3	4.97 ± 0.03^{b}

Note: ^{a,b} Several superscripts in the same column showed a very significant difference ($P < 0.01$) from the pH value of mayonnaise.

The pH of mayonnaise plays an important role in the structure and stability of the emulsion formed. The pH of mayonnaise is affected by the use of egg yolk, vegetable oil, mustard, and vinegar in the mayonnaise-making process. The pH of mayonnaise is affected by the isoelectric point pH of the yolk protein, which plays an important role in the structure and stability of the mayonnaise emulsion [7]. Increasing the percentage of addition of pumpkin flour will increase the pH value of mayonnaise. This is because the pH of pumpkin flour is 6.6.

3.2. Moisture content

The results of ANOVA show that the addition of pumpkin flour in different rations had a highly significant effect on the moisture content of mayonnaise ($P < 0.01$). Table 3 shows the average moisture content of mayonnaise with pumpkin flour added.

Table 3. The average moisture content of reduced-fat mayonnaise

Treatments	Moisture content (%)
Control	21.34 ± 0.09^{d}
T1	20.60 ± 0.07^{c}
T2	19.43 ± 0.08^{b}
T3	18.32 ± 0.05^{a}

Notes: ^{a,b,c,d} Several superscripts in the same column showed a very significant difference ($P < 0.01$) from the moisture content of mayonnaise.

Moisture plays an important role in the end result of the emulsion product, which affects the balance of protein content and fat content. The moisture content of reduced-fat mayonnaise comes from the ingredients used in its production, namely vinegar as an acidulant. Adding water and vinegar to reduce the oil concentration may increase the moisture content of the mayonnaise. [8]. The addition of pumpkin flour to mayonnaise causes a decrease in moisture content. Pumpkin contains pectin which has high anhydrogalacturonic acid and methoxyl [9]. Pectin in pumpkin flour has anhydrogalacturonic acid which is higher than the ash content. The heating treatment of pumpkin flour for too long causes the hydrolysis of pectin into galacturonic acid. Galacturonic acid plays an important role in the formation of the structure and texture of the pectin gel. The texture of reduced-fat mayonnaise with the use of pumpkin flour has a good consistency. Pectin is a hydrocolloid, which can bind water when added to food products. Another important role of hydrocolloids is that they can be used as food stabilizers. Another property of pectin is that it can form a gel at very low concentrations [10]. The content of carotenoids and vitamin C in pumpkin is higher than the content of sugar, starch, and total protein [11].

3.3. Emulsion stability

The results of ANOVA show that the addition of pumpkin flour in different rations had a highly significant effect on the stability of mayonnaise emulsion ($P < 0.01$). Table 4 shows the average stability emulsion of mayonnaise with pumpkin flour added.

The stability of the mayonnaise emulsion is influenced by the use of egg yolk, vegetable oil, acidifier, and the addition of stabilizer. Adding pumpkin flour increases the stability of the mayonnaise emulsion. Pumpkin flour has hydrophilic colloidal properties can be used as a stabilizer in reduced-fat mayonnaise. Destabilization of emulsion in mayonnaise is characterized by the separation of oil and water. The use of pumpkin flour will stabilize the oil phase and the water phase. Reduced-fat products are made by decreasing the dispersed phase and increasing the dispersing phase. The oil phase serves to achieve emulsion stability and prevent destabilization, but low-fat products can use a stabilizer.

Table 4. Average emulsion stability of reduced-fat mayonnaise

Treatments	Emulsion stability (%)
Control	89.29±0.06[c]
T1	86.60±0.05[a]
T2	88.53±0.05[b]
T3	90.70±0.05[d]

Notes: [a,b,c,d] Several superscripts in the same column showed a very significant difference ($P < 0.01$) from the stability emulsion of mayonnaise.

3.4. Sensory evaluation

Table 5 shows the results of the comprehensive evaluation of aroma, texture, tastem color, and mayonnaise overall acceptance.

Table 5. Average mayonnaise sensation assessment

Treatments	Texture	Aroma	Color	Taste	Overall Acceptance
Control	3.55±0.96	3.40±0.82	4.45±0.96	3.45±0.96	3.55±0.96
T1	3.75±0.50	3.75±0.96	4.10±0.58	3.70±0.56	3.88±0.82
T2	4.25±0.50	4.30±0.58	3.90±0.56	4.15±0.96	4.20±0.82
T3	4.55±0.96	4.45±0.96	3.65±0.50	4.55±0.96	4.50±0.58

3.4.1. Texture

The highest score for the texture evaluation by the panelists was obtained with reduced-fat mayonnaise with 6% pumpkin flour added. The crude fiber content in pumpkin flour is 5.8%. The addition of starch-containing materials can affect the resulting texture [12]. Reduced-fat mayonnaise with the addition of pumpkin flour has a very soft and very thick texture. The viscosity formed in mayonnaise is due to the presence of starch and pectin in pumpkin flour which functions as a stabilizer [10]. The addition of pumpkin flour with a higher percentage causes the mayonnaise to be more stable and thick. The addition of 6% pumpkin flour had a thick texture that was almost the same as full-fat mayonnaise. This addition can stabilize and replace the reduced oil composition of reduced-fat mayonnaise.

3.4.2. Aroma

Highest score of aroma given by the panelists was in the treatment of 6% addition of pumpkin flour. T3 treatment of reduced-fat mayonnaise has a distinctive aroma of pumpkin. Pumpkin flour has a distinctive caramel and fragrant aroma that is different from other flours [13]. The more addition of pumpkin flour concentration causes the sour aroma of the mayonnaise decrease. The aroma of mayonnaise is caused by the use of ingredients such as oil, vinegar, eggs, and mustard [14]. Panelists like mayonnaise which has a less acidic aroma, thus giving the highest score in treatment 3 compared to the control treatment.

3.4.3. Color

Participants gave the mayonnaise the highest color rating without the addition of pumpkin flour. The higher pumpkin flour added concentration causes the color of the mayonnaise to be more yellow to brownish-yellow. This is because pumpkin flour has a yellow-orange color [15].

3.4.4. Taste

The control treatment had a sour taste, while the addition of 6% pumpkin flour had a sweet and sour taste produced by pumpkin flour. Pumpkin flour has a sweet taste [15]. The flavor-forming factors in mayonnaise are salt, sugar, and mustard and other optional ingredients are added to add flavor, color, and maintain the stability of the mayonnaise [16]. Mayonnaise is a portion of food that sour taste.

3.4.5. Overall acceptance

The overall acceptance score of the product is based on the texture, aroma, color, and taste formed. The addition of pumpkin flour to 6% reduced-fat mayonnaise was favored by the panelists. The resulting mayonnaise has a very soft and thick texture, a fragrant aroma typical of pumpkin, a brownish yellow color that can still be accepted by panelists, and a sour and sweet taste of mayonnaise and pumpkin flour.

4 Conclusion

Adding 6% pumpkin flour to mayonnaise improves the quality of reduced-fat mayonnaise and is accepted by panelists.

References

[1] M. P. Gaikwad, H. M. Syed, D. D. Shinde, Journal of Pharmacognosy and Phytochemistry, **6**, 06-09 (2017)
[2] C. Marchelina, H. Sinaga, L. M. Lubis, Indonesia Journal of Agricultural Research, **3**, 10-22 (2020)
[3] M. Z. Amin, T. Islam, M. R. Uddin, M. J. Uddin, M. M. Rahman, M. A. Satter, Heliyon, **5**, 1-5 (2019)
[4] W. Ashraf, A. Shehzad, H. R. Sharif, R. M. Aadil, M. R. Khan, L. Zhang, Journal of Food Processing and Preservation, **44**, 1-11 (2020)

[5] A. A. Torkova, K. V. Lisitskaya, I. S. Filimonov, O. A. Glazunova, G. S. Kachalova, V. N. Golubev, T. V. Fedorova,*PloS one*,**13** (2018)
[6] S. El-Khatib, M. Muhieddine, *The Health Benefit of Foods–current Knowledge and Further Development* (Intechopen, London, UK, 2020).
[7] T. P. Keerthirathne, K. Ross, H. Fallowfield, H. Whiley, *Pathogens*, **5**(4), 1-11 (2016)
[8] M. H. H. Amin, A. E. Elbeltagy, M. Mustafa, A. H. Khalil, Journal Agroaliment Proc Technology,**20**(1), 54-63 (2014)
[9] P. Ramachandran, A. K. Dhiman, S. Attri, Journal of Ecology, **44**(6) 685-689 (2017)
[10] Y. Bai, M. Zhang, M, S. C. Atluri, J. Chen, R. G. Gilbert, Food Hydrocolloids, **106,** 105894 (2020)
[11] S. Sharma, T. R. Rao, International Food Research Journal, **20**(5), 2309 (2013)
[12] Đ. Ačkar, A. Jozinović, J. Babić, B. Miličević, J. P. Balentić, D. Šubarić, Innovative food science & emerging technologies, **47**, 517-524 (2018)
[13] Y. B. Pramono, D. Handayani, S. Mulyani, C. H. Wibowo, IOP Conference Series*: Earth and Environmental Science*, **803**, 012039 (2021)
[14] D. Rahmawati, N. Andarwulan, H. N. Lioe, Jurnal Mutu Pangan: Indonesian Journal of Food Quality, **2**, 80-86 (2015)
[15] S. H. Ghaboos, S. M. Ardabili, M. Kashaninejad, International Food Research Journal, **25** (2018)
[16] L. Golchoobi, M. Alimi, S. Shokoohi, H. Yousefi. Journal Texture Studies. **47**, 1-10 (2016)
[17] H. Evanuarini, A. Susilo, IOP Conference Series: Earth and Environmental Science, **478**, 1-6 (2020)
[18] H. Evanuarini, D. Amertaningtyas, D. T. Utama. IOP Conference Series: Earth and Environmental Science, **788**, 1-8 (2021)

Production Performance of Layer Strain Hy-Line Brown in Different Cage Locations in Closed House

Edhy Sudjarwo[1], *Muharlien* Muharlien [1], *Osfar* Sjofjan, and *Dyah* Lestari Yulianti[1*]

[1]Animal of Animal Science, Universitas Brawijaya, Malang 65145, Indonesia

Abstract. The aim of this study was to determine the production performance of layer hen strain Hy-Line Brown in different cage locations in closed house. The research material was 5,051 laying hens of the Hy-Line strain aged 28 weeks. House system was closed house, cage arranged to form frame A. Hy-Line Brown layer were housed at cage densities of 500 cm^2 per hen. Research feed is complete feed code 324 KJ produced by PT. Charoen Pokphand Indonesia. The research method is comparative, comparing the production performance on the location of the cage in zone 1 (inlet), zone 2 (middle), and zone 3 (outlet) of a closed cage. The research sample is 10% of the total population. Observations were carried out on 6 cages which were randomly selected in each cage zone. The average density of cages is 20 birds/m^2. Tabulated research data are Hen Day Production (%), egg weight (g), egg mass (kg), and feed conversion. Data analysis is analysis of variance analysis (One-Way ANOVA). Based on the results of the study, there were significant differences in the the production performance of layer hen strain Hy-Line Brown in different cage locations in closed house. The highest production performance is the cage located close to the inlet (Zone 1).

1 Introduction

The house is an important part of the management of laying hens, because it is a place for all birds' activities so that the comfort of livestock must be guaranteed in order to obtain healthy and productive livestock. If kept in poor environmental conditions, birds cannot show performance that is in accordance with their genetic potential, which means a loss [1]. To anticipate unstable climatic factors, currently in Indonesia, closed house are widely used. A closed house system is a type of cage that has good air ventilation settings with the help of an automatic control panel.

The closed house system has ventilation that uses the pressure generated by the fan (exhaust fan) and a cooling pad as the cooling system [2]. The closed house has zoning inside, i.e. the zone close to the cooling pad/inlet, the middle zone and the zone close to the exhaust fan/outlet. The distribution of air flow, temperature, and humidity in the house will

* Corresponding author: dyahlestariyulianti@gmail.com

be different from the point of inlet to outlet (exhaust fan) [3]. Where the zone near the cooling pad has a lower temperature than the zone near the exhaust fan, the house zone near the inlet has a cooler temperature [4]. Differences in the house environment which include temperature, humidity, and air velocity result in differences in the effective temperature of laying hens at the inlet to outlet points [4].

Environmental temperature is the main non-genetic factor that affects the genetic potential of the layer [5]. Temperature is one of the most important environmental factors that can affect the health, behaviour and productivity of poultry [6]. Poultry behaviours can significantly affect its growth [7]. Different temperatures will affect the production activity and the quality of the eggs produced. Egg production is very susceptible to heat stress problems because of concentrated feeding and laying hens are mostly raised in cages with high stocking densities [8]. Killic and Simsek [9] stated that heat stress significantly reduces productivity due to reduced feed intake, reduced feed conversion, higher mortality, reduced egg production, lower egg shell quality, and smaller egg size.

The aim of this study was to determine the production performance of layer hen strain Hy-Line Brown in different cage locations in closed house.

2 Materials and methods

2.1 Animal, hen house, and feed

Animal in this study were 5,051 laying hens of the Hy-Line strain, aged 28 weeks. Research feed is complete feed code 324 KJ produced by PT. Charoen Pokphand Indonesia. House was a closed house with dimensions of 4 x 80 meters equipped with automatic feeding, drinking and egg collection equipment, located in Sumber Sekar Village, Dau District, Malang Regency. Birds are kept in cages/battery in house with a layout as shown in Figure 1. The research sample is 10% of the total population. Observations were carried out on 6 cages which were randomly selected in each cage zone. The average density of cages is 20 birds/m^2.

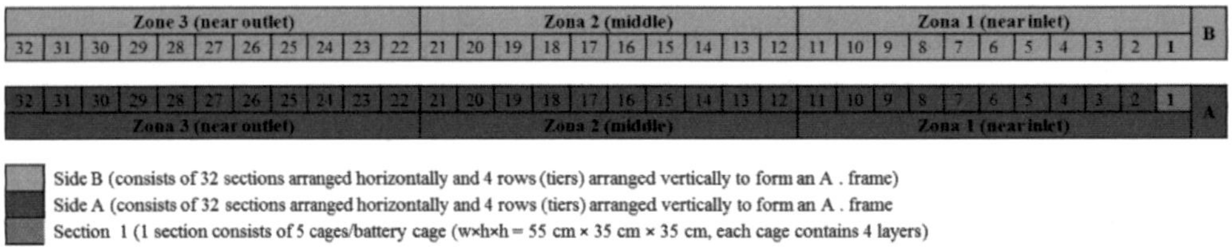

Fig. 1. Lay out cage/battery in closed house

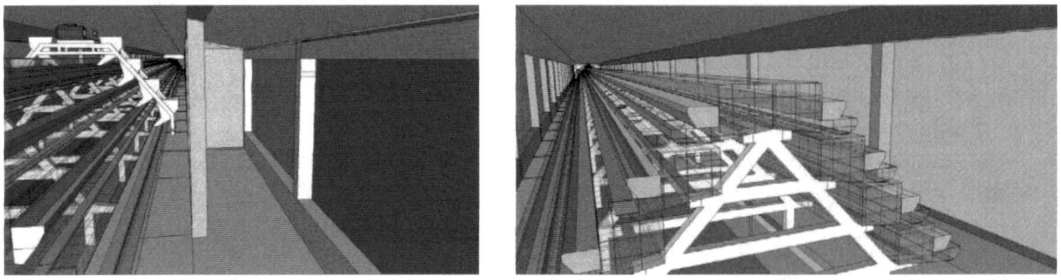

(a) (b)

Fig. 2. Cage layout (a) cages layer section A, (b) cage layer frame A

2.2 Method

Data was collected through observation of birds samples in each experimental unit. Microclimate data obtained by direct measurement using a thermohygrometer. Data that is recorded every day is climatic data including: temperature (°C) and humidity (%).

Layer productivity parameters include HDP (%), egg mass (kg), and observed for 1 month. Hen Day Production (HDP) is the calculation of egg production based on the number of layers per day (daily production = Hen day egg production), number of eggs (n) of layer groups on that day (m) in percentage. The HDP formula is as follows:

$$\text{HDP} = \frac{\text{n}}{\text{m}} \times 100\% \quad (1)$$

Egg quality is egg yolk weight (g) and Haugh Units.

Albumin consistance are calculated base on HU (Haugh Unit). The HU formula was as follow:

HU = 100 log $(H+7.57-1.7W^{0.37})$
H =Albumin height (mm)
W = Egg weight (g)

The data were subjected to an analysis of variance (ANOVA) using SPSS 16.0 software for Windows (SPSS Inc. Chicago,IL.USA). The differences between group were determined by Duncan's multiple-range tests. All value were presented as means and standard error of the mean, and significance level were set as $P<0.05$ and $P<0.01$.

3 Results and Discussion

Microclimate data in different house zones are presented in Table 1.

Table 1. Microclimate data in different house zones

Parameters	Letak Cage			*Sig.*
	Zone 1	Zone 2	Zone 3	
Temperature (°C)	24.83±0.70	25.53±1.14	25.85±1.17	0.247
RH (%)	73.83±4.83	72.00±5.51	68.00±7.20	0.252

Note: Mean± std. Deviation

Based on the results of statistical analysis, the microclimate in the three closed house zones laying hens strain Hy-Line strain rearing showed no difference, although numerically there was an increase in temperature from zone 1, zone 2, and zone 3. In line with research [2] which states that measurements from morning to evening show that the temperature will increase from the location of the house in front (inlet) to the location of the house behind (outlet). Yani et al. [3] stated that the position near the cooling pad is cooler because it is the first area to receive cold air that enters through the cooling pad. The cold air that enters through the cooling pad due to being sucked in by the fan at the end of the cage causes the air to move directly to the fan because it has a greater pressure. This condition causes the cold air that enters through the cooling pad to move continuously towards the fan and cools the cage so that the closer to the fan the air is higher than the position near the cooling pad.

According to Primaditya [10], closed house type have advantages such as birds that are not easily stressed due to extreme temperature changes from outside the house and humidity and temperature inside the cage that can be adjusted. The results showed that the highest humidity was in zone 1 and the lowest was in zone 3, but the relative humidity value was still in the standard range of layer period laying hens. According to Yuwanta [14], the ideal humidity for laying hens is 60-70%.

Based on the results of statistical analysis, the effect of different cage locations on the performance of laying hens of the Hy-Line strain, including Hen Day Production (%), egg weight (g), egg mass (kg), and feed conversion is presented in Table 2.

Table 2. The effect of different cage locations on the performance of laying hens of the Hy-Line strain, including Hen Day Production (%), egg weight (g), egg mass (kg), and feed conversion

Parameters	Cage Location			*Sig.*
	Zone 1	Zone 2	Zone 3	
HDP (%)	87.80±1.05^{c}	82.73±0.40^{b}	80.37±3.90^{a}	0.000
Egg weight (g)	63.50±3.16^{b}	64.50±3.25^{b}	59.75±2.30^{a}	0.010
Egg mass (kg)	82.27±2.28^{c}	73.38±0.90^{b}	69.22±5.81^{a}	0.000
FCR	2.63±0.08^{a}	2.97±0.05^{b}	3.17±0.08^{a}	0.000

Note: Week of age 28
Mean± std. deviation
The superscript on the same line shows the difference between treatments with a 95% and 99% significance level ($P<0.01$ and $P<0.05$)

The production parameters includes hen day production and egg weight. The average results of the effect of different cage locations on egg production performance are presented in Table 2. The results of statistical analysis showed a very significant difference in cage location on hen day production, egg mass, egg weight, and feed conversion ($P<0.000$). Based on the performance standard of Hy-Line Brown [11], laying hens of the Hy-Line Brown strain aged 28-29 weeks had an average egg production of 94%. Hen day production is one of the variables measured to determine the size of the level of layer egg production. The results of statistical analysis showed that the HDP at the location of each cage was significantly different. The average egg production at the front house (Zone 1) was the highest when compared to other treatments.

The productivity of modern laying hens varies greatly depending on breed, environmental factors, and disease. The factor that greatly affects production is heat stress. The heat stress will have an impact on the efficiency in converting feed into poultry products [16]. Heat stress is the most detrimental thing in a poultry business, especially for poultry in the high growth and production phase because in that phase the poultry has a high metabolic rate. The main effect of heat stress is a decrease in feed consumption. This is a consequence of the inability of birds to tolerate heat stress. Heat stress can be anticipated through several mechanisms, including sweating, panting, increasing the body area for the evaporation process [17].

Based on the results of the study, the average environmental temperature of the cages in zones 1, 2, and 3 was 24.83 °C, respectively; 25.53 °C; and 25.85 °C. The further away from the inlet, the higher the temperature of the cage. The temperature of the cage in the appropriate layer period is 18-20 °C, then it is increased by 1 °C every 2 weeks until it reaches 25°C. The temperature regulation in the cage must be equipped with a ventilation system that functions to maintain air quality at a certain temperature [11]. Lower temperatures after the peak production phase will trigger an increase in feed consumption so that it can fail to achieve production targets including egg weight, feed efficiency, and adult body weight. Meanwhile, high environmental temperatures will have an impact on decreasing feed consumption and increasing drinking water intake. This resulted in non-fulfillment of food intake so that the production target was not achieved.

High ambient temperature causes birds to carry out effective regulatory mechanisms, such as increased respiration rate. Optimal and efficient production is a combination of low feed intake and optimal egg production (total production). Generally, if the temperature increases, there will be a decrease in feed consumption and egg weight. The parallel impact caused by a decrease in feed consumption is a change in drinking water intake. The increase in environmental temperature will increase the consumption of drinking water.

The hormonal system and metabolic processes change due to heat stress. Heat stress is a sign of the start of changes in chemical reactions in the body. Changes that occur due to an increase in environmental temperature are a decrease in thyroid hormone secretion (although feed intake is maintained through a force-feeding mechanism) and a change in insulin and growth hormone secretion occurs. During prolonged heat stress, the secretion of hormones that play a role in anabolism and catabolism processes will decrease. The decrease in hormone secretion that plays a role in the anabolism process will reduce the efficiency and utilization of food substances that can be digested. This can result in production targets not being achieved.

Based on the results of statistical analysis, the effect of different cage locations on the egg quality of Hy-Line strain layer, including egg yolk weight (g) and Haugh Units is presented in Table 3.

Table 3. The effect of different cage locations on the egg quality of Hy-Line strain layer, including egg yolk weight (g) and Haugh Units (HU)

Parameters	Cage Location			*Sig.*
	Zone 1	Zone 2	Zone 3	
Yolk weight (g)	16.63±0.92	16.25±0.89	16.12±0.64	0.463
HU	70.04±2.83^{a}	77.14±5.76^{b}	76.20±7.78^{b}	0.048

Note: Mean± std. deviation
($P<0.05$)

The superscript on the same line shows the difference between treatments with a 95% and 99% significance level ($P<0.01$ and $P<0.05$)

The microclimate environment of the house can affect the quality of the eggs produced. The average results that there weren't significant difference of cage location on egg quality which are presented in Table 3. measurement of egg quality will show interactions between treatments on egg white height, haugh unit, and shell thickness [12].

Haugh unit (HU) is a parameter of egg interior quality which is calculated based on albumen height and egg weight [13]. The higher the HU value, the higher the egg quality and the better the albumin quality. The difference in the microclimate of each cage with a different location affects the Haugh unit value. The value of HU at each cage location is good in zones 2 and 3.

With the right supply of feed in quality and quantity, poultry is not infected with disease, and neutral environmental temperature conditions will produce eggs with relatively constant physical composition and size. When the nutrient content does not meet the needs, the layer will make adjustments in several ways, namely reducing the size and or some eggs produced but the chemical composition of the product is maintained. Another adjustment method is through the mechanism of modifying the chemical composition of the product while maintaining production or changing the size, number, and composition of eggs. The evoked response depends on the intake of nutrients that are related to the physiology of egg formation [17].

Albumin quality is related to consistency, addition, and functional properties. Albumin quality is measured by Haugh Units (HU) which is calculated from albumin height and egg weight [18]. The minimum value that can be accepted by consumers is 60. Most eggs produced from animal husbandry should have a HU value of 75-85 [18]. Egg white consistency is influenced by poultry age, genetics, age and shelf life of eggs, vanadium mineral content in a feed, and disease. Poultry with high egg production tends to produce fewer eggs with thick egg whites [19]. This statement explains the results of the study which showed that there was a negative correlation the higher the egg production, the lower the HU value.

4 Conclusions

Based on the results of the study, there were significant differences in the performance of production and egg quality of laying hens strain Hy-Line at different cage locations in closed house. The highest production performance is the cage located close to the inlet (Zone 1).

Authors wishing to acknowledge encouragement from Animal Science Faculty, Universitas Brawijaya, special work by Poultry Production Laboratory technical staff and financial support from Animal Science Faculty Universitas Brawijaya.

References

1. J. J. R. Feddes, E. J. Emanuel, M. J. Zuidhof, Poult. Sci. **81,** 774-779 (2002)
2. D. T. Amijaya, A. Yani, Rukmiasih, Jurnal Ilmu Produksi dan Teknologi Hasil Peternakan. **6**(3),98-103 (2018)
3. A. H. Yani, Suhardiyanto, Erizal, B. P. Purwanto, Med Pet. **37**, 87-94 (2014)
4. Renata, T. A. Sarjana, S. Kismiati, Jurnal Ilmu-Ilmu Peternakan **28**, 183-191 (2018)
5. A. O. Oguntuji, O. M. Alabi, World Poultry Science Journal. **66** (2010)
6. A. B. Webster, M. Czarick, Journal of Applied Poultry Research. **9**, 118-129 (2000)
7. D. P. Neves, T. M. Banhaji, I. A. Naas, Brazilian Journal of Poultry Sci. **16**, 1-6 (2014)
8. X. Tong, S. W. Hong, L. Zhao, Biosystems engineering **30**, 1-15 (2018)
9. I. Killic, E. Simsek, Journal of Animal Veterinary Advances. **12**, 42-47 (2013)
10. F. M. Primaditya, S. Hidanah, Soeharsono, Agroveteriner, **3**, 99-106 (2015)
11. Hy-Line, Management Guide Commercial Layers Hy-Line Brown (2018)

12. L. Komalasari, Dampak suhu tinggi terhadap respon fisiologi, profil darah dan performa produksi dua bangsa ayam berbeda, Tesis, Institut Pertanian Bogor, Bogor (2014)
13. K. M. Keener, K. C. McAvoy, J. B. Foegeding, P. A. Curtis, K. E. Anderson, J. A. Osborne, D. J. Bush. Poultry Science. **85**, 550-5 (2006)
14. SPSS 16.0,. Statistical Package in Social Sciences for Windows (Statistical Innovations Inc., Chicago, USA, 2010)
15. T. Yuwanta, *Telur dan Kualitas Telur* (Universitas Gajah Mada Press, Yogyakarta, 2010)
16. L. G. Cavalchini, S. I. Cerolin, R. I. Marian, Environmental influences on laying hens production. In: Sauveur B. (Ed.). L'aviculture en Méditerranée. Montpellier: CIHEAM. Options Méditerranéennes. Série A. Séminaires Méditerranéens. 153-172 (1990)
17. J. A. F. Rook, P.C. Thomas, *Nutritional Physioloy of Farm Animals* (Longman Inc., New York, 1983)
18. J. A. Coutts, G. C. Wilson, *Egg Quality Handbook* (Queensland Department of Primary Industries, Australia, 1990)
19. J. E. Jones, J. Solis, B. L. Hughes, D. J. Castaldo, J. E. Toler, Poultry Science. **69**, 378-387 (1990)

Alabio ducks, germplasms of South Kalimantan, as protein and income sources

Eni Siti Rohaeni[1], *Ratna Ayu* Saptati[2*], and *Lintje* Hutahaean[1]

[1]Indonesian Center for Agricultural Technology Assessment and Development, Bogor 16114, Indonesia
[2]Indonesian Center for Animal Research and Development, Bogor 16151, Indonesia

Abstract. Apart from providing food originating from meat and eggs, Alabio ducks have another important role for farmers in South Kalimantan, namely as an income source, employment source, and preservation of local genetic source. This paper aimed to study the economic profile of Alabio ducks in several types of businesses in Hulu Sungai Utara Regency, South Kalimantan. A survey method was carried out on several groups of duck farmers. In order to obtain more in-depth information, a focus group discussion (FGD) was held with key insiders in the duck industry. The results showed that the duck farming business has good prospects and can be a source of income and employment for farmers and young generations. The business of hatchery, slaughter-duck producers, hatching egg producers, and consumes-egg producers, is economically feasible. Improving bio-security and handling and managing environmental sanitation have to carry out, among others, through pens, equipment, and environmental disinfection, as well as routine vaccinations with the escort and support of local and central government. In addition, to increase the welfare and income of farmers need to introduce technological innovations that can increase farming efficiency.

1 Introduction

Changes in consumption patterns certainly follow the increase in people's welfare, one of which is by increasing the need for animal protein. Poultry is a source of animal protein and is mainly raised by the community. In addition to being easy to maintain, the cost of poultry farming is also cheaper than other livestock. A common type of poultry that plays an important role in meeting the nutritional needs of the community is ducks. Alabio duck is one of the local livestock commodities developed as a germplasm resource in South Kalimantan. Alabio ducks have an important role for farmers in South Kalimantan, namely

* Corresponding author: ratnaayusaptati@yahoo.com

as a source of animal food from meat and eggs, a source of farmers' income, creating employment opportunities, and at the same time as a genetic source of local germplasm.

The Alabio duck development center area in the province of South Kalimantan is in Hulu Sungai Utara Regency. The population of ducks in this area reached 1,492,013 heads or about 34.98% of the total duck population (4,264,995 heads) in 2018. A high population of ducks in HSU Regency shows that ducks have an important role for farmers in HSU. The production of duck eggs and meat in HSU Regency in 2018 was 10,230,946 kg and 525,634 kg, respectively. This production contributes to the production of eggs and duck meat in South Kalimantan Province reached 34.98% and 39.72% respectively [1].

As an effort to protect local genetic resources, the results of the Arabio duck color pattern research (Anas platyrhynchos Borneo) in South Kalimantan showed that the dominant color pattern of Alabio male ducks from Hulu Sungai Utara (HSU) was grayish-brown fur on the neck (56%), grayish-brown on the chest (72%), black-gray on the hindquarters (60%), bluish-green on the wings (56%), and black on the tail (80%). The female Alabio duck from HSU has a grayish-brown color pattern around the neck, back, chest, and tail ranging from 46.29 to 82.85%, and grayish-black on the wings by 40.57%. The differences in the dominant color of feathers that occur in male and female ducks is due to the different selection systems carried out by farmers. Usually, farmers make selections based on the uniformity of coat color phenotype, without considering the pedigree [2].

HSU Regency's duck farming industry specializes in a number of businesses, namely consumer egg producers, hatching egg producers, DOD producers (hatcheries), slaughter duck producers (finishing), and breeding to produce pre-laying female duck [3]. This article aims to determine the business profile of Alabio Duck in several types of businesses, the germplasm as a source of protein, and the revenue source in Hulu Sungai Utara Regency in South Kalimantan.

2 Materials and methods

2.1 Tannery process of tilapia fish skin

The research was conducted in North Hulu Sungai Regency, South Kalimantan Province, from June to October 2018. The research method used was a survey of several groups of duck farmers who run businesses, namely slaughter-duck producers (fattening), breedings, hatching-egg producers, consumes-egg producers, hatchery, and duck traders. Focus group discussion with key informants was also conducted to obtain more in-depth information.

The data used include primary data and secondary data. Primary data were obtained through interviews with farmers and field observations. Secondary data in the form of related data published by BPS, Dinas, Pusdatin as well as theories and results of previous research obtained through library research. The number of respondents as many as 53 respondents.

In addition to descriptive analysis to explain other parameters and their relationships, net farm income analysis and R/C ratios are also used to analyse the commercial viability of each business understudy. According to [4], income is the difference between income and all costs. Meanwhile, farmers' income is the multiplication between the production obtained and the selling price. Analysis of the contribution of duck farming income to the Provincial Minimum Wage (UMP) was obtained by comparing the income to the UMP using one year period and multiplied by 100%. The UMP value used is the 2018 South Kalimantan UMP, which is IDR 2,454,671 (BPS, 2019). The formula used as follows:

$$T1 = TR - TC$$

where:

TI = Total income of duck business

UMPj = the provincial minimum wage

3 Results and discussion

3.1 Characteristics of duck farmers

The average age of farmers was 43.03 years, and was included in the productive age, so that it was assumed to have high motivation and business spirit. A person's age affects the ability to carry out an activity, including farming [5]. At productive age, a person's ability if healthy, is in optimal condition to have work motivation, abilities and hone skills. The average education of respondents was 7.44 years. Although the level of education was relatively low, the ability of reading and writing can be used as capital to supports running their business. Education level has something to do with information acceptance and technology adoption. A person's education will determine the views, mindset, and decision making on a problem [5]. Furthermore, [5] reported that education level and age had a positive but not significant effect on farmers' income. The results of another study reported by [6] that age has an effect on income while education has no effect. Household labor is the number of production workers owned by agricultural households that can participate in crop/plantation and livestock production activities, expressed in working days (HOK). 1 HOK is equivalent to 8 working hours per day, 25 days per month, and 300 days per year. Based on the data above, the average availability of family labor was 849 HOK/year, but only 690 HOK used to work in agriculture (Table 1).

Table 1. Characteristics of duck farmers in Hulu Sungai Utara

Description	Average
Age of the head of the family (year)	43,03
Education of the head of the family (year)	7,44
Number of household member (people)	4,64
Age < 15 years old	1,81
Age >15 years old	2,83
Number of family labor works in agriculture (people)	2,3
Experience in duck farming (year)	15,2

Table 2. Asset control by duck farmers in Hulu Sungai Utara, 2018

Description	Average
Land (ha)	
a. Paddy field	0.73
b. Tidal land	0.30
c. Others: garden, not utilized	0.65
Total	1.68
Poultry asset (head)	
a. Duck	241
b. Chicken	28
Business scale of duck farming based on business type:	
a. Breedings (head)	70
b. Slaughter-duck producer (head)	83
c. Hatching-eggs producer (head)	130
d. Consumes-eggs producer (head)	96
e. Hatchery (egg/week)	500-3,000

The existing duck business specialties in Hulu Sungai Utara are breeding, producing slaughter-duck (fattening), producing hatching-eggs, consumes-eggs, and hatchery. In breedings, farmers are raising ducks to produce duck seeds. Farmers sell ducks based on consumer demand in the form of ducklings with an age of 1 month to 5 months. The longer the selling time, the higher the production cost. The production system in breedings is usually uses extensive and semi-intensive systems.

Production of slaughter-ducks (fattening male ducks) have mostly used an intensive and semi-intensive system. Producers of hatching eggs, using the ratio of raising male and female ducks for 1:10-20. This production system ranged from extensive, semi-intensive, and intensive systems. The average scale of this business was around 130 heads per farmer, which was higher than the scale of consumes-egg producers. The higher scale of hatching egg production was due to the higher price of hatching eggs about IDR 200-300 / egg compared to the price of consumes-eggs. Hatching-egg price was also relatively stable at IDR 2,400-2,500/egg compared to consumes-eggs, which continue to fluctuate in the same one-year period due to season. The range of consumes-eggs price was between IDR 1,900/egg to IDR 2,500/egg. The production of consumes-eggs was mostly done with an intensive system. Duck hatchery was carried out per week, with a scale between 500-1,500 eggs/week and ducklings sell at the ages of 1-7 days.

[7] pointed out that for low-income countries with food shortages, poultry farming is still considered an important source of protein, and provide additional income for smallholder farmers (especially women) with limited resources. Although it requires low capital, poultry farming contributes significantly to food security, poverty alleviation, and environmentally friendly management of natural resources. Poultry farming is a source of employment for underprivileged groups in many local communities. However, the constraints faced by the poultry production system are poor management conditions and an underdeveloped marketing structure. The skills and abilities of smallholders in all aspects of the poultry business have to improve through training and education. Credit or loans have to give to smallholder farmers, taking into account the economic circumstances and socio-cultural context in which the beneficiaries live. More than that, to reach business sustainability, the marketing system must be improved. Barriers faced by small-scale poultry farming in the future, with a particular focus on poultry disease given the current outbreak of highly pathogenic avian influenza in Asia. Identification of geographic hot spots needs to be done so that it can be determined where the most urgent intervention is needed, and where the location of small-scale farming has the potential to generate large impacts [8].

3.2 DOD-producing hatchery

The average age of farmers in hatcheries was 40.75 years with an average of 12 years of education (high school graduation). This age is in the productive age category, and the education level is quite good. Farmers with a good level of education usually have a good level of understanding in receiving information and innovation. The average experience in hatching and trading ducks was 20 years, and the average number of hatching eggs was about 2014 eggs/week. Hatching uses a combination of a hatching machine and hatchery bales. The average number of hatching machines was 4.25 units/family with a capacity of between 500-1,000 eggs. The average price for hatching machines was IDR 2,800,000/unit. DOD was mainly sold in HSU Regency (60%), and the rest is outside HSU Regency and South Kalimantan, each accounting for 20% (Table 3).

Farmers sell DOD all at once, and if there are any leftovers, farmers will raise and sell them in the following week. The demand for DOD usually increases after the rice harvest season ends. This situation especially occurs in the HSU area, which is a lowland area during the dry season and may be suitable for semi-intensive duck farming. The type of duck will affect the price of the duck. There is no difference between the price of one-day-old male and female Peking ducks between 6000-7000 IDR/head. As for Alabio ducks, there was a price difference between male and female DOD. The price of female DOD is more expensive than that of male DOD. Female DOD are IDR 10,000-12,000 /head, and male DOD are only IDR 2,000-3,000 /head.

Table 3. Characteristics of hatchery farmer in HSU, 2018

Description	**Average**
Age (year old)	40.75
Education (year)	12
Number of household member (people)	
Age < 15 years old	2
Age >15 years old	3
Experience in hatching and trading ducks (year)	20
Number of duck egg hatched (egg/week)	2014
Number of hatching machine (unit)	4.25
Percentage of DOD sold (%):	
a. in HSU	60
b. Outside HSU	20
c. Outside province	20
Percentage of not hatched egg (%) from:	
a. Infertile egg	9.69
b. Fertile egg	90.31

The hatchery business scale ranged from 800-5,000 eggs/week/farmer with an average of 2,014 eggs/week. Hatchability was about 61.24%. The incubator system used is a mixture of an incubator and husk. Egg germination was 90.31%. Farmers did candle eggs 4 (four) times, about 8.68% of total eggs hatched did not grow, and about 7.77% of the number of eggs that entered the incubator were eggs with dead embryos in them. Farmers sold infertile eggs at a price of between IDR 1,900- IDR 2,000/egg, while dead embryo eggs were between IDR 1,000- IDR 1,600/egg. Respondent utilized dead embryo eggs for feed mixtures. Sometimes, it was difficult for farmers to obtain uniform hatching eggs. Farmers did not encounter any problems related to the ducklings sold. Generally, hatchery farmers already have regular suppliers of hatching eggs and guarantee their availability. However, when the demand for DOD increases, especially in March – September, it is sometimes difficult for farmers to obtain hatching eggs. DOD marketing is not difficult, because it already has a fixed market and customers. Farmers regulate the number of eggs hatched according to fluctuations in DOD demand. The problems that farmers often face are frequent power outages that affect the hatching rate, the availability of a sufficient number of hatching eggs, and limited funds.

3.3 Duck and egg trader

Duck products sold are in the form of edible/consumes-eggs, hatching eggs, DOD, slaughter ducks, pre-laying female ducks, and rejected ducks (duck that no longer produce). The duck product market is located in the North Hulu Sungai (HSU) Regency and is open every Wednesday. The average age of a duck trader was 51 years with a bachelor's education, both of which indicate that the quality of human resources for duck traders was quite good, so that they were actively involved in communication or information about duck development. Traders had 27.5 years of experience in duck trading. The average sales of ducklings were 3,000 heads/week, young ducks were 150 heads/week, and adult ducks were 100 heads/week.

Traders sold ducks to meet demand in South Kalimantan, Central Kalimantan, and East Kalimantan. For several years, with the exception of Central Kalimantan and East Kalimantan, almost no ducks have been delivered or sold outside the provinces. In addition to HSU itself, there are also HST, HSS, Banjar, Tapin, Tanah Laut, and Banjarmasin, who have purchased a large number of ducks in South Kalimantan.

Table 4. Characteristics of duck trader in Hulu Sungai Utara, 2018

Description	Average
Age (year old)	51
Education (year)	16
Number of household member (people)	3
Age < 15 years old	2
Age >15 years old	2
Experience in duck trading (year))	27.5
Scale of sale (head):	
Duckling (DOD)	3,000
Young duck	150
Adult duck	100

Demand for ducks has increased from March to April and August to September. This is because the transition season has passed, so the disease outbreak will not affect duck raising. In addition, August and September are the dry seasons, the harvest of rice has been completed, and duck raising can be extensively carried out in the paddy field. Due to the outbreak of many diseases, the demand for ducks is reduced during the transition seasons and the peak of the rainy season.

The price of ducks increased in line with the increase in demand. Depending on the age and quality of the ducks, the price of ducklings will increase by IDR 3,000-4,000 per head, while the prices of young and adult ducks will increase by IDR 10,000-30,000 per head. The sale of eggs or live ducks is relatively unproblematic because their demand determines their

prices. This fact is consistent with the views of [9] and [10] that duck products, whether eggs, DOD, slaughtered ducks, or laying ducks, are relatively easy to sell.
There were several routes in the marketing system of ducks, namely:
1. Hatchery – Trader – Farmer (ducklings)
2. Hatchery – Trader – External Trader – Farmer (ducklings)
3. Farmer – Trader – Farmer (young duck/adult duck)
4. Farmer – Trader – Consumer (young duck/adult duck)

The sales scale of duck eggs per week was 1,500 eggs/trader according to market prices. An average of 15 years of duck egg trading experience. The eggs sold are for consumption, so the average buyer likes Peking duck eggs because of their large size.

3.4 Duck hatchery

Table 5. Economic analysis of duck hatchery in Mamar Village, HSU Regency

Description	**Number**		**Price/unit (IDR)**	**Total (IDR)**	**%**
Expenditure (IDR):					
Hatching egg	1,000	eggs	2,500	2,500,000	76.34
Feed	1	zak	375,000	375,000	11.45
Medicine	1	unit	100,000	300,000	9.16
Electricity	1	month	50,000	50,000	1.53
Sexing wage	1	time	50,000	50,000	1.53
Total				3,275,000	
Revenue (IDR):					
Selling of:					
Female DOD	300	heads	12,000	3,600,000	
Male DOD	300	heads	3,000	900,000	
Candling egg 1	50	eggs	1,700	85,000	
Candling egg 2	50	eggs	1,700	85,000	
Candling egg 3	50	eggs	1,000	50,000	
Total				4,635,000	
Income (IDR)				1,360,000	
R/C				1.42	

Source: Author's survey, 2018

The results of the analysis on duck hatchery business on a scale of 1,000 eggs a week and in an R/C value of 1.42, which means that the hatchery was profitable and feasible to do. The prices and assumptions that apply were in accordance with the data in Table 5. Farmers did this business throughout the year, but in certain months will increase the scale of their business because demand increases. The increase in demand for ducklings occurs during the dry season. This condition is because farmers have a lot of time after harvesting the food crops, and the land is available. What's more, in the dry season, the air becomes hot and not humid, so the ducks become healthier. The research result of Purnomo [11] that hatching duck eggs using an incubator with a capacity of 200 eggs was profitable and feasible with an R/C value of 1.3. In order to obtain more profit and serve as the main income, the number of incubators needs to be increased to at least six units. The structure of the incubator needs to be further studied to increase the hatching rate.

The hatchery business of 1,000 duck eggs/week has created an average income of IDR 4,635,000/week. The largest expenditure of the hatchery business is 76.34% of the purchase of hatching eggs, and only 11.45% of the feed, because the ducklings produced between 1-7 days are sold immediately.

3.2.2 Duck breeding

The business of breeding ducks to mature female ducks (pre-laying duck) was feasible, with an R/C value of 1.65 and a business scale of 175. Feed cost was the largest cost incurred, accounting for 64%, and the cost of purchasing ducklings accounts for 30.34%. The weakness of duck breeding is that it takes a long time so that there turn of business capital takes time (Table 6).

Table 6. Economic Analysis of duck breeding to pre-laying duck, 2018

Description	Number		Price/unit (IDR)	Total (IDR)	%
Expenditure (IDR):					
Duckling	175	heads	15,000	2,625,000	30.34
Commercial feed	4	zak	380,000	1,520,000	17.62
Rice bran	250	kg	2,800	700,000	8.12
Sago	11	sticks	300,000	3,300,000	38.26
Others	1	package	480,000	48,000	5.57
Total				8,625,000	
Revenue (IDR):					
Selling of pre-laying duck	150	heads	95,000	14,250,000	
Income (IDR)				5,625,000	
R/C				1.65	

3.3 Egg producer (hatching-egg and consumes-egg)

The duck raising business that produces hatching eggs (Table 7) and consumes-eggs (Table 8) generated an R/C value higher than 1, which means that both businesses were feasible and profitable. In the consumes-eggs production, the biggest cost was the cost of feed, which reached 74.95%, while the cost of purchasing duck breeds was only 22.23%. The results of the analysis in the consumes-eggs production with a scale of 212 heads obtained an R/C value of 1.24. The monthly income generated was IDR 1,721,234 (Table 7).

The analysis of hatching egg producer was carried out on a business scale of 183 heads, with some value assumptions described in Table 8. The cost for feed was 78.70% and the cost of brood stock was 17.92%. The analysis results showed that the business was profitable and feasible (R/C 1.27) and the income was IDR 1,859,666/month.

In the business of raising ducks, it is feasible to produce both hatching eggs and consuming eggs. Pay attention that there are no fatal diseases and insect pests. This fact is not much different from the study reported by Sitorus *et al.* [12]. The traditional laying duck business is feasible, with an R/C value of 1.58.

Due to the intensive raising of ducks, farmers must provide sufficient feed, and the cost of feed used by farmers for production consumes-eggs and hatching-eggs is more than 70%. Gutterres *et al.*[13] also said the same thing, that the cost of feed is the largest component in duck farming, which reaches 70% of the total cost. For the production of ducks, the feed costs incurred are lower, which is only 64%, because there is a mixture of maintenance, namely when ducks aged from a day to 1 month are intensively raised and after that, the ducks are raised in a semi-intensive way so that there is a saving in feed costs. The results of research by Soekarwati [14], stated that there was no difference in the size and quality of Alabio Duck eggs with intensive and semi-intensive maintenance, this fact was the basis so that it was sufficient to carry out semi-intensive maintenance.

Table 7. Economic analysis of consumes-egg producer with a scale of 212 heads, 2018

Description	Number		Price/unit (IDR)	Total (IDR)	%
Expenditure (IDR):					
Fixed Cost (IDR):					
Duck brodstock	212	heads	90,000	19,080,000	22.23
Equipment	1	year	250,000	250,000	0.29
Electricity	1	year	600,000	600,000	0.70
Cage depreciation	1	year	1,000,000	1,000,000	1.17
Variable Cost (IDR):					
Feed	1	year	64,320,000	64,320,000	74.95
Medicine	1	year	222,000	222,000	0.26
Others	1	year	350,000	350,000	0.41
Total Costs (IDR)				85,822,000	
Revenue (IDR):					
Consumes-egg	46,428	eggs	2,079	96,523,812	
Rejected duck	175	heads	50,000	8,750,000	
Manure	1	year	1,203,000	1,203,000	
Total Revenue (IDR)				106,476,812	
Income (IDR)				20,654,812	
Montly income (IDR)				1,721,234	
R/C				1.24	

Table 8. Economic Analysis of harching-egg producer with a scale of 183 heads, 2018

Description	Number		Price/unit (IDR)	Total (IDR)	%
Expenditure (IDR):					
Fixed Cost (IDR)					
Duck brodstock	183	head	80,000	14,640,000	17.92
Equipment	1	year	200,000	200,000	0.24
Electricity	1	year	600,000	600,000	0.73
Cage depreciation	1	year	1,000,000	1,000,000	1.22
Variable Cost (IDR):					
Feed	1	year	64,284,000	64,284,000	78.70
Medicine	1	year	600,000	600,000	0.73
Others	1	year	360,000	360,000	0.44
Total Costs (IDR)				81,684,000	
Revenue (IDR):					
Hatching egg	40,000		2,400	96,000,000	
Rejected duck	160		50,000	8,000,000	
Total				104,000,000	
Pendapatan				22,316,000	
Pendapatan per bulan				1,859,666	
R/C				1.27	

3.4 Duck farming income to the Provincial Minimum Wage (PMW)

The duck hatchery business with a scale of 1,000 eggs/week can generate revenue of IDR 2,454,671 per month higher than the provincial minimum wage (UMP) applicable in South Kalimantan in 2018. This result is not much different from the research reported by Purnomo [11] that the duck hatchery business can meet the South Kalimantan UMP of (IDR 1,870,000) and can be a source of income with a scale of 1,200 points.

The production of hatching-egg can generate 75.76% of UMP's revenue. The breeding business to produce duck brood stock which carried out for 4 months provided the smallest income compared to the types of duck farming (Table 9). Duck hatchery business is feasible to be carried out as the main business because the income generates disgreater than the applicable UMP. For other types of duck business, it still needs diversification with other commodity farming or non-agricultural businesses. The reality found in the field is that most farmers carryout several types of commodities to be cultivated other than livestock, especially rice and vegetables. Research conducted in Tegal and reported by [15] that farmers who apply technology well can generate income above the minimum wage (IDR 975,000).

Table 9. Income of duck farming to the UMP of Kalimantan Selatan, 2018

No	Business	Montly income (IDR)	%
1	Hatchery	5,440,000	221.62
2	Breeding	1,406,250	57.29
3	Consumes-egg producer	1,721,234	70.12
4	Hatching-egg producer	1,859,667	75.76

Source: Author's survey,2018

3.5 Internal and external factors in duck farming

The strength is the long-term experience in raising livestock, with an average of more than 10 years, and strong motivation for breeding. Even if the livestock is affected by the disease, the farmer will not easily get discouraged and stand up again. What's more, HSU regency is a duck development center supported by institutions. The weakness is that the limited capital and technological mastery are still considered low. The external factor that influences is the opportunity in the form of a market for livestock products every Wednesday, and the opening of markets outside the district and province. Government support is an important factor, in addition to a supportive environment such as available land and local feed ingredients. Another important support is infrastructure in the form of roads and markets. Threats that are often faced in raising ducks are disease attacks that can cause their business to go out of business, the change of seasons that cause livestock to get sick.

Table 10. Identification of factors in the development of duck farming in Hulu Sungai Utara Regency, 2018

Internal	External
Strength: Experience Motivation Capital Central area Solid group	Opportunity: Market Government support Supportive environment Adequate infrastructure
Weakness: Funds/capital Technology applied	Threat: Disease Weather/Season transition Import Decreasing quality of duck broodstok

Since the avian influenza virus attack in 2010, it has caused some changes, including the entry and development of other duck species besides Alabio, such as Peking duck, Mojosari duck, MA duck, Tegal duck, and so on. This outbreak has led to a decline in the scale of farmers raising ducks, which in turn affected the decline in the Alabio duck population. Before the outbreak of duck disease, the scale of duck farming was generally more than 200 heads per household, and some even reached the scale of thousands of ducks.The important factor identified is the forecast related to the price fluctuations of duck eggs, which are expected to be high in the month of Ramadan (Ramadan) from the middle to the end of the month and in the month of Maulud (Islamic celebration). Low egg prices are predicted to occur at the time of the entry of eggs from Java, the fish season, the beginning of the month of Ramadan. Due to the sufficient local feed supply and the ease of raising ducks, the highest record for ducks occurred during the dry season, from July to September. The disease usually occurs in April and May (transition season).

Some of the problems identified from the FGD results include:

1. Disease
2. Seed quality
3. Feed price
4. Product selling price fluctuates (eggs and livestock)

Based on the problems faced, the recommended policies include:

- Vaccination of ducks routinely according to age
- Maintain the cleanliness of the cage and the environment by disinfection regularly,
- Conducting selection in the seedling area,
- Optimizing the use of local feed along with training for farmers to be able to produce a good quality of feed,
- Low-cost processing of animal products training,
- Livestock product processing promotion
- Government facilitation for the development of a mini-feed mill

4 Conclusion

Duck farming has good prospects as a source of income and employment for farmers and the younger generation. The results of the business analysis how that they are all feasible in terms of hatching, production, and slaughter of ducks, production of hatching eggs, or consumption of eggs. The duck hatchery is a reliable business and it can generate more revenue than the applicable UMP. Developing of duck business needs escorts from related agencies, especially in disease prevention and control as well as assistance in technological innovation. Another important thing is the need for handling and managing environmental sanitation to keep it clean and healthy, efforts to spray cages and cleanse the cage environment, as well as vaccinations, are important.

References

1. A. J. Farid, P. H. Riyadi, U. Amalia, Indones. J. Fish. Sci. Technol. **10**, 80-83 (2015)
2. S. Moroki, A.J. Masinambow, B. Kalangi. J. Berkala Ilm. Eff. **8,** 132-142 (2018)
3. R. Olivi, R. Qurniati Firdasari. J. Sylva. Les. **3,** 1-12 (2015)
4. F. F. Gueye, World. Poult. Sci. J. **58**, 54-57 (2002)
5. V. Sivakumar, R. P. Prakash, P. G. Rao, B. V. Ramabrahmam, G. Swaminathan, J. Clean. Prod. **16,** 549-553 (2008)
6. S. Mack, D. Hoffmann, J. Otte, World. Poult. Sci. J. **51,** 7-14 (2005)
7. M. S. Rajasa, D. Sumardjono, B.T Eddy, Anim Agri J. **1**, 331-341 (2012)
8. O. Sjofjan, D.N. Adli, M.H. Natsir, Y.F. Nuningtyas, T.S. Wardani, I. Sholichatunnisa, S.N. Ulpah, Firmansyah, O. JITV. **26**, 39-48 (2021)
9. A. Wakhid, *Buku Pintar Beternal Itik dan Bisnis Itik,* (Agro Media, Jakarta, Indonesia, 2010)
10. N. Haq, Yunizal, Suherman, *Teknologi Pengawetan dan Penyamakan Kulit Ikan* (Pusat Penelitian dan Pengembangan Eksplorasi Laut dan Perikanan, Jakarta, Indonesia, 2000)
11. E. Purnomo, *Penyamakan Kulit Reptil* (Kanisius, Yogyakarta, Indonesia, 2001)
12. P. A. Sitorus, P. H. Riyadi, E. Susanto, Jurnal Ilmu dan Teknologi Perikanan **2**, 57-64 (2020)
13. M. Gutterres, L. M. D. Santos, J. Soc. Leather Technol. Chem. **93**, 171-175 (2009)
14. Soekarwati, *Prinsip Dasar Ekonomi Pertanian – Teori dan Aplikasi,* (PT.Raja Grafindo, Jakarta, Indonesia, 2002).
15. S. N. Hastang, Sirajudin, Asnawi, S. Kadir, Mahyudin. IOP Conf. Series : Earth and Environmental Sci. **492** (2020)

Nutritional content, performance and ratooning ability of four sorghum genotypes

Harmini Harmini*, *Achmad* Fanindi, *Endang* Sutedi, *Iwan* Herdiawan, and *Sajimin* Sajimin

Indonesian Research Institute for Animal Production, Ciawi 16002, Indonesia

Abstract. Sorghum is a grain plant that is a source of food, biofuel and feed, which can be produced using the techniques of ratoon cultivation. The objective of this study is to assess the nutrition content, the performance of the four sorghum genotypes and the vegetative characteristics of the sorghum ratoon. The research was carried out from December 2020 to July 2021 at the Ciawi Research Garden, the Livestock Research Institute. The study employed four treatments (genotype) and three replications of random block design. The range of protein content, crude fat crude fiber and ash in the four genotypes of sorghum was 7.91-9.30, 1.91-2.69, 33.41-37.57 and 8.25-9.11%, respectively. Genotype significantly affected $P<0.05$ on plant height, stem height, leaf width, internode length, flag leaf width and fresh weight, but not significantly different $P>0.05$ on leaf length, stem diameter and flag leaf length on the main plant. Genotype had no significant effect ($P>0.05$) on plant height, number of tillers, leaf length, leaf width and fresh weight of ratoons. It was concluded that the ability of ratoon in sorghum was not affected by the genotype of sorghum.

1 Introduction

Sorghum is a cereal grain with multiple uses, including food, animal feed, and fuel. As a result of its high nutritional value and ability to grow on dry land, this plant has the potential to be developed as a ruminant feed source [1]. One potential crop to develop with the ratoon system is sorghum. A few reasons for using the ratoon system to cultivate plants include: 30% regrowth, 20% labor savings, excessive vegetative regrowth, and disease potential (18 percent) [2]. The dry matter production, drought tolerance, and better regrowth after grazing or cutting of sorghum are more promising when compared to the other forages [3]. The Agency for Agricultural Research and Development has developed several superior varieties of sorghum. Super sorghum 1 is the result of improving the White Hamu Watar population from the Cereal Research Institute's germplasm collection, with characteristics of aphis resistance, anthracnose resistance, leaf rust and leaf blight, suitable for dry land and dry climates and environmental adaptation [4]. While sorghum super 2 is an improvement over the 15021 strain from ICRISAT, the characteristics of producing ratoon [4]. Sorghum bioguma, on the other hand, is the product of 59 Gy gamma rays and in vitro culture on

* Corresponding author: hmini2011@gmail.com

parents of the Numbu variety [5], Numbu, on the other hand, is the name of the native sorghum variety.

Ratoons are used in the cultivation of sorghum as one of the methods. Ratoons are raised by harvesting the first season's crop, allowing it to regrow, and then cultivating it like sorghum grown from seed. The genotype has a significant impact on the main plant's vegetative, reproductive, physiology, the ratoon's ability, and the vegetative character. Plant height, stem diameter, wet and dry stover weight, harvest age, and harvest index were found to be significantly correlated with the ability to ratoon. Significantly correlated with ratoon plant performance were the main plant's height, total leaf chlorophyll, wet and dry stover weight, panicle weight, seed weight per panicle, 1,000 seed weight, harvest index, and transpiration rate. Scientific studies have been done on the effects of cultivating ratoons on crop yield, sugar production, and pest attack [6], evaluation of sweet sorghum growth and yield at various spacings [7], evaluation of sweet sorghum genotypes with high biomass production and high ratoon capacity. We conducted a study to determine the productivity of ratoons from four types of sorghum consisting of one national variety and three breeding varieties to be developed as forage for livestock based on the information above

2 Materials and methods

2.1 Material

This research was carried out in the experimental garden of Indonesian Research Institute for Animal Production (IRIAP), from December 2020 to June 2021. The materials used were Sorghum varieties Super 1, Super 2, Bioguma and Numbu which are the result of breeding at the Agricultural Research and Development Agency.

2.2 Method

The experimental design used was a one-factor randomized block design, namely genotype with 4 replications. Sorghum is planted with a spacing of 50 cm, on a plot of 2 m x 1 m, with planting without tillage. Sorghum plants grown from seeds are called primary crops. Sorghum is ratified after harvesting by cutting the main stem and leaving one segment from the soil surface, approximately 5 to 10 cm from planting. Ratoon plants are plants that grow from the remaining internodes of other plants.

The variable observed in the main plant was the vegetative character. Vegetative characters for the main plant consisted of plant height, stem height, leaf length, leaf width, internode length, stem diameter, flag leaf length, flag leaf width, fresh weight, while for ratoon included plant height, number of tillers, leaf length, width leaves and fresh weight. Analysis of nutrient content in forage sorghum includes water, fat, and crude and ash content by gravimetric method, while protein and energy content by calorimeter. The data were analyzed by analysis of variance and if there was a significant difference, Duncan further tested it Steel [8] with SPSS 16.

3 Results and discussion

3.1 The nutritional content of forage sorghum

The nutritional content of the main forage sorghum plant can be seen in Table 1. Based on Table 1 the four types of sorghum have almost the same crude protein (CP) content from 7.91

to 9.30%, with an average CP content of 8.38%, where Super 1 has the highest CP content. Crude fiber (CF) content in the range of 33.41 to 37.57%, and an average of 35.93%. crude fat (CFat) content, 1.91 to 2.69% and an average of 2.28%, energy content from 3958 to 40 45 kcal kg-1 with an average of 3988 kcal kg-1, and ash content from 8.25 to 9.11% with an average of 8.77% .

Research result Koten *et al.* [9] in local sorghum Rote got slightly different results, the content of CP, ash, CFat and CF were 9.90; 11.45; 2.7 and 20.32 %, respectively. Local varieties of sorghum Rote were added, which were harvested at the age of 90 days with a dose of 100 kg ha-1 urea fertilizer, producing the highest Dry Matter (DM), Organic Matter (OM) and CP[10]. Sorghum is rich in carbohydrates, micronutrients, rich in soluble fiber and a source of nutraceuticals such as phenols and tannins [11]. Plant density affects the accumulation of dry matter in sorghum ratoon, the higher the density, the lower the dry matter content [12]. There is a difference in biomass quality between the main crop and ratoon [13].

Table 1. Simply character of tilapia fish skin as a tannery material [8-9]

Varietas	Crude Protein (%)	Crude Fat (%)	Energi (kcal kg^{-1})	Crude Fiber (%)	Ash (%)
Super 1	9.30	2.28	3958	33.41	8.98
Super 2	8.01	1.91	3969	36.30	9.11
Bioguma	7.91	2.25	3980	37.57	8.77
Numbu	8.31	2.69	4045	36.46	8.25
Average	8.38	2.28	3988	35.93	8.77

In this study, we investigated the using of fresh palm oil and used palm oil as a fat liquoring agent in the tannery of tilapia skin. The study aims to find out the physical quality of leather through the using of fresh and used palm oil as fatliquoring agent in the vegetable tannery process. The physical quality of tilapia leather was observed and evaluated by examining parameters of tensile strength, elongation, and shrinkage temperature.

3.2 Effect of genotype on main plant performance

Based on Tables 2 and 3, it can be seen that plant height, stem height, leaf width, internode length, leaf width and fresh weight were not significantly different from flag leaf length. Sorghum Super 1 has the highest vegetative growth compared to other varieties. In Table 4 it can be seen that for sorghum ratoon plant Super 1 had the highest plant height, Super 2 had the highest number of tillers and the highest leaf length, Numba had the highest leaf width and highest fresh weight. Genotype significantly affected the performance of the main plant and ratoon, ratoon production in sorghum reached 48.9-65.7% of the main crop production [14]. One indicator of sorghum production as forage is plant height, stem diameter and number of leaves. Sorghum varieties affect plant height, stem diameter, number of leaves and nutritional quality (ash content, CF and CP) [15]. Stem diameter days of 50% flowering had a significant correlation and high positive direct effect with panicle weight [16]. The average fresh weight of sorghum ratoon decreased by 83.33%. This result is different from the results of research [17] which states that the productivity of ratoon sorghum will be higher if planted in the rainy season. Crops and ratoons differ in their ability to accumulate and allocate dry matter between forage and grain biomass [18]. Differences in sorghum varieties affect the growth and yield of sorghum ratoon except for stem diameter and wet stover weight [19]. Ratoon production is 30% of the main crop in the dry season [20]. The height of the ratoon plant is lower than the main crop, this is different from Vinutha *et al.* [13]. The average plant height was greater in the ratoon crop than in the main crop. Season affects the height of ratoon sorghum, where in winter it is higher than summer [21]. Ratooning is a cultural practice to stimulate regrowth of the basal or lower epigeal buds after removal of the photo-synthetically active material.

Table 2. Vegetative growth of four varieties of sorghum

Variety	Height of plants (cm)	Stem height (cm)	Leaf length (cm)	Leaf width (cm)	Segment length (cm)
Super 1	319.62[a]	267.82[a]	72.72	7.47[a]	31.12[a]
Super 2	275.81[b]	242.43[a]	83.29	6.12[b]	29.19[ab]
Bioguna	206.33[c]	195.08[b]	75.94	5.91[b]	22.10[bc]
Numba	204.37[c]	180.65[b]	72.72	5.50[b]	25.03[c]
Average	251.53	221.49	76.17	6.25	26.86

Note: Numbers followed by different letters in the same column show significantly different effects (P <0.05)

Table 3. Stem diameter, flag leaf length, flag leaf width, total fresh weight

Variety	Stem diameter (mm)	flag leaf length (mm)	flag leaf width (mm)	total fresh weight(gram)
Super 1	1289.30	43.54	6.11[a]	338.73[a]
Super 2	1115.22	36.21	4.82[b]	192.86[b]
Bioguna	969.90	33.71	5.23[b]	177.71[b]
Numba	969.74	32.18	4.92[b]	158.71[b]
Average	1086.04	36.41	5.27	217.00

Note: Numbers followed by different letters in the same column show significantly different effects (P <0.05)

Table 4. Ratoon ability and vegetative character of ratoon plant four genotypes of sorghum

Variety	Plant height (cm)	Number of tiller	Leaf length (cm)	Leaf width (cm)	total fresh weight (gram)
Super 1	111.05	2.09	50.59	2.89	38.00
Super 2	101.80	2.36	55.48	2.89	27.60
Bioguna	99.99	1.80	49.02	3.093	23.36
Numba	94.35	1.62	43.88	3.11	43.03
Average	101.80	1.91	49.74	2.99	33.00

4 Conclusion

Sorghum has the potential to be developed as an animal feed crop. One of the cultivations is by developing ratoon sorghum. The type of variety has an effect on the ability of the maturation of sorghum. The Numba ratoon variety produced the highest fresh weight compared to the Super 1, Super 2 and Bioguna varieties.

References

1. H. Harmini, Livest. Anim. Res. **19** (2021)
2. P. Roge, S Snap, M. N. Kakwera, L. Mungai, I. Jambo, B. Peter, Agron. Sustain. Dev. **36,** 1–18 (2016)
3. A. R. S. Shakeri, E. Y. Emam, S. A. Tabatabaei, Intl. J. Plant Prod. **11** (2017)
4. Badan Penelitian dan Pengembangan Pertanian, Kementrian Pertanian, Jakarta, Indonesia (2018)
5. E. Lestari, I. Dewi, R. Yunita, A. Nur, Azrai, K. Syahruddin, Bioguma 1 Agritan. (2019)
6. K. Wilson, University of Greenwich. London, (2011)
7. D. Puspitasari, D. Kastono, S. Waluyo, Vegetalika **1** (2013)
8. R. Steel, J Torrie, *Prinsip dan Prosedur Statistika Suatu Pendekatan Biometrik* (Gramedia, Jakarta, Indonesia, 1995)
9. B. B. Koten, R. D. Soetrisno, N. Ngadiyono, B. Suwignyo, Buletin Peternakan **36** (2012)
10. B. B. Koten, R. D. Soetrisno, N. Ngadiyono, B. Soewignyo, Jurnal Pastura **3** (2014)
11. P. Pontieri, L. Del Giudice, *Encycl. Food Heal.* **2** (2016).
12. B. Ferdian, S. Sunyoto, A. Karyanto, M. Kamal, Jurnal Agrotek Tropika **3** (2015)
13. K. S. Vinutha, G. S. Anil Kumar, M. Blümmel, P. Srinivasa Rao, Trop. Grasslands-Forrajes Trop **5,** 40-49 (2017).
14. M. G. TMeliala, Trikoesoemaningtyas, D. Sopandie, Jurnal Agronomi Indonesia **45** (2017)
15. Hajar H, IPB (2020)
16. S. E. Ardiyanti, D. Sopandie, D. Wirnas, Trikoesoemaningtyas, IOP Conference Series: Earth and Environmental Science **399** (2019)
17. N. Tsuchihashi, Y. Goto, Plant Prod. Sci. **11** (2008)
18. T. J. Gerik, W. D. Rosenthal, W. F. Seavey, F. Crop. Res. **23** (1990)
19. G. D. Cahyo, K. F. Hidayat, S. Sunyoto, M. Kamal, Jurnal Agrotek Tropika **2** (2014)
20. H. Gnanagobal, J. Sinniah, Trop. Grasslands-Forrajes Trop. **6** (2018)
21. S. A. Hassan, M. I. Mohammed, S. O. Yagoub, J. Plant Breed. Crop Sci. **7** (2015)

E3S Web of Conferences **335**, 00021 (2022) https://doi.org/10.1051/e3sconf/202233500021
The 2nd ICESAI 2021

Physicochemical quality of low fat mayonnaise using whey protein concentrate

Tri Umar Satriawan[1], *Herly* Evanuarini[2,*], and *Imam* Thohari [2]
[1] Master Program of Animal Science, Faculty of Animal Science, Universitas Brawijaya, Jl. Veteran, Kota Malang, East Java, 65145, Indonesia
[2] Department of Animal Product and Technology, Faculty of Animal Science, Universitas Brawijaya, Jl. Veteran, Kota Malang, East Java, 65145, Indonesia

Abstract. Low fat mayonnaise as a low-calorie product modification has low emulsion stability. Whey Protein Concentrate (WPC) has a high protein content and can act as a good emulsifier for emulsion of mayonnaise. The purpose of this study was to evaluate of a low-fat mayonnaise using WPC based on pH, moisture content, viscosity, and protein content. The materials used are egg yolk, sunflower seed oil, vinegar, WPC, and other complementary materials. This study used a laboratory experiment with a completely randomized design with 4 treatments and 4 replications. The treatment using control treatment without additional use of WPC and treatments using WPC as much as 5%, 10%, and 15% of the total oil use. The variables measured were pH, moisture content, viscosity, and protein. The results showed that the use of WPC in mayonnaise gave significantly different on pH, moisture content, viscosity, and protein content. The conclusion of this research that the use of WPC as much as 15% produces the best low fat mayonnaise.

1 Introduction

Mayonnaise an oil-in-water emulsion product consisting of a dispersed phase and a dispersing phase and is stabilized with egg yolk as an emulsifier. Mayonnaise on the market is classified as a high-fat food because lipid as the main components. Lipid contributes 9 kcal per g.

Currently, consumers are more selective towards foods that contain excess lipid. Restrictions on lipid consumption are influenced by the increasing development of science and human awareness of a healthy lifestyle. So that some processed foods are designed to contain low fat.

Some products that reduce the oil content in food include mayonnaise products, namely low-fat mayonnaise, reduce fat mayonnaise, and light fat mayonnaise. The manufacture of low-fat products is done by lowering the oil phase and increasing the water phase. This will cause instability of the emulsion, so it is necessary to use an emulsifier and stabilizer to improve the quality of mayonnaise.

* Corresponding author: herlyfptub@ub.ac.id

The stabilizer serves to reduce the interfacial interaction on the oil surface. The materials that are often used to improve emulsions such as protein ingredients. This material is used because it has the ability as a good emulsifier, so it can reduce the interfacial tension of oil and water. Emulsifiers are used to help bind oil and water because they have two sides, which are hydrophobic and hydrophilic. Herald *et al.* [1] stated that emulsifiers can use ingredients such as modified corn flour, wheat protein, whey protein, and chewing gum. The utilization of whey protein as a processing ingredient is very helpful in the food industry. According to Sun *et al.* [2] revealed that the use of whey protein concentrate and pectin can be a fat mimetic, the use of 20% fat mimetic in low-fat mayonnaise displays good gel formation.

Cheese-making by-products such as whey need to be reused. Environmental pollution can also occur in whey disposal sites due to the uncontrolled decomposition process. The good nutritional content of whey if discarded can grow bacteria easily. So the use of whey needs to be increased for commercial purposes and to protect the environment where cheese products are processed. The use of whey protein in mayonnaise needs to be known as an effort to improve the quality of low-fat mayonnaise.

2 Material and methods

2.1 Research Material

The research materials used in mayonnaise production are sunflower seed oil and commercial whey protein concentrate which has 80% protein content in powder form purchased from an e-commerce store. The ingredients such as egg yolks of Isa Brown chicken were obtained from the Dau Malang Regional Livestock. Other ingredients used are vinegar or vinegar brand S&W, mustard brand Maestro, sugar, salt, and white pepper.

2.2 Research procedure

The research method used is a laboratory experiment method using a completely randomized design (CRD) with 4 treatments and 4 replications. Control treatment (P0) was made using sunflower seed oil as much as 70%. All treatment samples used 30% sunflower seed oil. Treatment includes the use of whey protein percentage. P1 uses a whey protein percentage of 5%. P2 uses a whey protein percentage of 10%. P3 uses a whey protein percentage of 15%.

2.3 Variable

The variables observed were pH, water content, viscosity, protein content.

2.3.1 pH

pH measurements were carried out using a pH meter. The pH meter tool was calibrated using a solution pH 4 and pH 7. The pH meter is dipped into a pH 7 solution, then pH 4, and dipped into a sample that has been diluted with aquades in a ratio of 1: 10. After each dip, make sure it is dry before moving on to the next step [3].

2.3.2 Moisture content

Moisture content is determined by applying heat to the sample at 110°C. Rusalim *et al.* [4] stated that determining the moisture content of feed using the proximate analysis oven method took 4 hours to produce constant dry matter.

2.3.3 Viscosity

Viscosity is measured based on the viscosity of the mayonnaise flow in the tester glass. The next step is to install the spindle on the viscometer and lower it into the mayonnaise to the spindle line. The spindle head is positioned in the center of the sample. Turn on the viscometer and read the viscosity on the tool, the calculation is carried out. It was repeated 3 times for each sample [5].

2.3.4 Protein content

Protein content measured in crude protein mayonnaise samples determined by the stages of destruction, distillation and titration [3].

2.4 Data analysis

The data obtained were analyzed using Microsoft Excel software and then searched for the average and standard deviation. Data were analyzed using Analysis of Variance (ANOVA) and if the results were significantly different or very significant, then further tested using Duncan's Multiple Distance Test (DMRT).

3. Results and discussion

3.1 The effect of whey protein addition on pH of low-fat mayonnaise

The results of the analysis of variance showed that the addition of whey protein with different percentages in low-fat mayonnaise had a significant effect on the pH value. The average value of pH can be seen in Table 1.

Table 1. Average pH value

Treatment	pH ± SD
P_0	4.23 ± 0.05^a
P_1	4.68 ± 0.10^b
P_2	4.73 ± 0.05^b
P_3	4.83 ± 0.05^b

The average pH value ranged from 4.23 to 4.83. The highest mean value of 4.83 was obtained from the treatment using the addition of whey protein P3 and the lowest average value of 4.23 was obtained from the treatment without the addition of whey protein P0. The results showed that the higher the percentage of addition of whey protein, the pH value of low-fat mayonnaise will increase.

The results of the study at P0 were control treatments such as commercial mayonnaise which had an acidic atmosphere with an average pH value of 4.23. This is in accordance with the opinion of Rizkyyani *et al.* [6] that the addition of banana peel mayonnaise to the control sample has a pH of 4.43. According to Badan Standarisasi Nasional [7] stated that the mayonnaise control sample had an average of 4.1. The average pH value of the mayonnaise control is not much different because it uses almost the same ingredient formulation.

The use of whey protein makes the mayonnaise sample higher based on the average pH value. Whey protein comes from the separation of milk fat, so the pH value of whey protein is almost the same as the pH value of milk. The pH of milk is according to Guo *et al.* [8] from 6.3 to 6.8. According to Badan Standarisasi Nasional [9], sweet whey is obtained from the cheese-making process using an inoculant with a pH above 5.6 during ripening (Cheddar and Mozzarella) while sour whey is obtained from a cheese-making process using an inoculant with a pH below 5.6 (Cottage).

3.2 The effect of whey protein addition on moisture content of low-fat mayonnaise

The addition of whey protein with different percentages in low-fat mayonnaise has a significant effect on moisture content. The average value of moisture content can be seen in table 2.

Table 2. The average value of water content %

Treatment	Water Content ± SD
P_0	19.05 ± 0.48^a
P_1	18.53 ± 0.28^b
P_2	17.10 ± 0.50^c
P_3	15.81 ± 0.40^c

The moisture content values range from 15.81 to 19.05. The highest mean value of 19.05 was obtained from the treatment without the addition of whey protein (P0) and the lowest average value of moisture content of 15.81 was obtained from the treatment using the addition of 15% whey protein. The results showed that the higher the percentage of addition of whey protein, the lower the moisture content of low-fat mayonnaise produced.

The standard of moisture content in mayonnaise products is set by Widodo *et al.* [10], a maximum of 30%. Low-fat mayonnaise product with the addition of whey protein according to standards.

Badan Standarisasi Nasional [7] state that the control sample of mayonnaise products has an average value of 18.17. In the control sample, the low-fat mayonnaise product has an average value of 19.05. This is because low-fat mayonnaise reduces the oil and vinegar phases as the water phase of full-fat mayonnaise, thereby increasing the water phase.

The use of whey protein in low-fat mayonnaise can reduce moisture content. This is because the whey protein used is in the form of a powder (solid), so it can bind water. Whey protein absorbs water more easily because it is made from milk that has been separated from the fat component. This is in accordance with the opinion of Evanuarini *et al.* [11] which states that the cause of water-insoluble in powdered milk is because there is a non-polar fat content. The lower the mixture of fat components, the more easily the whey dissolves in water.

3.3 The effect of whey protein addition on viscosity of low-fat mayonnaise

The results of the analysis of variance showed that the addition of whey protein with different percentages in low-fat mayonnaise had a significant effect on the viscosity value. The average value of viscosity can be seen in Table 3.

Table 3. The average value of cPs. viscosity

Treatment	Viscosity ± SD
P_0	1070 ± 29[a]
P_1	1350 ± 22[b]
P_2	1580 ± 29[c]
P_3	1700 ± 26[d]

Table 1. This shows that the viscosity values range from 1070 to 1700. The highest mean value of 1700 was obtained from the treatment using the addition of whey protein P3 and the lowest average value of 1070 was obtained from the treatment without the addition of whey protein P0. The results showed that the higher the percentage of addition of whey protein, the higher the viscosity value of low-fat mayonnaise produced.

Reduction of oil in low-fat mayonnaise which results in a decrease in viscosity modified by the addition of carbohydrates in the form of pectin from watermelon rind flour can produce good viscosity because carbohydrates have the potential to bind water [12]. Azizah *et al.* [13] added that the addition of surfactants to mayonnaise can increase the viscosity of the product. Whey protein has a protein content of 80% capable of acting as a surfactant for oil and water.

3.4 The effect of whey protein addition on low-fat mayonnaise protein levels

The results of the analysis of variance showed that the addition of whey protein with different concentrations in low-fat mayonnaise had a significant effect on the value of protein content. The average value of water content can be seen in table 4.

Table 4. The average value of protein content %

Treatment	Protein content ± SD
P_0	1.46 ± 0.05[a]
P_1	1.78 ± 0.10[b]
P_2	2.03 ± 0.17[c]
P_3	2.47 ± 0.08[d]

The value of protein content ranges from 1.46 to 2.47. The highest mean value of 2.47 was obtained from the treatment using the addition of whey protein P3 and the lowest average value of 1.46 was obtained from the treatment without the addition of whey protein P0. The results showed that the higher the percentage of addition of whey protein, the higher the protein content of low-fat mayonnaise produced.

The results showed that the addition of whey protein can increase the average protein content of low-fat mayonnaise because the addition of ingredients containing whey protein concentrate is 80%. The whey protein component is dominated by -lactoglobulin, -lactalbumin, serum albumin, lactoferrin, immunoglobulin, and several other proteins [9].

The results of the addition of whey protein increased significantly beyond the standard set by the Indonesian National Standardization Agency. The highest value of low-fat mayonnaise is 1.80. while the SNI mayonnaise protein content according to [10].

4. Conclusion

The conclusion of this study is that the addition of 15% whey protein will improve the physicochemical quality of low-fat mayonnaise.

5. Reference

1. T. J. Herald, M. Abugoush, F. Aramouni, J. Texture Stud. **40**, 692-709 (2009)
2. C. Sun, R. Liu, B. Liang, T. Wu, W. Sui, M. Zhang, Food Res Int. **108**, 151-160 (2018)
3. B. D. Ahadi, M. Y. Effendi, Jurnal Ilmu Peternakan Terapan **2**, 34-38 (2019)
4. M. M. Rusalim, T. Gusnawaty, Jurnal Sains dan Teknologi Pangan **2**, 770-778 (2017)
5. H. Evanuarini, A. Susilo, IOP Conf. Ser. Earth Environ. Sci. **478**, 1-5 (2020)
6. P. Rizkyyani, A. Khusna, M. Hilmi, M. H. Khirzin, D. Triasih, Jurnal Ilmu dan Teknologi Peternakan Tropis **7**, 52-58 (2018)
7. Badan Standarisasi Nasional, SNI 3141.1:2011, Susu Segar-Bagian 1: Sapi (Badan Standarisasi Nasional, Jakarta, Indonesia, 2011)
8. M. Guo, G. Wang, in Whey Protein Production, Chemistry, Functionality, and Applications, edited by M. Guo (John Wiley & Sons Ltd, New Jersey, USA, 2019), pp. 1-12
9. Badan Standar Nasional, Standar Mutu Mayonnaise (Badan Standarisasi Nasional, Jakarta, Indonesia, 1998)
10. Widodo, A. V. Rahmawati, R. Chulaila, I. G. S. Budisatria, Jurnal Teknologi dan Industri Pangan **13**, 132-139 (2012)
11. H. Evanuarini, D. Amertaningtyas, D. T. Utama, IOP Conf. Ser. Earth Environ. Sci **788**, 1-7 (2021)
12. K. Wynne, J. Phys. Chem. Lett. 8, 6189-6192 (2017)
13. N. Azizah, K. Suradi, J. Gumilar, Jurnal Ilmu Ternak Universitas Padjadjaran **18**, 79-85 (2018)

Supplementation of vitamin E and selenium in different diets on performance of IPB D-2 chicken during developer-layer period

Beby Murba Ningsih Saragih[1], *Sumiati* Sumiati[2,*], and *Rita* Mutia[2]
[1]Student of Department of Nutrition and Feed Technology, Faculty of Animal Science, IPB University, Bogor 16680, Indonesia.
[2]Department of Nutrition and Feed Technology, Faculty of Animal Science, IPB University, Bogor 16680, Indonesia

Abstract. Supplementation of vitamin E and selenium improved the productivity of livestock. This study aims to evaluate the effect of vitamin E and selenium supplementation on different diets of IPB D-2 chickens during 15-21 weeks. Completely randomized factorial 2 x 2 (factor A was diet type and factor B was vitamin E supplementation) and five replications was used in this experiment. The variables observed were performance, blood profile, organ immunity and percentage of carcass weight. The results showed that diet type had a significant effect ($P < 0.05$) on feed consumption in the developer and layer periods and had a highly significant effect ($P < 0.01$) on feed consumption in the pre-layer period and body weight gain in the layer period. Vitamin E and selenium supplementation had a significant effect ($P > 0.05$) on pre-layer feed consumption. The blood profile was in normal range with optimal stress levels. There was significantly interaction ($P < 0.05$) between diet type and vitamin E supplementation on carcass percentage and highly significantly interaction ($P < 0.01$) on layer feed conversion. Parameters of immune organ and egg production were not affected by treatment. The diet type was reduced by 5% from the standard with supplementation vitamin E dan selenium showed the best result for IPB D-2 chicken performance.

1 Introduction

Poultry products, especially native chickens, make a significant contribution to meeting the food and nutritional needs of the community. Consumers prefer the taste of free-range chicken meat and eggs. The increase in consumption of native chicken meat and eggs is not balanced with their availability because the development of domestic chicken itself is still relatively low. In addition to the low development, domestic chickens themselves have a slow growth rate. Efforts have been made to overcome this problem by crossing. One of the crossbreed chickens that have almost the same taste as native chicken is IPB D-1 chicken.

IPB D-1 chickens result from a cross from three chicken families, Pelung chickens, native chickens, and Sentul chickens with broilers. IPB D-1 chickens were divided into two

[*] Corresponding author: y_sumiati@yahoo.com

lines, male candidate line (IPB D-3) and female line candidate (IPB D-2). IPB D-2 chickens are genetically the same as IPB D-1 chickens. IPB D-2 chickens are prospective female strains (female) that have been selected with an IgY indicator of at least 10 mg/ml blood, body weight at 12 weeks of age 1.3 kg for males, and 1.1 kg for females. The purpose of establishing this IPB chicken is to increase the participation of local chickens, which have advantages in the form of meat taste and resistance to disease and can utilize local feed but have a weakness in the form of slow growth [1].

Chicken growth is influenced by feed quality and environmental factors. IPB's chicken performance and productivity can be improved by the quality and quantity of feed provided. Quality feed must contain the nutrients needed according to age development and maintenance purposes. The provision of the local chicken feed itself is still based on various sources, including the recommendation of Scott et al. (1982), NRC (1994), and Lesson and Summers (2005). The standard is actually for broilers, while the standard for local chickens' energy metabolism and protein requirements does not yet exist.

Another thing that can be done to increase the productivity of IPB chickens is to do vitamin and mineral supplementation. Vitamin and mineral supplementation play an essential role for livestock even though they are needed in small amounts. Lack of vitamins and minerals will have an impact on the growth of livestock production. One of the vitamins and minerals that can synergize well is vitamin E and selenium. So, this synergistic combination can act and protect tissues against oxidative damage and enhance the immune response [2]. Based on the description above, it is necessary to conduct research that discusses the effect of vitamin E and selenium supplementation on different diet to increase the performance and productivity of IPB D-2 chickens.

2 Materials and methods

2.1 Research time

This research was from February 2021 to April 2021. Raising IPB D-2 Chicken was carried out at the Poultry Animal Nutrition Field Laboratory. The analysis was carried out at the Poultry Livestock Nutrition Laboratory and the Meat and Work Livestock Laboratory, Faculty of Animal Husbandry, Bogor Agricultural University.

2.1.1 Tools and materials

The tools used in this study include battery cages, Thermo hygrometers, digital scales, feed and drink containers, lights, and fans. The materials used in this study were 80 IPB D-2 chickens aged 15 weeks, vitastress, disinfectant, rice husk, and limestone. Chickens were divided into four treatments consisting of 5 replications with a total of 20 experimental units, and each replication consisted of 4 chickens.

2.1.2 Work procedures

Before use, the cage was sanitized first by liming the floor and walls of the cage. After sanitation, the cage was left for a week before being used for maintenance. Then the equipment for feeding and drinking water were washed by using detergent. Then, the research ration formulation was prepared based on the nutritional needs of Lesson and Summers [3]. The treatment ratio consisted of R1S1, which was the standard ratio of Lesson and Summers [3] without supplementation, R1S2 was the standard ratio of Lesson and Summers [3] + Vitamin E (100 ppm) and Selenium (1 ppm), R2S1 was the ratio that was reduced by 5% of energy requirements, and the protein was from the R1S1 ration, and R2S2 was the R2S1 + vitamin E (100 ppm) and selenium (1 ppm) rations. The ingredients used in the formulation and composition of feed ingredients are shown in Table 1.

Table 1. Composition and nutrients content of dietary treatments.

Feed Ingredients (%)	Treatment ration					
	R1			R2		
	Developer	Pre-layer	Layer	Developer	Pre-layer	Layer
Yellow corn	55.00	60.00	53.56	44.50	48.45	49.20
Rice bran	22.55	13.00	0.00	37.75	30.00	11.00
Soybean meal	15.00	10.00	22.05	10.00	10.50	19.00
Meat Bone Meal	2.00	9.00	9.90	3.00	3.50	7.58
Palm oil	2.00	2.00	4.50	1.00	1.00	2.60
DCP	0.65	1.00	1.98	0.65	0.93	1.65
CaCO3	1.80	4.00	7.65	1.80	4.50	8.40
Nacl	0.20	0.20	0.16	0.20	0.20	0.25
Premix	0.50	0.50	0.01	0.50	0.50	0.01
L-Lysine	0.10	0.12	0.00	0.30	0.17	0.01
DL-Methionine	0.20	0.18	0.19	0.30	0.25	0.30
Total	100.00	100.00	100.00	100.00	100.00	100.00

R1: Standard diet Lesson and Summers [3], R2: diet which energy metabolizable energy and protein content 5% below the standard, Developer (15-17 weeks), Pre-layer (17-19 weeks), Layer (19-21 weeks)

2.1.3 Maintenance

IPB D-2 chickens aged 15 weeks were placed in a battery cage and raised for six weeks, with feeding done two times a day, 50% in the morning and 50% in the afternoon. While the provision of drinking water *ad libitum*. Feed consumption data was measured every day, and body weight gain was carried out every week. Disease control is done by regularly replacing the husks when they are wet.

2.1.4 Observed variables

The variables observed in this study: performances (feed consumption, body weight gain, FCR, and egg production), percentage of immune organ weight, blood profile, and percentage of carcass weight.

2.1.5 Data analysis and experimental design

Completely randomized factorial 2 x 2 pattern with factor A was the diets type including R1: Standard diet [3] and R2: <5% Standard diet [3]. Factor B was S1: Without supplementation of vitamin E and Selenium and S2: Supplementation of vitamin E (100 ppm) and selenium (1 ppm) with five replications. Analysis data used analysis of variance (ANOVA). If the analysis shows a significantly different effect, it is continued using the Duncan multiple range test.

3 Results and discussion

3.1 IPB D-2 chicken performance developer period

Data on the performance of IPB D-2 chickens during the developer period are presented in Table 3. There was no interaction between diet type and vitamin E and selenium supplementation on performance IPB D-2 chickens at developer period. The diet type for IPB D-2 chickens had a significant effect ($P<0.05$) on feed consumption. The diet which energy metabolizable energy and protein content 5% below the standard resulted higher feed consumption. The occurrence of differences in feed consumption in this study was caused by differences in metabolizable energy and protein between R1 and R2. The high consumption indicated that the ME and protein contents did not meet the needs of IPB D-2 chickens. Rasyaf [4] stated that energy requirements and ration energy levels influence ration consumption. If the energy level can meet the needs of life, the ration consumed is less and vice versa.

Table 3. The performance of IPB D-2 chicken during the developer (15-17 weeks) period with different diet type and vitamin E and selenium supplementation.

Parameters	Supplementation Vit. E dan Se	Treatment		Mean
		R1	R2	
Feed consumption (g/bird/day)	S1	87.29±1.40	91.60±2.08	89.69±2.80
	S2	90.03±5.31	93.13±0.91	91.58±4.11
	Mean	88.81±4.29^b	92.36±1.77^a	
	P-Value	0.033		
BWG (g/bird)	S1	98.85±21.08	92.47±15.80	96.01±19.18
	S2	110.28±18.67	90.30±14.39	100.29±19.43
	Mean	104.56±20.72	91.26±15.07	
	P-Value	0.173		
FCR	S1	6.32±0.81	7.51±0.88	6.85±1.03
	S2	6.08±0.96	7.91±2.04	7.00±1.84
	Mean	6.20±0.90^a	7.74±1.64^b	

	P-Value	0.039	

S1: Without vitamin E and Se supplementation, S2: vitamin E supplementation (100 ppm) and Se (1 ppm), R1: Standard ration Lesson and Summers [3], R2: R1 <5% of energy metabolism and protein requirements

The results in Table 3 showed that the diet type and vitamin E and selenium supplementation in IPB D-2 chickens had no significant effect (P>0.05) on body weight gain. The ration conversion was significantly affected (P<0.05) by the diet type. The conversion of the ration with the diet which metabolizable energy and protein content 5% below the standard (R2), was more significant than the standard ration (R1). The difference in the value of the ration conversion between the R2 and R1 diet type was caused by the consumption of more rations and due to relatively lower bodyweight gain. The resulting ration conversion was more incredible.

Table 4. The performance of IPB D-2 chicken during the pre-layer (17-19 weeks) period with different diet type and vitamin E and selenium supplementation.

Parameters	Supplementation Vit. E dan Se	Treatment		Mean	P-Value
		R1	R2		
Feed consumption (g/bird/day)	S1	90.59±0.62	96.45±3.82	93.19±3.90[b]	0.004
	S2	95.12±2.49	99.56±0.28	97.34±2.84[a]	
	Mean	92.85±2.90[b]	98.18±2.99[a]		
	P-Value	0.0004			0.543
BWG (g/bird)	S1	80.66±5.62	80.21±7.26	80.46±6.40	0.832
	S2	85.89±15.19	77.00±3.65	80.95±11.38	
	Mean	82.89±11.26	78.43±5.77		
	P-Value	0.335			0.382
FCR	S1	8.04±0.77	8.72±1.49	8.43±1.20	0.888
	S2	7.85±2.05	9.10±0.46	8.48±1.61	
	Mean	7.95±1.55	8.93±1.07		
	P-Value	0.184			0.692

S1: Without vitamin E and Se supplementation, S2: vitamin E supplementation (100 ppm) and Se (1 ppm), R1: Standard ration Lesson and Summers [3], R2: R1 <5% metabolism energy and protein requirements

Data on the performance of IPB D-2 chickens during the pre-layer period are presented in Table 4. There was no interaction between diet type and vitamin E and selenium supplementation on performance IPB D-2 chickens pre-layer period. The diet type with vitamin E and selenium supplementation had a significant effect (P<0.05) on feed consumption in the pre-layer period of IPB D-2 chickens. The high feed consumption of

IPB D-2 chickens with the diet of metabolizable energy and protein content 5% below the standard indicated that the R2 diet type was not meeting the needs. The IPB D-2 chickens consumed more rations than the standard R1 diet type.

Table 5. The performance of IPB D-2 chicken during the layer (19-21 weeks) period with different diet type and vitamin E and selenium supplementation.

Parameters	Supplementation Vit. E dan Se	Treatment		Mean	P-Value
		R1	R2		
Feed consumption (g/bird)	S1	101.15±0.94	107.41±7.19	104.28±6.01	0.389
	S2	103.62±3.85	108.58±0.15	106.10±3.68	
	Mean	102.39±3.06^{b}	107.99±5.12^{a}		
	P-Value	0.015			0.753
BWG (g/bird)	S1	88.39±3.22	80.43±7.74	84.41±7.14^{b}	0.012
	S2	100.04±3.89	83.88±5.23	91.96±9.30^{a}	
	Mean	94.22±6.83^{a}	82.15±6.82^{b}		
	P-Value	0.0003			0.142
FCR	S1	8.03±0.27^{b}	5.42±0.26^{a}	8.74±0.90^{b}	0.0002
	S2	5.42±0.26^{a}	9.13±0.58^{c}	7.28±1.91^{a}	
	Mean	6.72±1.33^{a}	9.29±0.68^{b}		
	P-Value	0.0002			0.0003

S1: Without vitamin E and Se supplementation, S2: vitamin E supplementation (100 ppm) and Se (1 ppm), R1: Standard ration Lesson and Summers [3], R2: R1 <5% metabolic energy and protein requirements

Data on the performance of IPB D-2 chickens during the layer period are presented in Table 5. There was no interaction between diet type and vitamin E and selenium supplementation on performance IPB D-2 chickens at layer period. The diet type for IPB D-2 chickens had a significant effect ($P<0.05$) on ration consumption. The increase in body weight of IPB D-2 chickens during the layer period was significantly affected ($P<0.01$) by diet type and vitamin E and selenium supplementation. As for feed conversion, there was a very significant interaction ($P<0.01$) between different ratio treatments with vitamin E and selenium supplementation.

3.2 Percentage of immune organ weights

Table 6. Percentage of immune organ weights of IPB D-2 chickens with different diet type and vitamin E and selenium supplementation

Parameters	Supplementation	Treatment	Mean	P-Value

	Vit. E dan Se	R1	R2		
Spleen	S1	0.38±0.12	0.35±0.15	0.37±0.13	0.982
	S2	0.38±0.13	0.35±0.11	0.37±0.12	
	Mean	0.38±0.13	0.35±0.13		
	P-Value	0.717			0.994
Thymus	S1	0.31±0.09	0.32±0.10	0.32±0.10	0.07
	S2	0.23±0.05	0.22±0.11	0.23±0.09	
	Mean	0.27±0.08	0.27±0.12		
	P-Value	1.000			0.898
Bursa	S1	0.05±0.03	0.05±0.03	0.05±0.03	0.791
	S2	0.04±0.01	0.06±0.02	0.05±0.02	
	Mean	0.05±0.02	0.06±0.03		
	P-Value	0.289			0.239

S1: Without vitamin E and Se supplementation, S2: vitamin E supplementation (100 ppm) and Se (1 ppm), R1: Standard ration Lesson and Summers [3], R2: R1 <5% metabolic energy and protein requirements

The study results in Table 6 showed that there was no interaction between diet type and vitamin E and selenium supplementation on percentage of immune organ IPB D-2 chickens. The diet type and vitamin E and selenium supplementation had no significant effect (P>0.05) on the percentage of immune organ weights. The weight of the spleen is affected by foreign substances such as toxins, anti-nutritional substances, and diseases that cause the spleen to form lymphoid cells and increase the weight of the spleen. The percentage of spleen weight in this study ranged from 0.35%-0.38%. According to Swito, *et al.* [5], the percentage of spleen weight ranged from 0.33%-0.41% in native chickens was still in the normal range. The percentage of thymus weight was not affected by the treatment. It showed that the ration treatment and the dose of vitamin E and selenium supplementation were not toxic to the body, so the thymus did not atrophy to disease. The percentage of thymus weight in this study ranged from 0.22%-0.32%. According to Harmonis [6], the percentage of average thymus weight ranges from 0.21% to -0.28% in native chickens. Different ration treatments and vitamin E and selenium supplementation were not able to increase the bursa weight. This is due to the growth rate and the regression of the bursa depending on the type, breed, condition of the chicks, and sex hormones [7]. The results of the observations show that the percentage of immune organ weights ranges from 0.05%-0.06%. According to Widia, *et al.* [8], the organ weights weight is around 0.05% of body weight.

3.3 Blood profile

The results of observing the blood profile of IPB D-2 chickens in this study showed that the blood profiles other than erythrocytes and leukocytes were still in the normal range. The average number of erythrocytes in this study ranged from 1.94-2.10 million mm-3, so that it was slightly below the normal range. Smith and Mangkoewidjojo [9] stated that normal chicken erythrocytes ranged from 2.0-3.2 million mm-3. While the research conducted by Wahyudi, *et al.* [10], the average number of erythrocytes ranged from 1.62 to 3.87 x $10^6/\mu L$, which was still in the normal range. The number of erythrocytes will be constant in a relatively normal environment. In this study, the low value of erythrocytes produced can be influenced by the adequacy of nutrition and environmental conditions.

Table 7. Profile of IPB D-2 chicken blood with different diet type and vitamin E and selenium supplementation.

Parameters	Supplementation Vit. E dan Se	Treatment		Mean
		R1	R2	
Hematocrit	S1	35.80±2.93	39.00±2.76	37.40±3.26
	S2	39.60±2.42	36.00±0.71	38.00±2.58
	Mean	37.70±3.29	37.67±2.58	
Hemoglobin	S1	9.16±1.44	8.60±0.82	8.91±1.24
	S2	9.94±1.08	7.82±0.78	8.88±1.42
	Mean	9.55±1.33	8.17±0.89	
Lymphocyte	S1	57.67±4.48	58.86±4.78	58.26±4.67
	S2	59.87±1.47	60.28±1.62	60.10±1.57
	Mean	58.65±3.65	59.57±3.64	
Heterophil	S1	34.62±4.12	34.40±4.55	34.51±4.34
	S2	32.08±1.46	31.32±4.49	31.61±3.51
	Mean	33.49±3.46	32.82±4.79	
Erythrocyte	S1	1.82±0.43	1.85±0.27	1.83±0.37
	S2	2.06±0.54	2.14±0.40	2.10±0.48
	Mean	1.94±0.50	2.01±0.38	
Leucocyte	S1	31.77±10.1 1	53.65±39.37	42.71±13.36
	S2	29.40±7.61	59.58±38.73	44.40±16.25
	Mean	30.58±9.03	56.61±49.03	
Eosinophile	S1	2.62±1.24	3.09±0.37	2.85±0.95
	S2	3.38±0.58	3.81±0.73	3.57±0.68
	Mean	3.00±1.04	3.41±0.66	
Basophile	S1	2.27±0.96	2.36±0.43	2.32±0.72
	S2	3.26±0.61	1.54±0.47	2.30±1.01
	Mean	2.76±0.94	1.95±0.61	
Monocyte	S1	2.06±0.94	1.29±0.50	1.68±0.84
	S2	1.24±0.38	1.42±0.46	1.32±0.43
	Mean	1.65±0.83	1.35±0.49	
H/L	S1	0.61±0.11	0.53±0.04	0.57±0.10
	S2	0.54±0.03	0.52±0.09	0.53±0.07
	Mean	0.58±0.09	0.52±0.07	

S1: Without vitamin E and Se supplementation, S2: vitamin E supplementation (100 ppm) and Se (1 ppm), R1: Standard ration Lesson and Summers [3], R2: R1 <5% metabolic energy and protein requirements

The average number of leukocytes of IPB D-2 chickens obtained was 30.58 thousand mm-3-56.61 thousand mm-3. The results showed that the leukocyte count was higher than the normal range. Smith and Mangkoewidjojo [9] stated that the normal range of leukocyte count in chickens is 16-40 thousand mm-3. Leukocytes play a role in fighting germs that cause infection and foreign objects that enter the body [11]. Infection will cause an increase in the number of leukocytes due to abnormal physiological conditions that affect the hormonal balance in the chicken body. In this study, the rearing environment had high temperatures and high humidity during the day, causing stress to the chickens. The increased number of leukocytes due to environmental stress will increase the production of corticosteroids and glucocorticoids; this causes a decrease in the chicken's body defence system.

In this study, the H/L ratio in IPB D-2 chickens ranged from 0.52 to 0.61. According to Swenson [12], the normal range of the H/L ratio for poultry is 0.45-0.50. The H/L ratio in this study was outside the normal range. According to Gross and Siegeh [13], the heterophile/lymphocyte ratio of 0.2, 0.5, and 0.8, respectively, is a characteristic of low, optimal, and high-stress levels. Therefore, it can be concluded that the H/L ratio of IPB D-2 chickens obtained from this study has not shown that chickens experience high levels of stress but are still in optimal condition. This result indicates that vitamin E and selenium can help to improve the immune system in IPB D-2 chickens. This shows that the provision of different rations and vitamin and selenium supplementation can maintain the H/L ratio of IPB D-2 chickens in a relatively normal condition. The supplementation of vitamin E 200 mg kg-1 shows the best results in maintaining the health of broilers based on blood profile [14].

3.4 Percentage of carcass weight

Table 8. IPB D-2 chicken percentage of carcass weight with different diet type and vitamin E and selenium supplementation.

Parameters	**Supplementation Vit. E dan Se**	**Treatment**		**Mean**	**P-Value**
		R1	**R2**		
Percentage of carcass weight (%)	S1	63.80±1.58[ab]	60.88±0.92[b]	62.50±1.97	0.324
	S2	62.31±1.77[ab]	64.34±2.56[a]	63.21±2.38	
	Mean	63.05±1.84	62.61±2.59		
	P-Value	0.657			0.022

S1: Without vitamin E and Se supplementation, S2: vitamin E supplementation (100 ppm) and Se (1 ppm), R1: Standard ration Lesson and Summers [3], R2: R1 <5% metabolic energy and protein requirements

The study results in Table 8 showed that there was significant interaction ($P<0.05$) between diet type and vitamin E and selenium supplementation on the percentage of chicken carcasses of IPB D-2. The highest carcass percentage was produced by R2S2 treatment. It was found that higher consumption and body weight gain in the R2S2 treatment than other treatments before the chicken was slaughtered. Pahlepi, *et al.* [15] stated that the higher the feed consumption, the higher the body weight gain produced so that the carcass weight obtained was also high and vice versa. The higher carcass percentage in the R2S2 treatment also indicated that vitamin E and selenium as antioxidants could reduce stress and maintain the performance of IPB D-2 chickens even though the ration formula was reduced from the standard requirement.

3.5 Egg production

Table 9. IPB D-2 chicken egg production in the player period with different diet type and vitamin E and selenium supplementation.

Parameters	Supplementation Vit. E dan Se	Treatment		Mean
		R1	R2	
Hen Day Production pre-layer	S1	0.40±0.80	0.00±0.00	0.20±0.60
	S2	0.50±0.87	0.00±0.00	0.22±0.63
	Mean	0.44±0.83	0.00±0.00	
Hen Day Production layer	S1	0.80±0.60	0.00±0.00	0.40±1.20
	S2	0.00±0.00	0.80±1.60	0.44±1.26
	Mean	0.44±1.26	0.40±1.20	

S1: Without vitamin E and Se supplementation, S2: vitamin E supplementation (100 ppm) and Se (1 ppm), R1: Standard ration Lesson and Summers [3], R2: R1 <5% metabolic energy and protein requirements

The results in the table show that different ration treatments and vitamin E and selenium supplementation had no significant effect (P>0.05) on Hen Day Production (%). The average Hen Day Production obtained during this study ranged from 0.00% - 0.44%. According to Brickman [16], the factors affecting egg production are the seeds used, age, housing, lighting, feed, and environmental temperature. The production period also affects egg production [17]. In this study, IPB D-2 chickens aged 17-19 weeks in the player period and 19-21 weeks in the layer period. So that the resulting egg production is not optimal because the age is not ready to lay eggs. According to Habiburahman, *et al.* [18] the age of 24 weeks was the age of laying eggs in IPB D-1 G7 chickens.

4 Conclusion

The diet type with a 5% reduce from the standard with supplementation vitamin E (100 ppm) dan selenium (1ppm) showed the best result for IPB D-2 chickens performance, maintenance of immune organ and blood profile.

References

1. C. Sumantri, S. Darwati, Proceedings of the National Livestock Industry Seminar, **1**, 29-30 (2017)
2. V. Shinde's, K. Dhalwal, A. R. Paradkar, K. R. Mahadik, Pharmacologyonline, **1**, 252-261 (2007)
3. S. Lesson, J. D. Summers, *Commercial Poultry Nutrition,* **3** (2005)
4. M. Rasyaf, *Beternak ayam kampung. (*Panebar Swadaya, Jakarta, Indonesia, 2005)
5. Swito, D. Sudrajat, R. Handarini, Journal of Animal Husbandry Nusantara, **1**, 25–32 (2015)
6. M. Harmonis (2019)
7. B. Glick, *Sturkie's Avian Physiology*, **5**, 658-659 (2000)
8. E. Widia, E. Widiastuti E, A. S. Teysar, Prosiding STAP FAPET UNSOED, **5**, 297-301 (2017)
9. J. B. Smith, S. Mangkoewidjojo, *Maintenance, Breeding, and Use of Experimental Animals in the Tropics* (1988)
10. Wahyudi, A. Indi, M. A. Pagala, Halu Oleo Animal Husbandry Scientific Journal, **3**, 137-142 (2021)
11. Isroli, S. Susanti, E. Widiastuti, T. Yudiarti, Sugiharto, Prosiding Seminar Nasional Kebangkitan Peternakan, 548-557 (2009)
12. M. J. Swenson, *Duke's Physiology of Domestic Animals*, **10** (1984)
13. W. B. Gross, H. S. Siegel, Avian Dis, **27**, 972-979 (1983)
14. F. Meidita, Sumiati, S. Tuti, AIP Conference Proceedings, **2296** (2020)
15. R. Pahlepi, H. Hapid, A. Indi, JITRO, **2**, 1-7 (2015)
16. W. L. Brickman, *Poultry Production in Tropical Areas*, **11** (1989)
17. S. Baktiningsih, S. Mugiyono, D. M. Saleh, JEEP, **1**, 993-1000 (2013)
18. R. Habibburahman, S. Darwati, C. Sumantri, Rukmiasih, Journal of Animal Production Science and Technology, **8,** 97-101 (2020)

Growth performance and prediction of carcass production of captive javan porcupine (*Hystrix javanica*)

Wartika Rosa Farida[1*], *Andri Permata* Sari[1], *Umar* Sofyani[1], and *R. Lia Rahadian* Amalia[1]

[1]Research Center for Biology, Indonesian Institute of Sciences, Bogor 16911, Indonesia

Abstract. Porcupines are unique mammals, their upper body is covered with hard hair resembling sharp cylindrical spines and belongs to the order Rodentia. It has long been used as a source of animal protein. Javan porcupine (*Hystrix javanica*) is an endemic animal on Java and its population is estimated to continue, due to habitat destruction, poaching, and its use without a conservation. Observations on the growth of young javan porcupines were carried out for 12 months in Small Mammal Captivity, Research Center Biology - Indonesian Institute of Sciences. This study aims to determine the growth performance of young porcupines in captivity based on number of births, namely singles, twins, and triplets. The young will be with their mother for 3 months, they will be placed in individual cages. Weighing porcupines is done once a week. The results suggest that porcupines continue grow until they are 12 months old and probably continued. The average weight gain of single born was 30.31 g/day; birth of twins 29.65 g/day, 29.72 g/day; and birth of triplets were 29.57 g/day, 28.14 g/day, and 29.79 g/day. Prediction of 12-month-old porcupine carcass production for single, twins, and triplets, respectively, was 6740.86 g; 6593.15 g; and 6470.50 g.

1 Introduction

The porcupine is the prickliest of rodents, with needle-like quills on its sides, back, and tail. The Javan porcupine (*Hystix javanica*) is one of the 11 species of Old World Porcupines (Family: Hystricidae). This animal is active at night, while throughout the day it stays in its nest inside the ground Nowak [1], the mating system is monogamous, and a pair becomes long-life partners and breeds several times a year [2]. In its natural habitat, porcupines like to eat tubers, rhizomes, forest fruits, leaves, young stems, roots, buds, internal stems, roots, bark, and bamboo shoots [3, 4].

The Javan porcupine is endemic to Indonesia, its distribution covering Java, Madura, Bali, Lombok, Sumbawa, and Flores Tanah Djampa, Rintja, and Sulawesi [5, 6]. These terrestrial animals can be found mainly in lowland areas, secondary forests, and degraded lands. Porcupines are omnivores that eat leaves, fruit and insects.

* Corresponding author: wrosafarida@gmail.com

Porcupines have a role as a seed dispersal agent [7]. Hystrix is a polyestrus and the gestation period for Hystrix is about 100 to 112 days [8]. A Hystrix usually has 2 to 4 youngs per birth. Female porcupine can give birth twice a year [9, 10]. According to Van Aarde [8], in their habitat the female porcupine gives birth to one to three young per birth, 58.8% single births, 32.1% twins, and 9.1% triplets. Although young porcupines begin to eat solid food after 2 weeks of birth, the mother porcupine still has to nurse it for 13 to 19 weeks postpartum [8]. Young porcupines will live in colonies until they reach two years of age. Before they reach the age of 2 years, they will live with their mother in the nest [9].

H. Javanica has protected status based on Minister of Environment and Forestry Regulation Number: P.92/MENLHK/SETJEN/KUM.1/8/2018, and It is currently classified as Least Concern by the IUCN Red List [11]. Although *H. javanica* has protected status, in reality, illegal hunting of porcupines continues to increase. People in several regions in Indonesia believe that porcupine meat and other parts of the body contain medicinal properties. Porcupines are being illegally hunted and exploited throughout their range in Indonesia facilitated by poor enforcement and legislative weakness. Porcupines are in decline due to habitat loss, retaliatory killings and uncontrolled poaching [12]. Their meat is consumed as an alternative and important source of protein in parts of their range and their quills are used for decorative purposes. Food and medicine derived from wildlife have important nutritional values especially in times of crisis, although the wildlife food often regarded as supplementary to local peoples' diet [13]. Mardiastuti, *et al.* [14] reported that the Javan porcupine (*H. javanica*) and the Malayan porcupine (*H. brachyura*) are used by local people in Indonesia as traditional medicine.

Based on the facts above, conservation and protection efforts for the Javan porcupine should be carried out before this animal is declared extinct. One of the efforts to save it is through captivity, namely the maintenance of wildlife outside their habitat (ex situ conservation). The use of protected wildlife should be followed by breeding efforts through captivity, so that harvesting can be carried out starting from the second generation (F2) which legally can be used commercially. Harvesting is no longer done by taking directly from nature. The purpose of this study was to observe the growth of young Javan porcupine born in captivity in the number of singles, twins, and triplets, and predicting the production of porcupines carcass at the age of 52 weeks.

2 Materials and methods

This experiment was conducted in the Small Mammal Captivity of The Research Center for Biology, Indonesian Institute of Sciences (LIPI), Cibinong, Bogor Regency. The material used was 6 young Javan porcupines which were born from 3 females, consisting of a single, twins, and triplets. From birth to 12 weeks of age, the young porcupines are still with their mothers in the cage. Starting at 13 weeks of age, they were separated from their mothers and placed in three individual cages, each measuring length x width x height (2.25 m x 2.00 m x 2.75 m) with two layers of wire walls and a concrete floor. Single young in cage A, twins in cage B, and triplets in cage C. The feed given is tubers, vegetables, fruit, koi fish pellets as a source of protein and minerals. Feeding was carried out 2 times a day, namely at 08.30 a.m. and 04.30 p.m. The amount of feed given is about 15% of the body weight of each porcupine. Drinking water is available ad libitum. No calculation of consumption and feed conversion was carried out during the study.

The method of measuring phenotypic variables is the body live weight of the young porcupines through weighing at birth and then the young is weighed every week until the 52nd week or 1 year old. Weighing of young was carried out in the morning at 08.00 a.m before the feed was served. Weighing youngs aged 0-12 weeks using a digital scale with a capacity of 5 kg with an accuracy of 1 gram, and weighing porcupines aged 13-52 weeks using a digital scale with a capacity of 25 kg with a scale of 0.1 kg. Daily weight gain was obtained from the different between final weights with initial weight which divided with total days of experiment (gram/porcupine/day) [15]. Meanwhile, the prediction of porcupine carcass weight at the age of 52 weeks was based on the percentage of carcass weight from previous studies, which was 59.68% Farida *et al.* [16] multiplied by the final body weight. Carcass is the body part of livestock without head, legs, skin, digestive organs, blood and kidneys. Therefore, the components called carcass are meat (muscle), bone and fat [17].

The data obtained from the research results were analyzed descriptively by describing the data in the form of tables or curves into a sentence as well as concluding the research results obtained [18].

3 Results and discussion

The average temperature around the captivity during the study was 25.52°C (morning), 30.78°C (noon) and 30.11°C (afternoon). Average air humidity in the morning, afternoon and evening, respectively 81.83%; 64.26 % and 70.15%. According to Bartos [2] the ideal temperature for tropical porcupines is 70-85°F or 21-29.4°C. The factors of low temperature and high humidity in the morning and high temperature and low humidity during the day will affect the condition and productivity of the porcupine.

From Table 1, it can be seen that the birth weight of young Sunda porcupine ranged from 206 - 280 g, the litter size did not have much effect on birth weight. Reported by Roth [19] and Van Aarde [8], individual mass at birth equaling ca. 2.1% of mean adult female mass. Male and female neonates do not differ in body mass [8] and sex ratio at parturition does not differ from 1:1 [20]. Females average ca. 1 litter per year with a mean litter interval of 385 days. Litter size ranges from 1 to 3, with 58.8% of litters being singletons, 32.1% twins, and the remainder triplets [8].

The porcupine's body weight continued to increase until the age of 52 weeks although there was a slight fluctuation in body weight gain. Early weaning performed at the age of 12 weeks (84 days) did not affect the increase in body weight of single, twins, or triplets, while according to Van Aarde [8] mean length of lactation in 9 females porcupines was 101 days. Next Van Aarde [21] reported Postnatal growth rates do not differ between males and females and are nearly linear for the first 20 weeks of life, reaching an asymptotic point (and presumably full adult size) at ca. 52 weeks and an average mass of 11.7 kg. Table 1 shows the average final weight of single, twins , and triplets ranging from 10.84 – 11.30 kg. Reported by Barthelmess [22], the Cape porcupine (*Hystrix africaeaustralis*) or South African porcupine approached the age of 20 weeks and was close to linear until it reached the asymptotic point (adult age) reaching approximately 52 weeks.

The birth weight of single, twins, and triplets varied between 206 – 280 grams (Table 1). In singletons and twins, the birth weight of the young female was greater than that of the young male, on the contrary in the triplets, the birth weight of the young male was greater than that of the young female. At week 12, the young is separated or weaned from its mother. Reported by Bartos [2] young porcupine should remain with their mother in the cage for 24 - 52 weeks. Early weaning is intended so that the porcupine mother can mate and get pregnant again. In addition, based on observations in captivity, since the age of 8 weeks, the youngs have started to learn to eat solid food that is usually given to their mother and the youngs already appear strong and healthy. Early weaning did not affect the growth of young porcupines, because their body weightshould continued to increase (Table 1). Nowak [1] reported that young porcupines have the fastest growth period, which is around the age of 4-5 months, then will go up and down, and will reach adult body weight at the age of 1-2 years.

Table 1. Body weight development of young porcupines born single, twins, and triplets

Week	Body weight					
	Single	Twins		Triplets		
	♀ (g)	♂ (g)	♀ (g)	♂ (g)	♀ (g)	♀ (g)
0 (birth)	261	206	280	254	212	210
1	388	325	430	292	323	279
2	408	472	647	470	460	370
3	454	602	793	587	595	529
4	774	750	954	676	702	646
5	903	879	1068	829	860	821
6	1126	1016	1183	1052	1105	914
7	1338	1205	1335	1176	1273	1133
8	1497	1285	1435	1380	1352	1376
9	1589	1439	1560	1455	1468	1384
10	1705	1622	1719	1653	1552	1500
11	2004	2102	2142	1826	1707	1625
12	2339	2235	2330	1904	1915	1907
13	2980	3140	3355	2112	2204	2211
14	3467	3213	3620	2305	2401	2399
15	3602	3340	3710	2462	2510	2521
16	3877	3510	3830	2611	2778	2764
17	4005	3550	3925	3026	3123	3017
18	4158	4110	4035	3312	3502	3398
19	4230	4220	4150	3616	3836	3792
20	4369	4340	4260	4037	4200	4102
21	4412	4355	4335	4508	4756	4613
22	4568	4410	4400	4765	5007	4822
23	4673	4600	4550	4839	5110	4938
24	4788	4635	4640	4910	5204	5108

25	4876	4710	4714	5132	5386	5307
26	5075	4730	4810	5348	5501	5435
27	5287	5170	5275	5698	5799	5797
28	5690	5610	5740	6005	6189	6101
29	6003	6025	6205	6414	6215	6416
30	6668	6615	6700	6799	6304	6811
31	7100	6985	7105	7100	6412	7069
32	7513	7490	7595	7255	6578	7236
33	7941	7901	8010	7430	6866	7504
34	8337	8433	8500	7646	7005	7709
35	8790	8965	9087	7854	7268	7942
36	9025	9307	9305	8038	7455	8094
37	9398	9411	9491	8224	7680	8235
38	9506	9521	9584	8467	7897	8444
39	9868	9651	10011	8659	8075	8652
40	10036	9725	10206	8891	8269	8847
41	10201	9775	10288	9065	8497	9006
42	10365	9795	10316	9287	8723	9244
43	10453	9823	10504	9465	8986	9406
44	10521	9902	10527	9577	9018	9539
45	10626	9936	10554	9605	9112	9657
46	10707	9949	10608	9799	9268	9706
47	10789	9973	10615	9895	9413	9968
48	10828	10022	10691	10112	9689	10254
49	10879	10042	10674	10487	9790	10586
50	10901	10056	10744	10698	9824	10820
51	10989	10610	10889	10895	9951	10915
52	11295	10997	11098	11018	10455	11053
Average of final body weight	11295	11047.5		10842		

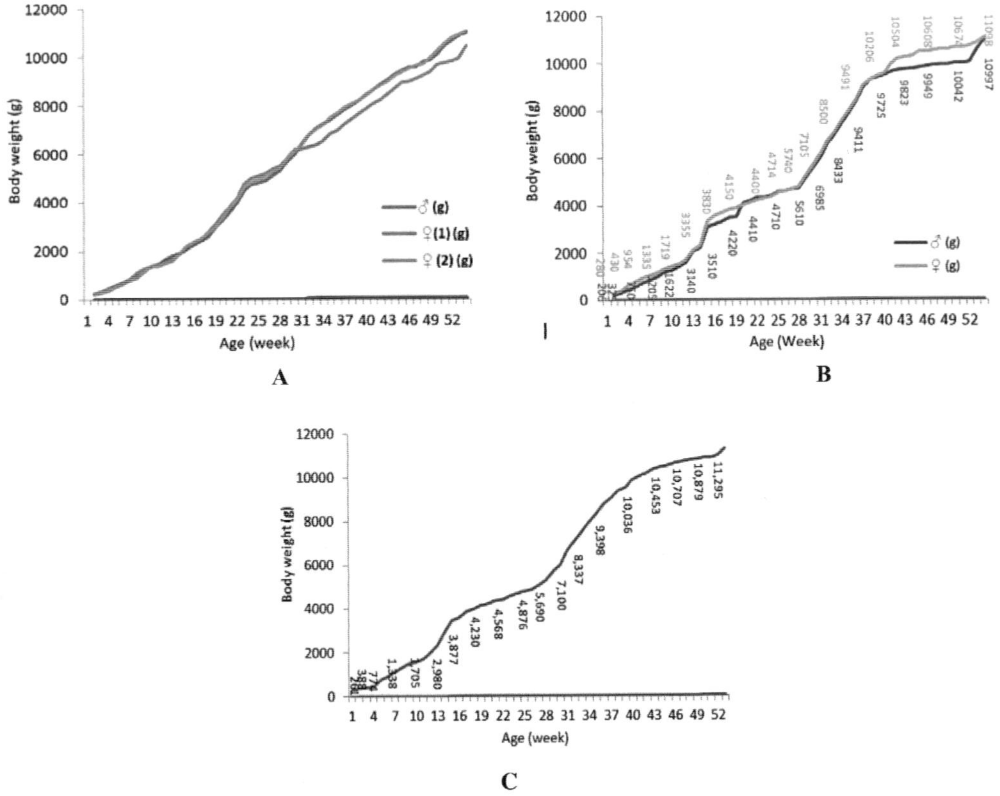

Fig 1. Body weight of development of sunda porcupine of litters being singletons (A), twins (B), and triplets (C)

In zootechnical studies, there are two main aspects in the growth process. The first aspect is the initial growth as an increase in body size (weight and length) per unit time, most of which have been achieved in the embryonic stage, including in the process. The second aspect is growth as a morphogenetic process, namely the increase in different body parts with changes in the size and shape of body components caused by differences in the intensity of growth of various organs and tissues. This differential growth can be called development. The part-to-whole relationship is considered stepwise, as carcass is related to live weight, major components are related to carcass weight, and individual bone, muscle, and fat depots are related to their respective total tissue weights. If animal height or weight are measured from conception to senescence, the data usually follow a flattened "S" shape called the sigmoid curve. The sigmoidal growth curve is formed because age does not cause an increase in body weight, but provides an opportunity for animals to grow, reach maturity and interact with the environment [23]. Growth initially took place very quickly (acceleration) then decreased and tended to be constant. But the growth curves of meat animals raised under commercial conditions may appear as relatively flat slopes (the middle segment of the flat "S", and the sigmoid shape may only become apparent if the data include very young animals or animal beyond a typical market weight. In other words, growth velocity is approximately constant during the commercial growing period.

From Figure 1, it can be seen that the three curves A dan B form like the letter S, while the C curve looks straighter because the body weight development of the three young porcupines from birth to 52 weeks of age is not much different. As reported by Von

Bertalanffy [24], curves of animal mass growth over time are shaped in S, which is also known as sigmoid or sigmoidal curves, while according to Silva *et al*. [25] and Rodrigues, *et al*. 26] that the body growth of most animal species can be described by a sigmoid curve and, therefore, fit by nonlinear models which are widely used for a large number of regression applications. Animal development is related to metabolic rate, which directly influences mass, shape, size, and other body traits [24]. The difference in the shape of the C curve is thought to be due to the number of triplets births, so that during lactation there is competition for mother's milk and after weaning there is competition for feed. This is in accordance with Smith and Mangkoewidjojo [27], stated that adult body weight was influenced by litter size, birth weight (initial weight), mother's milk production and feeding. Cahpman [28] defined maternal effect as the influence, contribution or impact on the phenotype of an individual caused directly by the phenotype of the parent. Body weight can be influenced directly by genetic and maternal effects as well as environmental factors [29, 30, 31]. Maternal effects on offspring phenotype can be caused by genetic or environmental differences between parents, or they can also be caused by genetic and environmental interactions. Body weight gain can be used as a variable to determine the growth of porcupines. Body weight gain is the difference between body weight (harvest) and initial body weight at a certain time. The average weight gain of porcupines for single, twins, and triplets are presented in Table 2.

Table 2. Body weight gain and prediction of carcass production of young porcupine, 1 year old (52 months)

Item	Single	Twins		Triplets		
	♀	♂	♀	♂	♀	♀
Initial BW (g)	261	206	280	254	212	210
Final BW (g)	11,295	10,997	11,098	11,018	10,455	11,053
BW gain per week (g)	212.19	207.52	208.04	207.00	196.98	208.52
BW gain per day (g)	30.31	29.65	29.72	29.57	28.14	29.79
Estimated carcass weight (g) based on carcass percentage = 59.68%*)	6740.86	6563.01	6623.29	6575.54	6239.54	6596.43
Average carcass weight	6740.86	6593.15		6470.50		

Source : Farida *et al*. [16]

Table 3. Slaughter weight, weight, and carcass percentage of Javan porcupine compared to other animals*)

Item	Sex	Slaughter weight (kg)	Carcas weight (kg)	Carcas (%)
Sunda porcupine	Female	6.40	3.79	59.68
	Male	8.70	5.16	59.66
Java mouse-deer	Female	1.92	0.99	50.65
	Male	1.53	0.75	50.48
Rex rabbit	Female	3.02	1.54	51.19

	Male	2.71	1.41	51.95
Goat	Female	24.27	9.77	39.39
	Male	24.23	11.03	42.48
Sheep	Female	25.13	11.70	43.01
	Male	25.80	12.53	44.18
Cattle	Female	214.32	110.60	51.18
	Male	226.14	115.51	51.02

Source: Farida *et al*. [16]

From Table 2, it can be seen that the average daily body weight gain of porcupines were 30.31 g/head (single), 29.65 and 29.72 g/head (twins) and 29.57; 28.14; and 29.79 g/head (triplets). Growth is generally expressed by increasing body weight and body dimensions as a reflection of the adequacy of food intake for body metabolism [32]. Table 2 shows that the average daily weight gain for single young is higher than that of twins and triplets. This is presumably because single-born get enough mother's milk intake, while twins and triplets have to compete for various milks from their mothers in the early stages of their growth. Body weight can be influenced directly by genetic and maternal effects as well as environmental factors [29, 31]. The increase in body weight of young porcupines after being weaned from their mother will depend on the adequacy of feed and nutritional intake. The increase in body weight between male and female was not affected by birth weight, it was proven that young born with low body weights could achieve high daily body weight gain. This may be the effect of the response of the porcupine to environmental conditions such as differences in response to the feed intake. It was stated by Soeparno [33] that among individuals within a breed or between animal races there are differences in responses to the environment, including nutritional, physical, and microbiological. The difference in response causes a difference in the rate of growth. In addition, there are still similarities in the breed or type of porcupine used in this study, namely the Javan porcupine. As reported by Tillman, *et al*. [34], food is not the only factor influencing body composition, as breeding and sex are also strong determinants, especially when food is sufficient to support rapid growth. The same thing was also stated by Soeparno [33] also Everitt and Jury [35], that sex, hormones and castration and genotype also affect animal growth.

Increased use of the Javan porcupine for meat and medicinal purposes in some areas of Indonesia [35]. It has also increased illegal hunting and overexploitation of these animals Gomes [12]. In order to maintain the preservation of protected wild animals, through captive efforts (ex situ) namely breeding wild animals outside their habitat in a controlled manner, harvesting for commercial purposes can be carried out starting from the second generation (F2). Based on the results of research Alves *et al*. [13], that the percentage of porcupine carcasses was higher than that of deer, rex rabbits, goats, sheep, and cattle (Table 3). Prediction of carcass weight calculation of captive Javan porcupine at the age of 52 weeks, respectively 6740.86 g (single), 6563.01 g and 6623.29 g (twins), and 6575.54g; 6239.54 g; and 6596.43 g (triplets). From the prediction of carcass weight (Table 2), singletons produced higher carcass weights than twins and triplets, while the carcass weight produced by twins was not much different from triplets.

The high carcas percentage of the Javan porcupine indicates that this animal has the potential to be bred as a meat-producing animal and can support food diversification programs, especially in Central Java and East Java, where most of the population has long consumed porcupines meat [35]. Porcupines are relatively undemanding to look after in captivity and acclimatise quickly to the captive environment. They become conditioned to people and are of a large enough size to make a memorable experience for captive visitors, as well as being quite engaging. They can also be used as education/outreach animals, easy to be tame, particularly if they have been hand-reared.

4 Conclusion

The growth curves of porcupines born single, twins, and triplets continued to increase until the observation limit of 52 weeks. Number of births, birth weight, and gender had no effect on final body weight and daily weight gain. The carcass weight of the porcupine is quite high at the age of 52 weeks, so that the porcupine has the potential to be bred as a meat-producing animal to meet protein needs and support food diversification.

References

1. R. M. Nowak, *Walker's Mammals of the World*, **5**, 643-1629 (1991)
2. C. Bartos, Husbandry Standards for Keeping Porcupines in Captivity (Baltimore Zoo, Druid Hill Park, Baltimore, MD 21217. (1998)
3. W. R. Farida, R. Ridwan, Jurnal Biologi Indonesia, **7**, 157–170 (2011)
4. A. Y. F. Khan, F. A. Asuhaimi, T. K. Jalal, F. O. Roheem, H. A. Natto, M. F. Johan, Q. U. Ahmed, R. A. Wahab, Preliminary Study Antioxid, **8**, 39 (2019)
5. D. van Weers, Beaufortia, **29**, 215–272 (1979)
6. H. Michael, M. McDade, D. G. Kleiman, V. Geist, *Grzimek's Animal Life Encyclopedia* (2004)
7. S. Timóteo, M. Correia1, S. Rodríguez-Echeverría, H. Freitas, R. Heleno, Nature Communications, **9**, 140 (2018)
8. R. J. Van Aarde, Journal of Reproduction and Fertility, **75**, 577-582 (1985)
9. A. H. Norsuhana, S. M. D. Nor, A. Aminah, Z. Z. Zahar, Sains Malaysiana, 38, 595-600 (2009)
10. D. Lunde, K, Aplin, S, Molur, Hystrix brachyura. The IUCN Red List of Threatened Species (2016)
11. K. Aplin. Hystrix javanica. The IUCN Red List of Threatened Species (2016)
12. L. Gomez, Nature Conservation, **43**, 109–122 (2021)
13. R. R. N. Alves, T. P. R. Oliveira, I. L. Rosa, Complementary and Alternative Medicine (2013)
14. A. Mardiastuti, B. Masy'ud, L.N. Ginoga, H. Sastranegara, Sutopo, Biodiversitas, **22**, 329-337 (2021)
15. W. A. T. Bowker, R. G. Dumsdey, J. E. Frisch, R. A. Swan, N. M. Tulloh, *Beef cattle Management and Economics* (1978)
16. W. R. Farida, A. S. Tjakradidjaja, A. P. Sari, Jurnal Biologi Indonesia, **8**, 381-398 (2012)
17. A. Lawrie, D. A. Ledward, Lawrie's Meat Science, **7** (2006)
18. R. G. D. Steel, J. H. Torrie, *Principles and Procedures of Statistics: A Biometrical Approach*, **2** (1986)
19. E, Mohr, M. Rohrs, Zeitschrif fur saugetierkunde, 313–316 (1964)
20. R. J. Van Aarde, Journal of Zoology, 213, 205-212 (1987a)
21. R. J. Van Aarde, Journal of Zoology, 211, 25-33 (1987c)
22. E. L. Barthelmess. Hystrix Africaeaustralis. Mammalian Species, **788**, 1-7 (2006)

23. H. L. Davies, *A Course Manual Nutrition and Growth* (1982)
24. L. Von Bertalanffy, The Quarterly Review of Biology, **32**, 217-231 (1957)
25. F. de L. Silva, M. M. de Alencar, A. R. de Freitas, I. U. Packer, G. B. Mourão, Pesquisa Agropecuária Brasileira, **46**, 262-271 (2011) A. Rodrigues, L. M. Chaves, F. F. Silva, I. P. Garcia, D. A. S. Duarte, H. T. Entura, Revista Ceres, **65**, 24-27 (2018)
26. J. B. Smith, S. Mangkoewidjojo, *Pemeliharaan, Pembiakan, dan Penggunaan Hewan Percobaan di Daerah Tropis*, **1** (1988)
27. A. B. Chapman, *General and Quantitative Genetics* (1985)
28. E. L. Bihan-Duval, C. Berri, E. Baeza, N. Millet, C. Beaumont, Poult. Sci., **80**, 839-843 (2001)
29. A. N. M. Koerhuis, R. Thompson, Genetics Selection Evolution, **29**, 225–249 (1997)
30. S. G. Vellemen, J. Anderson, K. E. Nestor, Poultry Science, **82**, 1479-1484 (2003)
31. R. Herman, Suwartono, Kadarman, Media Peternakan, **10**, 1-11 (1985)
32. Soeparno, *Ilmu dan Teknologi Daging*, **1** (1992)
33. A. D. H. Tillman, S. Hartadi, S. Reksohadiprojo, S. Prawirokusumo, Lebdosoekojo, *Ilmu Makanan Ternak Dasar* (1991)
34. G. C. Everitt, K. E. Jury, Journal of Agricultural Science, **66**, 1-14 (1966)
35. W. R. Farida, Proceedings of the National Biodiversity Seminar, **2**, 167-174 (2013)

Exploration of potential regional resources for beef cattle farming development in West Java, Indonesia

Rini Widiati[1,*], *Bambang* Suwignyo[2], and *Ahmad Romadhoni Surya* Putra[1]

[1]Department of Livestock Socio-Economics, Faculty of Animal Science, Gadjah Mada University, Yogyakarta, Jl. Fauna No.03, Karang Gayam, Caturtunggal, Kec. Depok, Kabupaten Sleman, Daerah Istimewa Yogyakarta, 55281, Indonesia
[2]Department of Animal Nutrition and Feed Science, Faculty of Animal Science, Gadjah Mada University, Yogyakarta, Jl. Fauna No.03, Karang Gayam, Caturtunggal, Kec. Depok, Kabupaten Sleman, Daerah Istimewa Yogyakarta, 55281, Indonesia

Abstract. More than 90% of cattle in Indonesia are raised by smallholder farmer. They are facing limited land resources so that the input of feed mostly depend on the agriculture waste and its environment. The aimed of this study was to determine the potential agricultural resources which influence to the development of beef cattle. The sample locations were 3 districts from the eastern part of West Java province. Panel data of variables that were thought have an effect on beef cattle development were taken from 2012 to 2019 in three regencies. These data were analyses by Multiple Regression Model using the OLS method. The stationary test was done before using time series data by unit root tests of Augmented Dickey-Fuller. The results of this research showed that the corn and Cassava production, as well as the population as a source of agriculture labor in a region was potential resources to increase of beef cattle population. The integration of cattle and corn farming on suitable land in a region needs to be considered as the strategy in beef cattle farming development, because the corn production per unit of land can produce the highest feed compared to the other food crop residues.

1 Introduction

Indonesia is an agricultural country. Most of its population live in rural areas and work in the agricultural sector. Unfortunately 57% of farmer households in Indonesia or 74.43% of them in Java controlled narrow areas of less than 0.5 ha (The Central Bureau of Statistics of the Republic of Indonesia, 2018). In addition to this, 96.41% of them grow food crops and raise livestock. Small farmers generally raise beef cattle with 1 to 3 Animal Units (AU) to support and complement the crop farming to generate the farmer household income [1]. Food crop farming produces crop residue for animal feed, and the livestock can produce manure to fertilize the agricultural land. These overall activities can increase the agricultural production. Small-scale livestock production activities in developing countries, including

* Corresponding author: rini_w@ugm.ac.id

Indonesia, can be an important route to address poverty for rural communities [2]. Cattle can both generate income for farmers and produce meat for highly nutritious food needed to increase human intelligence. It has been recorded that beef consumption in Indonesia continues to increase. In 2014, it was 101,416 tons per year, and in 2018, it became 125,551 tons per year, there was an increase in beef consumption by 4.76% per year on average [3].

Beef cattle development is a prospective business opportunity for farmers in rural areas in Indonesia. Research shows that local beef cattle fattening is profitable and has comparative advantages [4, 5]. This opportunity is supported by the condition in Indonesia, which geographically, has abundant natural resources to support the production of the ruminant beef cattle farming [6]. However, there are still various kinds of obstacles in beef cattle farming development in Indonesia, especially in Java, including the availability of forage, especially during the dry season [7]. In addition, there are constraints of capital and technology, poor land, and incompetent human resources [6, 8, 9]. These constraints are part of the socio-economic dynamics of farmers and agricultural resources, such as types of soil, climatic conditions, and rainfall, so that many breeders experience difficulties in providing feed, especially forage. Cattles are large ruminants that require large volumes of feed. This study is to confirm the previous research that one of the strategies to grow the beef cattle industry in Indonesia is by encouraging people to invest on the forage industry as a feed supplier, the product of which is processed with technology [9]. This way, farmers will be facilitated to access large quantities of feed throughout the year with a low price. Administratively, an area of a village, sub-district, regency, and city can have extensive agricultural land resources, workforce, and consumers, and they can create market demand. Furthermore, they are also theowners of the capital in the region.

An agricultural production process including beef cattle farming in an area can be described in the following function [10]:

$$Y = f(A, L, l, M) \quad (1)$$

Where: Y = output or product is the function of A = area of land for various household activities, L = labor, I = investment/capital, and M = management or technology. Agricultural resources such as crops and beef cattle are interrelated, dependent, and mutually supporting. The availability of resources in an area is important to be researched to increase the population and production of beef cattle. The researchers have also conducted a similar study in East Java Widiati *et al.* [11], and on this occasion, the researchers conducted a study in West Java. This study tries to find alternatives to develop smallholder beef cattle farming based on the potential of regional resources, especially plant resources that can support beef cattle farming.

2 Materials and Methods

West Java was chosen as the research location based on some considerations. West Java is a buffer area for the capital city of Jakarta in terms of the provision of livestock products including beef. However, the balance of local beef supply and beef demand in West Java was in deficit throughout 2012-2017 [12]. The eastern part of West Java has also developed new facilities, such as the new Kertajati Majalengka airport and the toll road to the capital city of Jakarta. Besides, the beef culinary in Cirebon City has been well known for a long time, so that the eastern part of West Java was chosen as the research location. Furthermore, three (3) regencies with low population density per area of agricultural land whose residents were possible to develop beef cattle are Majalengka, Indramayu, and Ciamis Regencies BPS Jawa Barat [13] were chosen as the samples.

West Java is a buffer area for the capital city of Jakarta in terms of providing livestock products including beef. However, the balance of local beef supply and beef demandin West

Java experienced a deficit throughout 2012-2017 [12]. The eastern part of West Java has also developed new facilities, such as the new Kertajati Majalengka airport and a toll road to the capital city of Jakarta. Besides that, beef culinary in Cirebon City has been famous for along time, so that the eastern part of West Java was chosen as the research location. Furthermore, three (3) districts with low population density per area of agricultural land were selected whose citizens allow for the development of beef cattle, namely Majalengka, Indramayu, and Ciamis Regencies BPS Jawa Barat [13] as samples.

This study used panel data collected during the period from 2012 to 2019 (8 years) for each sample regency. The beef cattle development in an area could be measured from the increase in the population of beef cattle from year to year. The population of beef cattle could be affected by the area of land for livestock, the production of various types of plants that can produce animal feed, and the number of people who provide labor. The data sources were the Central Bureau of Statistics (BPS), the Livestock Service, the Agricultural Service in each sample regency, and other related agencies.

The existence of cattle development in an area can be measured from the increase in cattle population from year to year as dependent variable that was incl uded on the multiple linear regression model as shown as Equation 2 [14, 15].

$$BCP_{it} = \beta_0 + \beta_1 Pop_{it} + \beta_2 FL_{it} + \beta_3 RProd_{it} + \beta_4 CrProd_{it} + \beta_5 CsProd_{it} + \beta_6 SProd_{it} + \beta_7 MbProd_{it} + \beta_8 PnProd_{it} + \beta_9 SptProd_{it} + e_{it} \quad (2)$$

Where:

BCP_{it} = beef cattle population as the dependent variable in the region i in year t (head)
β_0 = intercept
$\beta_1, \beta_2, \dots \beta_5$ = regression coefficient of independent variable
Pop_{it} = population in region i, in the year of t (person)
FL_{it} = farm land area in region i in the year of t (ha)
$RProd_{it}$ = rice production in region i and year t (ton/year)
$CrProd_{it}$ = corn production in region i and year t (ton/year)
$CsProd_{it}$ = cassava production in region i and year t (ton/year)
$SProd_{it}$ = soybean production in region i and year t (ton/year)
$MbProd_{it}$ = mungbean production in region i and year t (ton/year)
$PnProd_{it}$ = peanut production in region i and year t (ton/year)
$SptProd_{it}$ = sweet potato production in region i and year t (ton/year)
e_{it} = error term
I = 1,2,3…8 : region to 1,2,3…8
t = 1,2,3… 8 : year to 2012-2018

Estimation and validation of the model started from the correction of the time series data by using tests such as stationer, co-integration, and residual tests. The stationer tests considered here in are unit root tests of Augmented Dickey-Fuller (ADF) [15]. Furthermore, the coefficient of Regression estimated by using Ordinary Least Square method (OLS). The accuracy of the model was tested using a hypothesis test consists of a coefficient of determination (R^2), Overall test (F test) and partial test (t test). All of the above tests was done with the help of a computer using *EViews* 9th.

3 Results and Discussion

3.1 The conditions of the agriculture, beef cattle farming, and its supporting resources

Java is one of the major islands with 126.7 million ha or 6.63% of Indonesia's land area [16]. Nevertheless, it is also inhabited by 57.49% of Indonesia's population [17]. Meanwhile, West Java Province, which is one of the six provinces in Java, has a fairly large population, namely 31.52% of the population in Java. The detailed conditions of the agriculture, beef cattle farming, as well as agricultural and livestock households are presented in Table 1.

Table 1. The conditions of the agriculture, beef cattle farming, and supporting resources inWest Java, Indonesia

No	Activities	Indonesia		Java		West Java	
		Total	**%**	**Total**	**%**	**Total**	**%**
1	Beef cattle population (000 head)[a)]	16,433	100	7,156	43.55*	406	5.67**
	Population (000 person)[b)]	237,641	100	136,611	57.49*	43,053	31.52**
2	The area of agricultural land (000 ha)	7,460	100	3,128	41.93*	928	29.67**
3	The number of land-users agricultural sector households (000 RT)	27,682	100	13,862	50.08*	3,251	23.45**
	The number of land ownerhouseholds <0.5 Ha (000 Household)	15,809	100	10,318	74.43*	2,499	24.21**
4	The number of food crop farming (Households_	20,284	100	11,367	56.04*	2,666	23.45**
	Rice	13,155	64.85	7,129	54,19	2,249	31.54**
	Secondary crops (dry land food crops_)	7,129	35.15	4,238		417	15.64**
5	The number of households inbreeding sectors (000 RT)	13,562	100	8,242	60.77*	1,435	17.41**
	- Livestock and food crops	13,346	98.41	8,113	98.43**	1,414	98.54**
	-Livestock and non-food crops	216	1.59	129	1.56**	21	1.46**

Note: *) percentage in Indonesia, **) percentage in Java

Sources: a) Directorate General of Livestock and Animal Health [3], b) Kartawisastra et al. [16], Central Bureau of Statistics [17]

The development of beef cattle farming cannot be separated from its agricultural conditions, especially the land area, land use, and types of plants that will produce crop residues. Since beef cattle are plant-eating ruminants or herbivores, the source of feed depends on theagricultural land to produce forage. From Table 1, it can be seen that regarding the supply of beef, West Java is a beef consumer area as seen from the total population of 31.52% of the population in Java. However, the whole population of beef cattle as a beef producer is only 5.67% of the population in Java. Based on the capacity of the agricultural land in West Java, which is 30% of the agricultural land in Java can produce forage, this condition gives an opportunity to increase the cattle population in West Java. Unfortunately, 74.43% of farmers only control 5,000 m2 of land in Java, and 24.21% of the farmers are in

West Java. Becauseof the dependence of ruminants on the agricultural land as a producer of forage and the narrow land, the integrated farming and livestock activities are generally carried out by the farmers, especially in Java. It can be proven that 98.41% of farmer households in Java have livestock and food crop activities (Table 1), including beef cattle farming. For that reason, itis necessary to explore local resources, especially what kind of food plants that can increase the population of beef cattle.

3.2 Factors that influence to the development of beef cattle population

Based on the root test using the Augmented Dickey-Fuller (ADF) method, at the stationary level, there are 5 stationary independent variables out of the 9 independent variables in the model, namely the population, corn production, cassava production, green bean production, and peanut production (ADF Prob <0.05). Furthermore, the multiple linear regression analysis was carried out with stationary independent variables using the OLS method. The results of the multiple linear regression analysis show that the population variable *(Pop)*, corn production (*CrProd*) and cassava production (*CsProd*) has a significant positive effect (p<0.01) on the beef cattle population, while the independent variables of green bean and peanut production have no significant effect on the beef cattle population (Table 2).

Table 2. The results of the regression analysis of the factors that influence the development of beef cattle population

Variable	Coefficient	Sig
(Coefficient)	-68951.99	0.0000***)
Population (Pop)	0.0469	0.0000***)
Corn Production (CrProd)	0.1775	0.0000***)
Cassava Production (CsProd)	0.1640	0.0005***)
Green Bean Production (Gb Prod)	-0.1248	0.8772
Peanut Production (PnProd)	1.7644	0.2643
R-square	0.9242	
Adjusted R^2	0.9032	
Prob (F-statistik)	0.0000	

Note: ***) Significant at p<0.01

Rice straw production has not significant effect on the increase of beef cattle population even though the rice plant produces potential rice straw for beef cattle feed. This is because the recorded rice straw data were fluctuated and not stationer, so that the data were not included in further analysis. The crop residue of corn and cassava has a significant positive effect, which means that the increase in the yields in West Java will increase the population of beef cattle. Nevertheless, the crop residue production of corn per hectare is greater than that of cassava as can be seen from the following research results (Table 3).

Table 3. The crop residue production of corn vs. cassava from some previous research

The crop residue production of corn*		
No.	Tons/ha/harvest	Reference
1	5.99	[18]
2	6.00	[19]
3	5.70	[20]
The crop residue production of cassava**		
1	4.90	[21]
2	5.00	[22]
3	6.3	[23]

Note: *harvested twice a year, ** harvested once a year

Table 3 shows that the crop residue production of corn per ha per harvest is almost the sameas that of cassava. The production of corn, however, is shorter than that of cassava. Corn can be harvested 2 to 3 times per year, while cassava can only be harvested once per year [21]. Thus, the crop residue production of corn per ha per year is higher than cassava. From the results of this study, it can be concluded that the integration of corn and beef cattle farming is an instrument for the development of beef cattle farming in West Java and can be used as the basis for policies, especially the development of beef cattle for smallholder farmers.

4 Conclusion

The development of beef cattle in West Java should be directed towards the areas whose land has the capacity to grow corn and cassava. In addition, it is necessary to consider areas where the population has the potential to work in the sector of agriculture-breeding. The integration of corn and beef cattle farming is an instrument for the development of beef cattle farming in West Java and can be used as the basis for policies, especially the development of beef cattle for smallholder farmers. This is because corn can produce crop residue as forage for beef cattle higher than that of cassava per unit of land per year.

References

1. R. Widiati, T. S. M. Widi, J. Anim. Prod. **18**, 51-58 (2016)
2. W. Stur, T. T. Khanh, A. Duncan, J. Agric. Sustain. **11,** 363-381 (2013)
3. Directorate General of Livestock and Animal Health, *Livestock and Animal Health Statistic 2019* (Ministry of Agriculture of the Republic of Indonesia, Jakarta, Indonesia, 2019).
4. Y. Dwi, D. Suryadi, Jurnal Agripet **11**, 32-38 (2017)
5. R. Widiati, S. Nurtini, T. A. Kusumastuti, S. P. Syahlani, M. A. U. Muzayyanah, Int. J. Bus. Soc. **20**, 417-431 (2019)
6. A. A. Rouf, A. Daryanto, A. Fariyanti, Wartazoa **24**, 97-107 (2014)

7. Amam, R. Yulianto, N. Widodo, S. Romadhona, Livest. Anim. Res. **18**, 160-170 (2020)
8. D. Priyanto, J. Agric. Res. Dev. **30**, 108-116 (2011)
9. R Widiati, Wartazoa **24**, 191-200 (2014)
10. J. B. Penson, O. Capps, C. P. Rosson III, *Introduction to Agricultural Economic,* 3rd edition (Prentice Hall, New Jersey, USA, 2005)
11. R. Widiati, T. A. Kusumastuti, M. A. U. Muzayanah, *Exploration of potential regional resources for beef cattle farming development in Java, Indonesia*, In The 6th International Seminar on Tropical Animal Production "Integrated Approach in Developing Sustainable Tropical Animal Production" October 2015, Yogyakarta, Indonesia (2015), pp. 571-576
12. A. Firman, M. M. Sulaeman, L. Herlina, M. S. Mimbar, Jurnal Pemikiran Masyarakat Ilmiah Berwawasan Agribisnis **4**, 98-108 (2018)
13. BPS Jawa Barat, Propinsi Jawa Barat Dalam Angka 2019 (Badan Pusat Statistik Provinsi Jawa Barat, Bandung, Indonesia, 2019).
14. R. L. Ott, M. T. Longnecker, *An Introduction to Statistical Methods and Data Analysis, Illustrated Edition* (Cengage Learning, Boston, USA, 2015)
15. W. H. Greene, *Econometric Analysis*, 5th edition (Prentice-Hall, Upper Saddle River, USA, 2003)
16. S. Kartawisastra, S. Ritung, E. Suryani, D. Subardja, H. Suparto, A. Mulyani, *Sumberdaya Lahan Pertanian Indonesia: Luas, Penyebaran dan Potensi Ketersediaan* (Indonesian Agency of Agricultural Research and Development Press, Jakarta, Indonesia, 2018)
17. Central Bureau of Statistics, *The Result of Inter-census Agricultural Survey 2018* (Central Bureau of Statistics, Jakarta, Indonesia, 2018)
18. I. W. Ardiana, Y. Widodo, Liman, Jurnal Ilmiah Peternakan Terpadu **3**, 170-174 (2015)
19. Supriadi, L. Izhar, O. I. Safitri, *Potensi ketersediaan hijauan pakan limbahtanaman jagung manis di Provinsi Kepulauan Riau*, In: Prosiding Seminar Nasional 2016 "Membangun Pertanian Modern dan Inovatif Berkelanjutan dalam Rangka Mendukung MEA", Mei 2016, Bogor, Indonesia (2016), pp. 710-716
20. D. Yuzaria, M. I. Rias, M. Zaki, *Potensi ketersediaan limbah tanaman jagung sebagai pakan alternatif untuk peningkatan populasi sapi potong di Kabupaten Pasaman Barat*, In: Prosiding Seminar Teknologi dan Agribisnis Peternakan VII-Webinar: Prospek Peternakan di Era Normal Baru Pasca Pandemi COVID-19, Fakultas Peternakan Universitas Jenderal Soedirman, Juni 2020, Purwokerto, Indonesia (2020), pp. 119-128
21. I. Hernaman, A. Budiman, S. Nurachmah, K. Hidayat. *Potensi limbah tanaman singkong sebagai pakan ruminansia*, In: Seminar Nasional Fakultas Peternakan UNPAD ke-2 "Sistem Produksi Berbasis Ekosistem Lokal", November 2010, Bandung, Indonesia (2010), pp. 553-556
22. T. P. Daru, Suhardi, R. Yusuf, A. Wibowo, P. Pujowati, Jurnal Teknologi Pertanian **8**, 30-36 (2012)
23. R. Antari, G. P. Ningrum, D. E. Mayberry, Marsetyo, D. Pamungkas, S. P. Quigley, D. P. Poppi, Anim. Prod. Sci. **54**, 1228-1232 (2014)

Productivity of Benggala grass (*Panicum maximum*) at different ages of oil palm

Achmad Fanindi[1*], *Sajimin* Sajimin[1], *Endang* Sutedi[1], *Iwan* Herdiawan[1], and *Harmini* Harmini[1]

[1]Research Institute of Animal Production, Ciawi, Bogor 16002, Indonesia

Abstract. The study aimed to observe the cultivars of Benggala grass grown at different ages of oil palms. The research was conducted at the location of oil palm plantations at the Sei Putih, Deli Serdang, North Sumatra. The study used a Factorial Completely Randomized Block Design, using 3 replications, the first treatment was 2 oil palm planting ages, namely the age of fewer than 5 years and more than 5 years plus control (open space), while the second treatment was 5 cultivars of Benggala grass consisting of the Hamil, Riversdale, Natsuyutaka, Gatton and Petrie. The observed variables were morphological and reproductive characters. The results showed that the fresh and dry weight of the Benggala grass cultivars decreased at the age of oil palms less or more than 5 years when compared to the control ($P > 0.05$). Hamil and Riversdale cultivars had better yield than other cultivars ($P > 0.05$). The agronomic and reproductive characteristics of Benggala grass planted in the shade of oil palm showed a decrease when compared to open land. The decrease in Benggala grass production in shade indicates the need to assemble Benggala grass cultivars in oil palm shade.

1 Introduction

The oil palm plantation area can be used as a source of forage for livestock [1]. Palm oil plantations have the main limiting factor, namely low light intensity as the age of oil palm plant increases, causing a decrease in forage production in oil palm plantations [2]. Light limitations, resulting in the acclimatization of morphological forage to compensate for the low rate of photosynthesis per unit of leaf area through an increase in editorial ratios: roots, rods: leaves, and weight ratio of leaf area [3].

Forage that will be developed in the plantation area, must be tolerant of shade [4]. One of the forages, which is tolerant of moderate shade conditions is Benggala Grass (*Panicum maximum*) [5]. Tanzania cultivar bengal grass production ranges from 1.4 -2.4 tons/ha/harvest [6], With the content of 9.2 - 18.7% and digestibility 58.6 - 66.3% [7]. Some known Benggala grass cultivars are: A). Large type with plant height between 3.6 - 4.2 m like a Hamil cultivar, b). Moderate type with 1.5 - 2.5 m plant height such as Gatton and Common cultivars, and C) short types with 1.0 m plant height as in Sabi and Trichoglume cultivars [8].
To introduce Benggala Grass in oil palm plantations, it is necessary to test and select the cultivars of Benggala Grass to grow well on oil palm plantations. This selection is part of the breeding program in order to increase the productivity of Benggala Grass, because with breeding research can increase productivity of up to 63% [9].

2 Methodology

The study was conducted at the location of oil palm plantations in North Sumatra. The study used a randomized design of the factorial group with 3 replications, where the first factor was 3 ages of palm oil (location of Benggala grass planting) and the second factor was 5 Benggala grass cultivars (*Panicum maximum*). The age of oil palm used is less than 5 years (light intensity of 30%) and more than 5 years (light intensity of 10%), and without shade control (light intensity of 100%). Benggala grass cultivar that was planted was 5 cultivars. Before being planted in the field, the pols were planted in a polybag for one month, then transferred to the field. Fertilization is done using urea fertilizer, TSP and KCL with a dose of 100 kg/ha/year and manure with a dose of 10 tons/ha/year. Cutting is done at the age of 45 days of plant. The parameters observed are the morphological characters of plants, reproductive properties, forage production and seeds per spikelet (on average in 3 spikelet / clumps).

* Corresponding author: afanindi@gmail.com

3 Results and Discussion

Grass productivity indicators are characterized by high forage weights. The weight of forage 5 Benggala grass cultivars, are presented in Table 1. The results of statistical analysis show that cultivars and palm oil agents have a significant effect (P> 0.05) in the weight of fresh and dry forage. The highest fresh and dry-dry weight is obtained on control (100% light intensity), while fresh-weighted weight on palm oil is less and more than 5 years in a row equivalent with a light intensity of 30% and 10%, not significantly different. Decreased fresh weight of 91.63% and 91.28% compared to fresh forage weight at the control site. The decrease in the weight of Benggala grass at low light intensity, indicating that the reduced light will cause a reduced photosynthesis rate, so it will inhibit the growth and production of plant biomass [10]. The decrease in forage weight at low light intensity also occurs in *Paspalum*, *Arachis pintoi* [11].

The highest fresh and dry forage at the control location and oil palm less than 5 years is found in Hamil cultivars with forage weight are 1,383.70 g / clumps and 101.63 g / clumps, while at the age of oil palm more than 5 years, fresh weight and the highest dry forage is obtained by the Riversdale cultivar.

Table1. Fresh weight (g/clump) 5 Benggala grass cultivars at different oil palm ages

No	Cultivar	Age of oil palm			Average
		control	< 5 years	>5 years	
1	Gatton	588.15	59.07	81.85	243.02^{bc}
2	Petrie	309.85	45.00	70.37	141.74^{c}
3	Natsuyutaka	996.67	73.06	83.33	384.27^{ab}
4	Riversdale	1145.56	91.54	92.04	443.05^{a}
5	Hamil	1383.70	101.63	58.01	514.45^{a}
	Average	884.79^{a}	74.06^{b}	77.12^{b}	

Table 2. Dry weight (g/clump) 5 Benggala grass cultivars at different oil palm ages

No	Cultivar	Age of oil palm			Average
		Control	< 5 years	>5 years	
1	Gatton	77.73	12.19	11.43	19.40^{c}
2	Petrie	39.13	7.45	11.62	33.78^{bc}
3	Natsuyutaka	121.57	11.91	6.75	46.74^{abc}
4	Riversdale	143.86	15.97	17.12	58.98^{ab}
5	Hamil	176.85	20.69	7.73	68.42^{a}
	average	111.83^{a}	13.64^{b}	10.93^{b}	

Benggala's grass plant height is influenced by the age of oil plam and cultivar (Table 3). Benggala Grass on the control location shows higher plant height if compared to other locations. Whereas at the location of oil palm more than 5 years it shows higher plant height compared to the location of the plant less than 5 years. The height of shade-tolerant Benggala grass plants increased in 50% (moderate) shade compared to no shade reported by Malaviya *et al.* [3] and Matta *et al.* [12] The highest high plant cultivars was Hamil cultivars. Hamil cultivars also have higher when compared to other cultivars in control, oil palm age less and over than 5 years.

Table 3. Height (cm) 5 Benggala grass cultivars at different oil palm ages

No	Cultivar	Age of oil palm			Average
		Control	< 5 years	>5 years	
1	Gatton	112.89	98.64	96.41	102.65^{b}
2	Petrie	102.41	52.04	75.07	76.51^{c}
3	Natsuyutaka	169.33	64.58	108.91	114.11^{ab}
4	Riversdale	169.48	95.07	120.63	128.39^{a}
5	Hamil	172.52	109.18	120.21	134.09^{a}
	Average	145.33^{a}	83.90^{c}	104.22^{b}	

Table 4. Number of tillers 5 Benggala grass cultivars at different oil palm ages

No	Cultivar	Age of oil palm			Average
		Control	< 5 years	>5 years	
1	Gatton	51.19	14.76	9.85	25.27^{a}
2	Petrie	46.89	13.30	9.81	23.33^{ab}
3	Natsuyutaka	61.11	10.36	13.22	28.18^{a}
4	Riversdale	50.74	17.14	7.93	25.27^{a}
5	Hamil	27.22	13.28	7.15	15.88^{b}
	Average	47.43^{a}	13.77^{b}	9.59^{b}	

Table 5. Booting time (day after planting) 5 Benggala grass cultivars at different oil palm ages

No	Cultivar	Age of oil palm			Average
		Control	< 5 years	>5 years	
1	Gatton	11.19	16.67	47.11	24.99^{a}
2	Petrie	10.70	25.44	12.33	12.83^{b}
3	Natsuyutaka	13.56	10.56	8.00	8.04^{b}
4	Riversdale	18.41	15.78	11.45	11.40^{b}
5	Hamil	40.44	40.09	12.65	31.06^{a}
	Average	18.86	21.71	12.41	

The number of tillers is influenced by the type of cultivar and the age of the oil palm. The number of tillers at the control location was higher than the number of tillers in oil palms aged less and more than 5 years (table 4). A decrease in the number of tillers in selected shaded Benggala grass cultivars was also reported by [13] and [14]. Shade changes the intensity and quality of radiation in plants, and affects the potential number of tillers [15].

Age of booting and early flowering showed no difference ($P<0.05$) at the control site and in oil palms aged less and more than 5 years. The type of cultivar has a significant effect on booting age and early flowering), where the Petrie, Natsuyutaka and Riversdale cultivars faster than the Gatton and Hamil cultivars.

Table 6. First flowering time (day after planting) 5 Benggala grass cultivars at different oil palm ages

No	Cultivar	Age of oil palm			Average
		Control	< 5 years	>5 years	
1	Gatton	14.46	19.65	49.28	27.80[b]
2	Petrie	13.63	28.63	14.56	14.94[c]
3	Natsuyutaka	17.67	9.44	9.04	9.04[c]
4	Riversdale	22.22	16.65	12.00	12.95[c]
5	Hamil	43.15	40.70	36.78	40.21[a]
	Average	22.23	23.01	17.72	

Flowering time in Benggala grass cultivars was strongly influenced by planting location (Table 7). Benggala grass planted in control locations took longer to flowering when compared to Benggala grass planted in oil palm plantations aged less than 5 years and more than 5 years. Faster flowering in shade also occurs in soybeans [16]. The earlier flowering process occurs due to the presence of soluble proteins (phytochromes), where shaded environmental conditions can change the pigments (phytochromes) in shaded soybean plants into forms that initiate flowering induction [17]. Meanwhile, the Petrie, Natsuyutaka and Riversdale cultivars have a faster flowering age than the Gatton and Hamil cultivars.

Table 7. Flowering time (DAP) 5 Benggala grass cultivars at different oil palm ages

No	Cultivar	Age of oil palm			Average
		Control	< 5 years	>5 years	
1	Gatton	19.30	25.93	27.70	34.31[a]
2	Petrie	19.48	24.57	15.00	14.69[b]
3	Natsuyutaka	21.63	10.78	11.00	10.80[b]
4	Riversdale	27.63	20.00	12.78	13.47[b]
5	Hamil	51.87	44.13	20.00	32.00 [a]
	Average	27.98[a]	21.08[b]	14.10[c]	

Flag leaf length and flag leaf width (Tables 8 & 9) are affected by planting location. The length of the flag leaf at the oil palm location is more than 5 years old, shorter than other locations, as well as the width of the flag leaf, which is narrower than other locations. Meanwhile, the length and width of the flag leaf of the Hamil cultivar are longer than other cultivars. The width of the flag leaf in each cultivar is more varied than the length of the flag leaf, this happens, because the width of the flag leaf is a physiological response of a plant to shade stress [18].

Leaf length and width (Tables 10 and 11) were strongly influenced by light intensity. One of the plant responses at low light intensity is leaf morphology which is wider and thinner when compared to plants at full light intensity. In this study, it was shown that planting location and cultivar had an effect on leaf length and width. The leaf length of the Benggala grass at the control location was longer than the other locations. Leaf length decreases with increasing age of oil palm. While the Hamil cultivar had longer leaf lengths when compared to other cultivars, the other 4 cultivars did not show any difference in leaf length.

Table 8. Flag leaf length (cm) 5 Benggala grass cultivars at different oil palm ages

No	Cultivar	Age of oil palm			Average
		control	< 5 years	>5 years	
1	Gatton	11.85	9.00	16.20	9.35[b]
2	Petrie	12.37	16.37	9.50	9.58[b]
3	Natsuyutaka	16.63	2.28	6.30	6.30[b]
4	Riversdale	23.04	10.00	8.83	10.62[b]
5	Hamil	33.15	23.67	18.00	18.94[a]
	Average	19.41[a]	8.46[b]	5.01[c]	

Table 9. Flag leaf width (cm) 5 Benggala grass cultivars at different oil palm ages

No	cultivar	age of oil palm			Average
		control	< 5 years	>5 years	
1	Gatton	0.89	0.60	1.18	0.69[b]
2	Petrie	1.05	1.24	0.70	0.79[b]
3	Natsuyutaka	0.89	0.18	0.40	0.36[c]
4	Riversdale	1.45	0.66	0.53	0.66[b]
5	Hamil	2.69	2.11	1.00	1.60[a]
	Average	1.39[a]	0.71[b]	0.35[c]	

Table 10. Leaf length (cm) 5 Benggala grass cultivars at different oil palm ages

No	Cultivar	age of oil palm			Average
		control	< 5 years	>5 years	
1	Gatton	20.07	30.89	6.54	19.17[b]
2	Petrie	19.11	2.44	27.00	16.19[b]
3	Natsuyutaka	31.74	11.00	10.00	10.58[b]
4	Riversdale	43.26	14.59	18.00	19.28[b]
5	Hamil	61.15	46.76	20.00	35.97[a]
	Average	35.067[a]	18.937[b]	6.707[c]	

Table 11. Leaf width (cm) 5 Benggala grass cultivars at different oil palm ages

No	Cultivar	Age of oil palm			Average
		Control	< 5 years	>5 years	
1	Gatton	1.12	0.90	1.44	0.95^{b}
2	Petrie	1.26	1.53	0.90	0.96^{b}
3	Natsuyutaka	1.28	0.50	0.22	0.50^{b}
4	Riversdale	1.91	0.80	0.64	0.85^{b}
5	Hamil	2.99	0.96	2.01	1.99^{a}
	Average	1.71^{a}	0.56^{b}	0.88^{b}	

Stem diameter and internode length were affected by the age of oil palm and Benggala grass cultivars (Tables 12 and 13). The stem diameter and internode length of the Benggala grass planted in the control locations were higher than other locations. Hamil cultivar, has the highest stem diameter when compared to other cultivars, which is 3.62 cm, while the Natsuyutaka cultivar has a smaller value when compared to other cultivars, larger stem diameter in the Hamil cultivar, because morphologically this cultivar has a larger stem diameter when compared to other cultivars under normal conditions

Table 12. Stem diameter (cm) 5 Benggala grass cultivars at different oil palm ages

No	Cultivar	Age of oil palm			Average
		Control	< 5 years	>5 years	
1	Gatton	1.94	1.69	1.70	1.78^{cb}
2	Petrie	1.70	0.17	1.59	1.16^{cd}
3	Natsuyutaka	1.99	0.29	0.67	0.76^{d}
4	Riversdale	1.93	1.20	1.21	1.96^{b}
5	Hamil	3.49	2.92	1.00	3.62^{a}
	Average	3.41^{a}	1.25^{b}	0.90^{b}	

Table 13. Internode length (cm) 5 Benggala grass cultivars at different oil palm ages

No	Cultivar	Age of oil palm			Average
		Control	< 5 years	>5 years	
1	Gatton	15.00	12.11	5.05	10.72^{ab}
2	Petrie	17.04	13.07	6.07	12.06^{ab}
3	Natsuyutaka	18.41	2.67	1.78	7.62^{b}
4	Riversdale	27.33	8.78	4.23	13.45^{ab}
5	Hamil	30.00	18.44	10.54	19.66^{a}
	Average	21.56^{a}	11.02^{b}	5.53^{c}	

Table 14. Seed weight (g/clump) 5 Benggala grass cultivars at different oil palm ages

No	Cultivar	Age of oil palm			Average
		Control	< 5 years	>5 years	
1	Gatton	0.25	0.18	0.02	0.15[ab]
2	Petrie	0.25	0.09	0.00	0.11[b]
3	Natsuyutaka	0.43	0.06	0.00	0.16[ab]
4	Riversdale	0.87	0.00	0.00	0.29[a]
5	Hamil	0.07	0.00	0.05	0.04[b]
	Average	0.37[a]	0.06[b]	0.02[b]	

The highest seed production of Benggala grass was obtained at the control site, while at the location where oil palms were more than 5 years old had the lowest seed production. The cultivar with the highest production of pithy seeds was obtained by the Riversdale cultivar, while the lowest was the Hamil cultivar.

4 Conclusion

Benggala grass has decreased productivity, which is indicated by the low number of agronomic characters that support growth and productivity such as forage weight, number of tillers and plant height at the age of oil palm more or less than 5 years. Locations that can still be planted with Benggala grass are oil palm plantations that are less than 5 years old. Cultivars that still show good productivity and agronomic characteristics in oil palm plantations are the Hamil and Riversdale cultivars. The decline of Benggala grass cultivars in oil palm shade indicates that there is an influence of light intensity on the growth and productivity of Benggala grass, so it is necessary to look for cultivars that are tolerant of low light intensity (shade) through breeding activities.

References

1. B. Suwignyo, E. Baliarti, B. Suhartanto, M. Hamdani, A. Agus, I. G. S. Budisatria, Panjono, B. Guntoro, H. Trisakti, S. Bintara, Yuriadi, B. A. Atmoko, G. T. Yuwono. *Pengembangan peternakan Berbasis Plasma Nutfah dan Kearifan Lof,al Mendukung Agroekologi Berkelaniutan* (Fakultas Peternakan Universitas Gadjah Mada, Yogyakarta, 2016)
2. I. Ezenwa, O. A. Aribisala, M. E. Aken'ova, Tropical Grasslands **30,** 414-417 (1996)
3. D. R. Malaviya, M. J. Baig, B. Kumar, and P. Kaushal, Revista de Biología Tropical **68**, 563-572 (2020)
4. N. D. Purwantari, Wartazoa **26**, 051-056 (2016)
5. C. C. Wong, *Shade tolerance of tropical forages* (Proceeding Workshop on Forages for Plantation Crops, Canberra (AUS), ACIAR, 1990)
6. M. D. Hare, S. Phengphet, T. Songsiri, N. Sutin, E. Stern, Tropical Grass-lands–Forrajes Tropicales **1**, 87-89 (2013)
7. M. E. Siregar, M. Martawijaya, T. Herawati, Bulletin Lembaga Penelitian Peternakan **26** (1980)
8. Sajimin, E. Sutedi, N. D. Purwantari, B. R. Prawiradiputra, *Prosiding Lokakarya Nasional Tanaman Pakan Ternak* (Pusat Penelitian dan Pengembangan Peternakan, Badan Penelitiaan dan Pengembangan Pertanian, 2005)
9. Subandriyo *et al, Prosiding Lokakarya Nasional Tanaman Pakan Ternak* (Pusat Penelitian dan Pengembangan Peternakan, Badan Penelitiaan dan Pengembangan Pertanian, 2005)
10. J. Sirait, A. Tarigan, K. Simanihuruk, Junjungan, *Produksi dan nilai nutrisi enam spesies hijauan pada tiga tingkat naungan di dataran tinggi beriklim kering* (Prosiding Seminar Nasional Teknologi Peternakan dan Veteriner, Pusat Penelitian dan Pengembangan Peternakan, Badan Libang Pertanian, 2007)
11. R. S. Barro, A. C. Varella, G. Lemaire, R. B. Medeiros, J. C. Saibro, C. Nabinger, F. V. Bangel, I. J. Carassai, Zootec **41**, 1589-1597 (2012)
12. P. M. Matta, S. M. Souto, P. F. Dias, A. A. Colombaria, B. C. Azevedo, M. S. Vieira, Archivos Latinoamericanos de Producción Animal **17,** 097-102 (2009)

13. D. M. Victor, L. Jank, B. Lempp, R. M. Simeão, M. D. V. Resende, M. C. Gonçalves, Rev. Ceres, Viçosa **62**, 199-207 (2015)
14. D. S. C. Paciullo, C. A. M. Gomide, C. R. T. Castro, P. B. Fernandes, M. D. Müller, M. F. A. Pires, E. N. Fernandes, D. F. Xavier, Pesquisa Agropecuária Brasileira **46,** 1176-1183 (2011)
15. O. Guenni, S. Seiter, R. Figueroa, Tropical Grasslands **42**, 75-87 (2008)
16. Chairudin, Efendi, Sabaruddin, J. Floratek **10**, 26 – 35 (2015)
17. L. T. Karamoy, Soil Environment **7**, 65-68 (2009)
18. C. Hu, J. Rao, Y. Song, S. A. Chan, T. Tohge, B. Cui, PLOS ONE **15**, 1-23 (2020)

Honey moisture reduction using several thermal methods and their effects on its quality

Firman Jaya[1,2], *Lilik Eka* Radiati[2*], *Teti* Estiasih[3], *Djalal* Rosyidi[2], *Anang* Lastriyanto[4] *Mochammad* Junus[5], *Jati* Batoro[6], *Erwan* Erwan[7], *Jacobus Suruka Aifaman* Lamerkabel[8], *Dewi* Masyithoh[9], *Ustadi Ustadi*[10], *Eggi Pur* Pinandita[11]

[1]Animal Science Doctoral Program, Faculty of Animal Science, Universitas Brawijaya, Malang 65145, East Java, Indonesia
[2]Department of Animal Products Technology, Faculty of Animal Science, Universitas Brawijaya, Malang 65145, East Java, Indonesia
[3]Department of Agricultural Product Technology, Faculty of Agricultural Technology, Universitas Brawijaya, Malang 65145, East Java, Indonesia
[4]Department of Agricultural Engineering, Faculty of Agricultural Technology, Universitas Brawijaya, Malang 65145, East Java, Indonesia
[5]Department of Animal Products, Faculty of Animal Science, Universitas Brawijaya, Malang 65145, East Java, Indonesia
[6]Department of Biology, Faculty of Mathematics and Natural Sciences, Universitas Brawijaya, Malang 65145, East Java, Indonesia.
[7]Department of Animal Husbandry, Faculty of Animal Science, University of Mataram, Mataram 83125, Indonesia
[8]Department of Agrotechnology, Faculty of Agriculture, University of Pattimura, Ambon 97233, Indonesia
[9]Department of Animal Husbandry, Faculty of Animal Husbandry, Islamic University of Malang, Malang 65144, Indonesia
[10]PT. Kembang Joyo Sriwijaya, Malang 65152, Indonesia
[11]Animal Science Magister Program, Faculty of Animal Science, Universitas Brawijaya, Malang 65145, East Java, Indonesia

Abstract. Honey is thermally heated at various methods to decrease the moisture content and prolong the shelf life. The heating methods might decrease the quality of honey's physicochemical and biochemical properties. The present study thermally treated the honey with a pasteurizer, evaporator, and dehumidifier. The most thermally affected to decrease the moisture content were treatment by dehumidifier (14.09%), subsequent evaporation (8.41%), and pasteurization (8.41%). After heating, significant differentiation was also observed in the variation of both HMF (Hydroxymethylfurfural) content and diastase activity according to the botanical origin of the honey sample. In line with the biochemical analysis of honey, total phenolic decreased significantly during the pasteurization treatment. Evaporation was the most resistant thermal treatment due to its ability to maintain the level of HMF and the enzyme diastase as a standard for honey quality. However, changes made

* Corresponding author: lilik.eka@ub.ac.id

in physicochemical and biochemical quality are still in compliance with national and international legal limits.

1 Introduction

Honey is sweet due to monosaccharides and viscous liquid, about 17-22% of moisture [1]. The amount of water found in honey determines its balance against fermentation and crystallization. Therefore, it is vital to procedure the honey by subjecting it to thermal remedy to prevent fermentation through yeasts [2]. Heating eases the bottling manner by reducing honey's viscosity, reducing honey's water content material to prevent fermentation, and delaying crystallization in honey [3].

Previous studies have shown that based on moisture reduction systems [4] such as, developed a small-scale honey dehydrator to reduce the moisture content of honey beneath 17%. Furthermore, Singh *et al.* [5] set a desiccant honey dehydrator that heats and dehumidifies air to reduce honey's moisture content using a silica gel desiccant bed. While studies on microwave-based honey moisture reduction, such as Cui *et al.* and Kowalski *et al.* [6, 7], have been investigated microwave vacuum as a capability approach for obtaining high-quality dried honey. Few studies are using the drying method and comparing it with other honey moisture reduction.

The present study aims to compare methods between heating and drying processes such as pasteurization, dehumidification, and evaporation to measure the effect of thermal heating on honey quality in Indonesia. In the literature, there is limited studies effect of thermals such as pasteurization, dehumidification, and evaporation on the quality of honey. The current study aims to determine the best method for honey moisture reduction using heating and drying processes based on physicochemical and biochemical analysis.

2 Materials and methods

2.1 Samples

Fresh honey samples from different botanical origins were obtained from a representative from three areas in Indonesia: four honey samples from Jawa Barat (west), two from Ambon (east), and two from Nusa Tenggara Timur (central). In total, eight samples were studied: six samples were from monofloral honey came from Jawa Barat (rubber and rambutan), Ambon (kosambi and jambu bol), and Nusa Tenggara Timur (kosambi and eucalyptus), and one from mango honeydew (Jawa Barat), while the other one was multiflora nectar honey from different plant species (Jawa Barat). Honey samples from Jawa Barat were provided from honey bee cultivation, while honey samples from Ambon and Nusa Tenggara Timur were from wild forest honey bees (*Apis dorsata*). Honey samples were stored at -18 °C until analysis.

2.2 Honey thermal treatment

2.2.1 Pasteurization

The pasteurization was evaluated according to Wibowo *et al.* [8]. It was a colonized form (3 double jacket pans) equipped with a control system to regulate the temperature and a temperature sensor (thermocouple type-K), as shown in Figure 1. Honey was heated at 65°C.

(A) **(B)**

Fig. 1. Pasteurization with the colonized method (A) and control system box (b)

2.2.2 Evaporator

According to Lastriyanto *et al.* [9], the evaporator was evaluated for this research. The prototype scale evaporator became geared up with a vacuum pump jet and an evaporator tube using a heat-resistant pyrex glass flask. The heating stove layout was positioned above the water tub. The vacuum stress used became 7.8 kPa, and the temperature set on the manage box became 45 °C. The stove flame will routinely decrease while the honey's temperature in the room exceeds 45 °C, consequently that the heat stays steady. The water content of honey samples was achieved each 10 min. The evaporator is shown in Figure 2.

Fig. 2. Evaporator

2.2.3 Dehumidifier

According to Lastriyanto *et al.* [9], the dehumidifier used in this study was evaluated. The capacity of the water jet vacuum evaporator was 1 L with a portable dehumidifier with dimensions of 35 x 32 x 27 cm (Figure 3). Furthermore, the remark changed into finished through weighing every pattern's weight every 1 hour (60 min) until the 12 hours (720 min). The RH value became the lowest condition (CE) on the dehumidifier. The temperature turned into measured the usage of a type k thermocouple linked to Omron Ecwl 5.

Fig. 3. Dehumidifier

2.3 Physicochemical analysis

Analysis for moisture content, acidity, HMF (Hydroxymethylfurfural) concentration, and diastase enzyme was estimated by the procedure of SNI 8664:2018 [10].

2.4 Biochemical analysis

Determination of reducing sugars and non-reducing sugar (sucrose) was carried out using SNI 8664:2018 [10]. At the same time, the total phenolic content was determined using the Folin-Ciocalteu method, and the results were expressed as mg gallic acid/kg honey. Moreover, the total flavonoid contents of the honey samples were determined using aluminum chloride, and the results were expressed as mg quercetin/kg honey developed by [11].

2.5 Statistical analysis

Statistical evaluation becomes finished on Microsoft excel for home windows. Furthermore, evaluation of variance turned into performed for the considerable distinction the usage of F-test ($P < 0.05$). All the assays had been done in triplicate, and outcomes have been expressed in mean ± SD.

3 Results and discussion

3.1 Moisture content

The moisture content of honey is a restricting thing for figuring out its excellent stability and spoilage resistance against yeast fermentation [12]. Eshete *et al.* [13] stated that honey's composition and properties rely, initially, on the floral origin concerning the geographic location and the weather regime. The initial honey samples from east Indonesia were significantly higher than the west area of Indonesia. Honey samples ranged from 19.38 ± 0.80 in rubber honey in Jawa Barat to 24.57 ± 0.54 in kosambi tree honey in Ambon (Table 1). This investigation confirmed that the moisture content of the nonthermally-processed honey sample is a concordance to the set values of The National Standardization Agency of Indonesia [10], in which moisture content is a maximum of 22% unless shown by three samples from the east area of Indonesia.

Heating resulted in slightly lower moisture content than raw honey, thus indicating a slight reduction in moisture content. However, no significant moisture content changed into located among honey processed at distinctive heating methods (Table 2).

This study has noticed that pasteurization did not cause a significant decrease in moisture content following the national limits. While the lowest moisture content was found using a dehumidifier. The moisture content of all types of honey was slightly decreased where pasteurization treatment of around 5.88%. While evaporation treatment was decreased around 8.41%, and dehumidifier is around 14.09%. These results are higher than Zarei *et al.* [12], which was achieved in a temperature-managed water bath, and samples had been fast cooled in the ice bath at 63°C. After 30 min thermal treatment, they found a lower of 0.4–0.56 % in all kinds of honey.

Table 1. The quality properties of the nonthermal-processed honey sample

Source of honey	Moisture content (%)	Acidity (mL NaOH/kg)	HMF (mg/kg)	Diastase activity (DN)	Reducing sugar (%)	Non-reducing sugar (sucrose) (%)	Total Phenolic (mg/kg)	Total flavonoid (mg/kg)
Rubber tree from Jawa Barat	19.38 ± 0.80	18.01 ± 1.43	5.41 ± 2.59	21.17 ± 0.78	73.09 ± 0.81	0.84 ± 0.32	372.40 ± 46.31	31.22 ± 1.21
Rambutan tree from Jawa Barat	20.72 ± 1.00	26.05 ± 1.76	5.00 ± 2.86	22.90 ± 0.80	69.54 ± 1.12	0.43 ± 0.41	533.62 ± 49.77	35.80 ± 0.98
Honeydew from Jawa Barat	20.37 ± 0.76	42.80 ± 1.98	5.09 ± 3.23	21.10 ± 0.92	57.83 ± 0.99	0.40 ± 0.26	723.36 ± 39.32	61.24 ± 1.16
Multiflora honey from Jawa Barat	19.92 ± 0.87	22.04 ± 1.02	5.20 ± 3.76	17.01 ± 1.01	72.79 ± 1.29	1.65 ± 1.19	450.88 ± 38.01	28.92 ± 0.76
Eucalyptus from Nusa Tenggara Timur	22.08 ± 0.98	41.45 ± 1.79	6.50 ± 2.99	1.74 ± 0.76	69.83 ± 0.82	1.99 ± 1.08	567.53 ± 32.33	36.24 ± 0.71
Kosambi tree from Nusa Tenggara Timur	23.86 ± 1.24	43.70 ± 2.04	7.90 ± 2.32	3.03 ± 0.82	50.21 ± 0.79	2.21 ± 1.83	951.29 ± 31.24	129.84 ± 1.58
Kosambi tree from Ambon	24.57 ± 0.54	48.00 ± 2.37	9.21 ± 3.43	3.02 ± 0.80	49.46 ± 1.35	2.30 ± 1.04	908.97 ± 38.83	125.45 ± 1.34
Jambu Bol from Ambon	23.98 ± 1.02	49.60 ± 1.56	10.34 ± 2.48	4.90 ± 0.95	57.50 ± 1.64	1.11 ± 0.62	635.14 ± 40.02	73.38 ± 0.97

HMF: Hydroxymethylfurfural

3.2 Acidity

Acidity. The initial ranges of free acidity for studied honey samples were 18.01 ± 1.43 mL NaOH/kg (rubber honey in Jawa Barat) to 49.60 ± 1.56 mL NaOH/kg (jambu bol honey in Ambon). According to The National Standardization Agency of Indonesia [10], these values are acceptable (below 50 mL NaOH/kg). The free acidity of honey can be defined with the aid of taking into account the presence of organic acids in equilibrium with their corresponding lactones, or internal esters, and a few inorganic ions, inclusive of sulfate, chloride, and phosphate [14]. Thermal treatment did not significantly affect the free acidity compared to Table 1 to Table 2 and 3.

3.3 HMF

HMF is identified as a marker of honey freshness and significant deterioration because of the motion of usual honey acidity on decreasing sugars. The HMF content material tends to grow at some point in the processing or growing old of the product. Several factors influence the levels of HMF. Considered one of them is temperature and time of heating. Thus, it indicates overheating and storage in poor situations [14].

In this study, the mean values of initial HMF content were ranged from 5-30 mg/kg [13], indicating the freshness of the honey and good storage condition. Jambu bol honey in Ambon showed the highest initial level of HMF (10.34 ± 2.48 mg/kg), and rambutan tree honey in Jawa Barat showed the lowest (5.00 ± 2.86 mg/kg). HMF content in this study ~~is~~ increased gradually with different thermal methods by 75.60 %, 94.82 %, and 107.02 % using pasteurization, evaporator, and dehumidifier, respectively. Nevertheless, the suggested values for HMF content in three types of thermal methods had been still below the perfect acceptable limit based on national and international legal limits (≤40 mg/kg). This increase following the report of Zarei *et al.*[12], HMF content increased ranged from 81.3% to 108.3% using temperature-controlled water bath at 63°C. Eshete *et al.* [13] stated that commercial pasteurization at 70-78°C and rapidly cooling at 32-40°C minimize heat damage. After pasteurization, diastase activity and HMF content remain almost unchanged because this temperature is similar to beehives and does not affect the honey.

3.4 Diastase Activity

A notable disparity turned into determined in the preliminary diastase activity among the fresh samples. For instance, rambutan tree honey in Jawa Barat had the highest diastase activity (22.90 ± 0.80 DN), while eucalyptus honey in Nusa Tenggara Timur showed the lowest value (1.74 ± 0.76 DN). These discrepancies are due to the honey from wild forest honey bees.

Blidi *et al.* [15] noted that these discrepancies are due to the honey's floral origin and the species of bees. However, all the honey samples complied with the SNI limits for diastase activity (min 1 DN for wild forest honey bee and min 3 DN for honey bee cultivation). The diastase activity decreased within the different thermal treatments attaining a maximum decrease after heating using a dehumidifier under the limit of about 58.26%. On the other hand, pasteurization and evaporator decreased the diastase activity by about 6.70% and 25.93%, respectively. Eshete *et al.* [13] pronounced the same thermal treatment that long heating periods of 60 to 90 s at thermal degrees of 30, 50, and 70 °C reduced the diastase activity of honey via almost 50% of its unique value. The heating of honey above 90 °C affects the caramelization of sugar.

3.5 Reducing sugar and non-reducing sugar (sucrose)

In general, sugars and different components of honey may also alternate during storage. While honey is heated, sugars decompose to form unwanted compounds (which include furfural and furan). They are related to non-enzymatic browning reactions (i.e., Maillard reaction, sugar degradation in an acidic medium caramelization) [16]. The sugar composition depends highly on bees' type of flower and regional and climatic conditions. Reducing sugar presents values of initial honey varying from 49.46 ± 1.35 % (kosambi tree honey in Ambon) to 73.09 ± 0.81% (rubber tree honey in Jawa Barat). This result is similar to Mouhoubi-Tafinine *et al.* [17] research, reducing sugar produce values varying from 62.75% to 76%.

The Sucrose level ranged from 0.40 ± 0.26 (Honewdew in Jawa Barat) to 2.30 ± 1.04% (Kosambi tree in Ambon). Sucrose content material for all samples fell inside the limits of the Codex Alimentarius (5%). The excessive sucrose content may also suggest a spread of adulterations, including cane sugar or subtle beet sugar. Moreover, early harvest indicated that the sucrose turned completely transformed into fructose and glucose; or extended artificial feeding of honeybees with sucrose syrups[18]. The presence of low sucrose levels suggests the opportunity that honey has been unadulterated with invert syrup.

3.6 Total phenolic and flavonoid

Folin–Ciocalteu's method evaluates overall total phenolics in many samples, along with honey. Phenolic compounds encompass distinct subclasses (usually flavonoids and phenolic), displaying many systems, some of which may interfere in the assay. Honey samples typically contain a number of these compounds, increasing absorbance values and overestimating the phenolic compounds [12].

The present study confirmed that the total phenolic content (mg tannic acid/kg honey) numerous many honey kinds. The lowest preliminary value was determined in rubber tree honey in Jawa Barat, where the average result of the samples was 372.40 ± 46.31 mg/kg, and the highest initial value was kosambi tree honey in Nusa Tenggara Timur (951.29 ± 31.24 mg/kg). Those data are similar to research by [12], where the average result of the multiflora honey became 462 ± 53 mg/kg, growing further in thyme (538 ± 41 mg/kg) and lotus (609 ± 60 mg/kg) honey.

Total phenolic content decreased after all thermal methods in rubber tree honey, kosambi tree in Ambon samples, and pasteurized method in rambutan tree honey in Jawa Barat and kosambi tree honey in Nusa Tenggara Timur, while in others it increased. Total phenolic content decreased after pasteurization, evaporation, and dehumidification were 4.00%, 1.18%, and 3.39%, respectively, in rubber tree honey in Jawa Barat. At the same time, kosambi tree honey in Ambon was 25.98%, 14.60%, and 14.47%, respectively. In rambutan tree honey in Jawa Barat and kosambi tree honey in Nusa Tenggara Timur, after the pasteurization process, total phenolic content decreased by 5.33% and 11.81%, respectively.

Table 2. The physicochemical quality properties of the thermal honey sample

SOURCE OF HONEY	PASTEURIZATION				EVAPORATION				DEHUMIDIFIER			
	Moisture content (%)	**Acidity (mL NaOH/kg)**	**HMF (mg/kg)**	**Diastase activity (DN)**	**Moisture content (%)**	**Acidity (mL NaOH/kg)**	**HMF (mg/kg)**	**Diastase activity (DN)**	**Moisture content (%)**	**Acidity (mL NaOH/kg)**	HMF (mg/kg)	Diastase activity (DN)
Rubber tree in Jawa Barat	18.24 ± 0.63	17.2 ± 0.88	11.54 ± 1.32	3.89 ± 1.32	17.75 ± 0.58	17.2 ± 1.43	13.54 ± 0.75	9.49 ± 1.43	16.65 ± 0.40	19.2 ± 0.63	17.66 ± 0.94	5.98 ± 1.76
Rambutan tree in Jawa Barat	19.85 ± 0.98	28.8 ± 1.34	12.07 ± 0.51	19.1 ± 0.56	18.34 ± 0.32	52.4 ± 0.52	15.21 ± 1.43	1.87 ± 2.43	17.42 ± 1.21	29.4 ± 0.71	19.32 ± 2.31	16.3 ± 1.32
Honeydew in Jawa Barat	19.81 ± 1.65	45.7 ±1. 62	11.87 ± 0.70	19.37 ± 0.81	18.10 ± 0.64	43.7 ± 0.76	13.32 ± 1.33	34.9 ± 0.24	17.36 ± 1.52	48.9 ± 2.42	18.22 ± 1.69	13.7 ± 1.93
Multiflora honey in Jawa Barat	20.79 ± 1.58	30.1 ± 0.71	12.22 ± 1.98	15.87 ± 0.64	18.66 ± 1.54	24 ± 0.37	14.21 ± 1.65	12.6 ± 1.33	17.58 ± 0.88	33.5 ± 0.86	19.22 ± 0.99	7.10 ± 2.01
Eucalyptus in Nusa Tenggara Timur	22.66 ± 0.86	46.4 ± 1.87	13.36 ± 2.59	2.24 ± 1.53	20.68 ± 0.87	42.1 ± 1.24	16.21 ± 0.88	3.31 ± 0.59	19.43 ± 0.92	46.5 ± 2.77	21.32 ± 1.22	3.0 ± 0.62
Kosambi tree in Nusa Tenggara Timur	22.98 ± 1.87	51.3 ± 2.41	15.98 ± 2.89	1.66 ± 2.43	22.26 ± 1.65	45.6 ± 2.31	17.32 ± 1.93	5.3 ± 1.02	21.01 ± 0.44	52.1 ± 3.61	20.21 ± 0.80	1.53 ± 1.39
Kosambi tree in Ambon	23.29 ±1.77	52.4 ± 1.97	17.75 ± 1.32	1.95 ± 2.89	22.97 ± 1.21	23.5 ± 1.63	20.30 ± 0.56	20.1 ± 0.73	21.53 ± 0.51	61.1 ± 3.09	24.22 ±.1.47	1.72 ± 1.66
Jambu Bol in Ambon	22.62 ±1.90	69.4 ± 2.43	20.56 ± 2.43	2.04 ± 2.08	22.13 ± 0.93	63.4 ± 0.54	22.21 ± 1.66	1.81 ± 2.31	21.09 ±1.72	51.2 ±3.47	25.32 ± 2.44	2.68 ± 1.39

HMF: Hydroxymethylfurfural

Table 2. The biochemical quality properties of the thermal honey sample

SOURCE OF HONEY	PASTEURIZATION				EVAPORATION				DEHUMIDIFIER			
	Reducing sugar	Non-reducing sugar (sucrose)	Total Phenolic (mg/kg)	Total flavonoid (mg/kg)	Reducing sugar	Non-reducing sugar (sucrose)	Total Phenolic (mg/kg)	Total flavonoid (mg/kg)	Reducing sugar	Non-reducing sugar (sucrose)	Total Phenolic (mg/kg)	Total flavonoid (mg/kg)
Rubber tree in Jawa Barat	63.8 ± 1.46	0.68 ± 0.65	357.49 ± 46.32	30.95 ± 2.34	73.1 ± 1.59	1.42 ± 1.19	368.02 ± 39.22	38.88 ± 1.23	67.9 ± 2.46	0.91± 0.55	359.80 ± 33.43	57.14 ± 1.37
Rambutan tree in Jawa Barat	62.4 ± 0.88	0.04 ± 0.03	505.18 ± 34.55	44.02 ± 1.48	60.6 ± 1.20	0.49 ± 0.41	882.82 ± 42.90	55.34 ± 1.61	62.8 ± 1.32	1.51 ± 0.65	787.70 ± 50.45	42.89 ± 1.88
Honeydew in Jawa Barat	58.5 ± 0.82	1.94 ± 1.77	846.02 ±52.23	86.80 ± 3.83	55.56 ± 0.42	0.88 ± 0.66	928.30 ± 25.71	92.66 ± 1.85	59.3 ± 1.38	0.15 ± 0.10	854.58 ± 41.34	81.80 ± 1.21
Multiflora honey in Jawa Barat	62.2 ± 1.00	2.0 ± 0.61	534.09 ± 30.88	42.66 ± 2.43	65.90 ± 0.93	0.08 ± 0.02	618.13 ± 30.70	52.29 ± 0.77	63.1 ± 1.33	0.44 ± 0.33	662.36 ± 47.77	68.38 ± 1.29
Eucalyptus in Nusa Tenggara Timur	59.8 ± 2.33	1.18 ± 0.45	592.26 ± 42.87	52.22 ± 1.51	57.90 ± 3.25	0.16 ± 0.08	596.52 ± 35.66	92.91 ± 1.68	63.4 ± 0.71	1.38 ± 0.53	796.39 ± 26.79	42.94 ± 0.76
Kosambi tree in Nusa Tenggara Timur	60.7 ± 1.46	1.29 ± 0.21	838.92 ± 29.54	96.44 ± 1.19	57.6 ± 0.29	0.1 ± 0.05	1068.36 ± 57.80	146.80 ± 2/55	56.8 ± 1.73	0.20 ± 0.03	1108.75 ± 57.22	125.09 ± 2.43
Kosambi tree in Ambon	57.3 ± 1.77	0.34 ± 0. 29	672.83 ± 53.76	76.70 ± 2.48	67.8 ± 0.73	4.21 ± 0.55	776.28 ± 44.09	109.60 ± 0.77	64.6 ± 2.68	0.40 ± 0.22	777.42 ± 34.67	93.86 ± 1.45
Jambu Bol in Ambon	57.6 ± 1.54	0.98 ± 0. 35	708.593 ± 48.62	88.37 ± 1.44	61.9 ± 0.79	0.62 ± 0.21	734.19 ± 37.31	104.60 ± 2.59	63.3 ± 1.37	0.61 ± 0.45	812.36 ± 54.384	83.19 ± 3.49

HMF: Hydroxymethylfurfural

In contrast, the other honey showed the thermal treatment increased the total phenolic content as liberated phenolic compounds and thus increased the number of active compounds in the honey. According to Šarić *et al.* [19], heating honey at 95°C for 5 min no longer affects its antioxidant activity and total phenolic content. Aydogan-Coskun *et al.* [20] commented that the simple phenolics could increase due to the breakdown of supramolecular structures containing phenolic groups.

4 Conclusion

The study has shown that the heating of honey reduces the moisture with a subsequent rise in HMF and acidity. The thermal treatment decreased moisture content by 5.88% using pasteurization and 8.41% using evaporation. The most affected was by the dehumidifier decreased around 14.09%. During the thermal treatment, physicochemical and biochemical analysis of different types of honey have been altered. Among the physicochemical characteristics, dehumidifier increased HMF content and decreased enzyme diastase were more prominent. In line with the biochemical properties of honey, total phenolic decreased significantly during the pasteurization treatment. Although the percentage decreased moisture content is lower than dehumidifiers, it is recommended to conduct thermal treatment by evaporation. It could maintain HMF and the diastase activity within the standard for honey quality.

This study was supported and funded by RISPRO LPDP (Indonesia Endowment Fund For Education), the Ministry of Finance of the Republic of Indonesia (PRJ-45/LPDP/2019)

References

1. I. Singh, S. Singh, J. Food Sci. Technol., **55**, 3861-3871 (2018)
2. B. Fallico, M. Zappala, E. Arena, A. Verzera, Food Chem., **85**, 305-313 (2004)
3. R. Subramanian, H. U. Hebbar, N. K. Rastogi, Int. J. Food Prop., **10**, 127-143 (2007)
4. R. S. Gill, V. S. Hans, S. Singh, P. Pal Singh, S. S. Dhaliwal, J. Food Sci. Technol., **52**, 6695-6702 (2015)
5. S. Singh, R. S. Gill, P. P. Singh, Int. J. Ambient Energy, **32**, 62-69 (2011)
6. Z. W. Cui, L. J. Sun, W. Chen, D. W. Sun, J. Food Eng., **84**, 582-590 (2008)
7. S. Kowalski, M. Lukasiewicz, S. Bednarz, M. Panus, Czech J. Food Sci., **30**, 21-26 (2012)
8. S. A. Wibowo, A. Lastriyanto, E. Erwan, F. Jaya, J. Batoro. Quantum Teknika, **2**, 66-74 (2021)
9. A. Lastriyanto, S. A. Wibowo, E. Erwan, F. Jaya, J. Batoro, D. Masyithoh, J. S. A. Lamerkabel, JMEST, **4**, 153-163 (2020)
10. The National Standardization Agency of Indonesia, *Honey (Madu)*, **SNI 8664** (2018)
11. J. A. Pontis, L. A. M. A. da Cota, S. J. R. da Silva, A. Flach, Food Sci. Technol. Campinas, **34**, 69-73 (2014)
12. M. Zarei, A. Fazlara, N. Alijani, FFHD, **9**, 593-605 (2019)
13. Y. Eshete, T. Eshete, Madridge J Food Technol., **4**, 158-162 (2019)
14. H. M. Habib, F.T. Al Meqbali, H. Kamal, U. D. Souka, W.H. Ibrahim, Food Chem., **153**, 35-43 (2014)

15. S. Blidi, P. Gotsiou, S. Loupassaki, S. Grigorakis, A. C. Calokerinos, Adv. Food Sci. Eng., **1**, 1-8 (2017)
16. C. Thamasopinkul, P. Ritthiruangdej, S. Kasemsumran, T. Suwonsichon, V. Haruthaithanasan, Y. Ozaki, JNIRS, **25**, 36-44 (2017)
17. Z. Mouhoubi-Tafinine, S. Ouchemoukh, M. Bachir bey, H. Louaileche, A. Tamendjari, IFRJ, **25**, 1044-1050 (2018)
18. P. M. Da Silva, C. Gauche, L. V. Gonzaga, A. C. O. Costa, R. Fett, Food Chem., **196**, 309–323 (2016)
19. G. Šarić, K. Marković, D. Vukičević, E. Lež, M. Hruškar, N. Vahčić, Czech J. Food Sci., **31**, 601-606 (2013)
20. B. Aydogan-Coskun, H. Coklar, M. Akbulut, Food Sci. Technol., **40**, 629-634 (2020)

Effect of energy and protein level in complete feed used elephant grass (*Pennisetum purpureum, Schum.*) and maize stover (*Zea mays.* L) silage on nutrient content, total digestible nutrient, and in vitro degradation

Hartutik[1*], *Marjuki*[1], *Asri Nurul* Huda[1], and *Roi Katul* Jannah[1]

[1]Faculty of Animal Science, University of Brawijaya, Jl. Veteran, Malang 65145, Indonesia

Abstract. The purpose of this research was to determine of ideal ration of energy and protein in complete feed used elephant grass and maize stover silage. The materials were use elephant grass, maize stover silage with 10% molasses and *Lactobacillus plantarum* 1x106 CFU/g and concentrates. The method used experimental laboratory, the data of nutrient and TDN content using descriptive analysis. In vitro degradation value was analysed by Analysis of Variance from a factorial randomized block design and followed by Duncan's Multiple Range Test. The complete feed was use 12.5% elephant grass + 37.5% maize stover silage + 50% concentrates with consist of energy level (E1 =12.5, E2 =13.5, E3 =14.5 MJ/kg DM) and protein level (P1 =10.5, P2= 13.5, P3= 16.5%). The results showed that in vitro DM and OM degradation respectively energy or protein level showed has significantly ($P<0.01$), while the interaction did not significant ($P>0.05$). The best treatment is E3P3 with energy 14.5 MJ/kg and protein 16.5% on nutrient content DM 92,51%., OM 90,33%., CP 16.57%, CF 19.29%, EE 1.77%, NFE 53.70%, TDN content 67.14%, In vitro DM degradation 66.14 % and in vitro OM degradation 70.01%.

1 Introduction

Forage as source of fiber is the main feed for ruminants and the fiber to be one of energy source. Elephant grass is a common perennial grass for farmers to be used as fiber source for their livestock. According to Rukmana [1], the nutritional content of elephant grass consists of dry matter 19.9%, crude protein 10.2%, crude fat 1.6%, crude fiber 34.2%, ash 11.7% and extracts without nitrogen 42.3%. Indonesia has two seasons, namely the rainy and dry seasons, during the rainy season forage will be available in abundance, while in the dry season it is very difficult to obtain so alternative feed is needed for substitution.

* Corresponding author: hartutik@ub.ac.id

Utilization of feed from agriculture or plantations can be used as a solution to overcome forage supplies and as an effort to reduce environmental pollution, for example, maize stover. In otherwise, the provision of feed for animals is a major contributor to land and water use greenhouse gas emission [5, 6]. Maize stover including stems, leaves, flowers and young corn fruit which are generally harvested at the age of 45-65 days contain nutrients, namely, Crude Protein 12.06%, Crude Fiber 25.2%, Ca 0.28%, P 0, 23% according to Koloud *et al* [2]. Based on the nutritional quality of maize stover above, it is very necessary to do preservation to maintain its nutritional content as feed, in the long term by using the silage method. Silage is a preservation technology with airtight (anaerobic) conditions that are specially placed, namely silos. Generally, alternative feeds from agriculture have low levels of Water-Soluble Carbohydrate (WSC), according to [3] corn WSC levels reach 11-16%, so additional sources of soluble carbohydrates such as molasses are needed to accelerate the formation of lactic acid which is useful in fermentation.

The addition of lactic acid bacteria can accelerate the ensilage phase to ensure the perfect lactic acid fermentation. Lactic acid bacteria such as *Lactobacillus plantarum* are producers of lactic acid in the manufacture of silage [4]. In addition to being given feed in the form of forage, ruminants are also given concentrate. Concentrate is feed that is given with the addition of other feeds to increase nutrition by being mixed together as a supplement (complete feed) [4]. Good quality feed has an optimal balance of nutrients for livestock, mainly the energy and protein content in a complete feed. According to Widodo [4] Complete Feed or Total Mixed Ration (TMR) is feed consisting of forage and concentrate in an optimal balance, with the aim of ensuring the intake of nutrients in daily feed so that it is able to supply fiber needs which are very important for stability rumen function. Good feed quality can be seen from the composition of nutrient content through nutrient content analysis which can describe in general the nutrients of feed ingredients. Nutrient content in feed will be useful if the Total Digestible Nutrient is known because in Indonesia it is used as a standard for nutrient substances in feed ingredients. Feed in the rumen will be degraded by rumen microbes, not all nutrient content of feed can be degraded, degradable feed will be used for nutritional needs of rumen microbial activity, this can be done in vitro. In vitro degradation is digestibility that mimics the condition of the body of livestock using laboratory equipment so that it will describe the amount of feed that is degraded in the rumen in vitro using the method from Sjofjan *et al* [7]. For this reason, further evaluation is needed regarding complete feed using elephant grass and corn silk silage with nutritional balance. Therefore, it is necessary to conduct research on the effect of different energy and protein levels in complete feed using elephant grass and corn husk silage on nutrient content, total digestible nutrient (TDN) and in vitro degradation.

2 Materials and methods

2.1 Materials

The material used for silage was hybrid maize stover aged 65 days from Sumber Sekar plantation Malang, molasses obtained from animal feed shops and *Lactobacilus plantarum* obtained from the Wiyasa Mandiri Singosari Laboratory, Malang. The next material was complete feed used 45 days old elephant grass which was obtained from the Sumber Sekar Field Laboratory, Faculty of Animal Husbandry, Universitas Brawijaya Malang.

Concentrates such as soybean meal, pile, coffee husk and bran were obtained from the animal feed shop Karangploso, Malang. Additional concentrate of "SAE Pujon Concentrate" obtained from Pujon Dairy Cooperative, Malang.

2.2 Data analysis

The research data were analysed using a Randomized Block Design (RBD) and continued with Duncan's Multiple Range Test (DMRT). TDN was based on complete feed including energy sources with CF > 18 and crude protein content < 20 with the following calculations. TDN (%) = 70.6 + 0.259 CP + 1.01 EE– 0.76 CF + 0.091 NFE

2.3 Method

The research method used is a laboratory experiment using descriptive analysis for nutrient content and TDN. The next research method is a laboratory experiment using analysis of variance from a Randomized Block Design (RBD) factorial pattern of two factors, 1st factor (energy level) and 2nd factor (protein level) which were grouped into three groups based on time of rumen fluid collection for degradation analysis. If there is a significant effect on the treatment, it will be continued with Duncan's Multiple Distance Test (DMRT). The Complete feed was added 12.5% EG (Elephant Grass) and 37.5% MS (Maize Stover Silage) with 50% concentrate. The metabolizable energy (ME) content of concentrate was a) 12.5 MJ/kg DM, b) 13.5 MJ/kg DM, c) 14.5 MJ/kg DM and the level of protein content 10.5%, 13.5%, 16.5%. The combination and composition of the treatment are as follows:

Table 1. Combination and composition of the treatment

Energy (Mj/kg)	Protein (%)	Code	Description
E1 (12.5)	P1 (10.5)	E1P1	12.5% EG + 37.5% MSS + 50% Concentrate (Energy 12.5 MJ/kg dan Protein 10.5%)
	P2 (13.5)	E1P2	12.5% EG + 37.5% MSS + 50% Concentrate (Energy 12.5 MJ/kg dan Protein 13.5%)
	P3 (16.5)	E1P3	12.5% EG + 37.5% MSS + 50% Concentrate (Energy 12.5 MJ/kg dan Protein 16,5%)
E2 (13.5)	P1 (10.5)	E2P1	12.5% EG + 37.5% MSS + 50% Concentrate (Energy 13.5 MJ/kg dan Protein 10.5%)
	P2 (13.5)	E2P2	12.5% EG + 37.5% MSS + 50% Concentrate (Energy 13.5 MJ/kg dan Protein 13.5%)
	P3 (16.5)	E2P3	12.5% EG + 37.5% MSS + 50% Concentrate (Energy 13.5 MJ/kg dan Protein 16.5%)
E3 (14.5)	P1 (10.5)	E3P1	12.5% EG + 37.5% MSS + 50% Concentrate (Energy 14.5 MJ/kg dan Protein 10.5%)
	P2 (13.5)	E3P2	12.5% EG + 37.5% MSS + 50% Concentrate (Energy 14.5 MJ/kg dan Protein 13.5%)
	P3 (16.5)	E3P3	12.5% EG + 37.5% MSS + 50% Concentrate (Energy 14.5 MJ/kg dan Protein 16.5%)

EG = Elephant grass; MSS = Maize Stover Silage

Table 2. Chemical Composition of Raw Materials

	Raw material	Chemical Composition (%DM)							
		DM	OM	Ash	CP	EE	CF	NFE	TDN
1	SAE Pujon Concentrate	97.60	90.06	9.94	18.38	4.70	14.97	52.00	70.18
2	Elephant grass	96.12	85.95	14.05	12.08	2.35	31.99	39.54	55.39
3	Silage 14 days	94.54	89.36	10.64	7.8	0.94	22.45	58.17	61.81
4	Soybean Meal	93.53	91.62	8.38	47.53	2.57	4.04	37.48	80.68
5	Coffee husk	94.14	89.42	10.58	10.11	1.49	34	43.82	52.87
6	Bran	90.63	87.40	12.60	10.15	13	16.20	48.06	71.43
7.	Pile	92.59	82.87	17.13	1.76	0.44	25.39	55.28	57.24

Information :
*) The results of the analysis at the Laboratory of Nutrition and Animal Feed, Faculty of Animal Science, Universitas Brawijaya (2021)
*) DM = Dry Matter OM = Organic Matter CP = Crude Protein EE = Ether Extract CF = Crude Fiber NFE = Nitrogen Free Extract TDN = Total Digestible Nutrient

Table 3. Feed Formulation

Energy Level (MJ/kg)	Protein Level (%)	Code	MSS (%)	EG (%)	SPC (%)	Bran (%)	Pile (%)	SBM (%)	CH (%)	Total (g/DM)
E1 (12.5)	P1 (10.5)	E1P1	37.5	12.5	18	12	4.5	-	15.5	100
	P2 (13.5)	E1P2	37.5	12.5	23	7	-	5.5	14.5	100
	P3 (16.5)	E1P3	37.5	12.5	18	-	-	14.5	17.5	100
E2 (13.5)	P1 (10.5)	E2P1	37.5	12.5	24	6	11.5	-	8.5	100
	P2 (13.5)	E2P2	37.5	12.5	50	-	-	-	-	100
	P3 (16.5)	E2P3	37.5	12.5	36.5	-	2.3	11.2	-	100
E3 (14.5)	P1 (10.5)	E3P1	37.5	12.5	30	-	19.5	0.5	-	100
	P2 (13.5)	E3P2	37.5	12.5	23	-	17.6	9.4	-	100
	P3 (16.5)	E3P3	37.5	12.5	13.5	-	17	19.5	-	100

MSS: Maize Stover Silage; EG: Elephant grass; SPC: SAE Pujon Concentrate; RB: Rice Bran; SBM: Soyabean Meal; CFH: Coffee Husk

3 Result and Discussion

3.1 Chemical Composition Profile

The quality of feedstuff is reflected on its chemical composition, higher content of nutritive value better quality of feedstuff. The complete feed chemical composition is presented in Table 3. Based on the chemical composition analysis of complete feed using elephant grass, maize stover silage with different energy and protein balances, showing differences in nutritional content, it because each treatment contains more than one concentrate with different nutrients and compositions that cause different chemical composition. In addition, the differences in nutrient quality may be due to the difference feed manufacturers, as well as the quality of raw materials and difference processes. The DM content in complete feeds with the highest different energy and protein balances was 93.09% in E2P3 (Energy 13.5 MJ/kg + Protein 16.5%).

Dry matter consists of organic matter and ash while organic matter consists of CP, CF, EE and NFE. The higher the CP content, the dry matter content can increase Riswandi, *et al.* [8]. The OM content in E3P3 (Energy 14.5 MJ/kg + Protein 16.5%) had the highest OM content of all energy and protein balance treatments, which was 90.33%. According to Riswandi, *et al.* [8] high organic matter content of other nutrients such as CP, CF, EE, NFE, nitrogen and vitamins are also high. According to Rukmana [1] the content of organic matter is a substance in feed ingredients that is related to other components such as dry matter and ash. The highest fat content in E2P2 (Energy 13.5 MJ/kg + Protein 13.5%) is 3.14%, this is presumably due to the use of 50% concentrate in the ration, while the EE content in SAE Pujon concentrate is high 4.70%. Ether Extract contains some dietary fat, namely triglycerides which are commonly found in grains, high fat content in feed has a negative effect on dry matter digestibility of feed. The type of fat that affects digestibility is saturated fat which can reduce the digestibility of dry matter, organic matter, NDF (fiber) in the rumen, the higher the fat content in the feed, the lower the digestibility [9].

The content of CF E1P3 (Energy 12.5 MJ/kg + Protein 16.5%) has a low crude fiber content of 18.15% this is due to the use of concentrated feed ingredients in the amount of 18 g/DM, soybean meal 14.5 g/ DM and coffee husk 17.5 g/DM. Crude fiber is composed of cellulose (polysaccharide), lignin and hemicellulose which functions as a protective plant cell wall. Low CF content does not necessarily mean high digestibility, CF with high lignin content makes feed difficult to digest. R. Rukmana [1] stated that CF is difficult to digest due to the cellulose, hemicellulose, and lignin content in the cell wall. The NFE content in E3P2 (Energy 14.5 MJ/kg + Protein 13.5%) is the highest value among all energy and protein balance treatments. This is presumably due to the high use of feed ingredients such as SAE Pujon concentrate and cassava bagasse because both feed ingredients are high in NFE content. According to Budiman *et al.* [7], the increasing of protein content affects the absorption and nutrient content of feed, so NFE can increase. Sutowo *et al.* [10] stated that the increase of NFE contributes to the increase the digestibility of dry matter, organic matter and TDN.

Table 4. The Average value of the chemical composition of complete feed consist of elephant grass, maize stover silage with a balance of energy and protein

Chemical Composition (%DM)	Protein Level (%)				
	Energy Level (MJ/kg)	P_1 (10.5)	P_2 (13.5)	P_3 (14.5)	Average
DM	E1 (12.5)	92.93	92.70	92.72	92.78
	E2 (13.5)	92.26	92.37	93.09	92.57
	E3 (14.5)	92.15	93.07	92.51	92.58
	Average	92.45	92.71	92.77	
OM	E1 (12.5)	88.18	88.52	89.21	88.64
	E2 (13.5)	89.16	88.06	88.75	88.65
	E3 (14.5)	89.86	90.21	90.33	90.13
	Average	89.06	88.93	89.43	
Ash	E1 (12.5)	11.83	11.48	10.79	11.37
	E2 (13.5)	10.84	11.94	11.25	11.34
	E3 (14.5)	10.14	9.79	9.67	9.86
	Average	10.94	11.07	10.57	
CP	E1 (12.5)	10,51	13.37	16.55	13.48
	E2 (13.5)	10.74	13.49	16.73	13.65
	E3 (14.5)	10.55	13.24	16.57	13.45
	Average	10.6	13.37	16.62	
EE	E1 (12.5)	2.83	2.77	2.32	2.64
	E2 (13.5)	2.87	3.14	2.48	2.83
	E3 (14.5)	2.28	2.24	1.77	2.09
	Average	2.66	2.72	2.19	
CF	E1 (12.5)	22.73	21.22	18.15	20.7
	E2 (13.5)	22.40	19.62	19.18	20.4
	E3 (14.5)	21.45	18.75	19.29	19.83
	Average	22.19	19.86	18.87	

Chemical Composition (%DM)	Energy Level (MJ/kg)	Protein Level (%)			
		P_1 (10.5)	P_2 (13.5)	P_3 (14.5)	Average
NFE	E1 (12.5)	52.09	51.17	52.19	51.82
	E2 (13.5)	53.15	51.81	50.36	51.77
	E3 (14.5)	55.58	55.98	53.70	55.09
	Average	53.61	52.99	52.08	

Information : *) The results of the analysis at the Laboratory of Nutrition and Animal Feed, Faculty of Animal Science, Universitas Brawijaya (2021).

3.2 Nutrient degradability profile

Table 5. TDN content of complete feed of elephant grass, maize stover silage with different energy and protein balances

TDN (%DM)				
Energy Level (MJ/kg)	Protein Level (%)			Average
	P1 (10.5)	P2 (13.5)	P3 (16.5)	
E1 (12.5)	63.30	65.16	67.81	65.42
E2 (13.5)	63.61	66.67	67.15	65.81
E3 (14.5)	63.65	66.37	67.14	65.72
Average	63.42	66.06	67.36	

Information: *) The results of the analysis at the Laboratory of Nutrition and Animal Feed, Faculty of Animal Science, Universitas Brawijaya (2020)

The higher the TDN value, the better the quality of the feed ingredients. E1P3 (Energy 12.5 MJ/kg + Protein 16.5%) showed the highest TDN value among all treatments, this was due to the high protein content of the feed ingredient, namely soybean meal. The high TDN is assumed due to the contribution of CP, EE, CF and NFE, so that the nutrients in the feed are easily digested. TDN describes the total energy in the feed consumed by livestock. The TDN value depends on the digestibility of organic matter, such as crude protein, crude fiber, ether extract and NFE [11]. TDN can be determined from the chemical composition of digested CP, EE, CF and NFE. According to Indah *et al* [12] TDN content will increase if the OM, CP and EE content increases, while TDN will decrease if the content of CF and fiber fractions such as ADF, NDF, lignin, cellulose and hemicellulose increases.

Table 6. Dry Matter Degradation (DMD) Value (%) used elephant grass, maize stover silage with different energy and protein balances.

Energy level	Dry Matter Degradation (%)			
	Protein Level (%)			Average
	P1 (10.5)	P2 (13.5)	P3 (16.5)	
E_1 (12.5)	46.05	51.66	61.21	53.88^{a}
E_2 (13.5)	53.34	55.49	63.33	56.74^{b}
E_3 (14.5)	62.26	63.07	66.14	63.56^{c}
Average	53.88^{a}	56.74^{ab}	63.56^{b}	

Information: a-c) Different superscripts in the same row and column show a very significant effect ($P<0.01$) on DMD

Table 7. Organic Matter Degradation (OMD) Value (%) used elephant grass, maize stover silage with different energy and protein balances.

Energy Level	Organic Matter Degradation (%)			
	Protein Level (%)			Average
	P1 (10.5)	P2 (13.5)	P3 (16.5)	
E1 (12.5)	52.92	58.79	65.06	58.23^{a}
E2 (13.5)	58.07	57.79	63.65	60.81^{ab}
E3 (14.5)	63.70	65.84	70.01	66.24^{b}
Average	58.92^{a}	59.84^{a}	66.52^{b}	

Information: a-c) Different superscripts in the same row and column show a very significant effect (<0.01) on OMD

The results of the analysis of variance DMD and OMD combinations of different energy and protein balances did not show any interaction (P>0.05), but in respectively treatment with an increase in energy or protein levels, there are high significant effect (P<0.01). E_3P_3 (Energy 14.5 MJ/kg + Protein 16.5%) showed the highest dry matter degradation value 66.14%. It can be concluded that with the addition of energy and protein levels will produce a high degradation value. It because the increase of feed balance, good quality of feedstuffs. According to Noorsati *et al* [14] the factors that affect dry matter digestibility are composition of feed ingredients, pH, good temperature, or air quality (aerobic/anaerobic), rumen fluid, buffer solution, sample particle size, and incubation time. The CF content is also very influential on dry matter digestibility, the lower CF in the feed will increase digestibility. It because the high fiber content causes the cell walls to become thick, so the feed is difficult to degrade. Dewi *et al* [15] Added that lignocellulose is the main component of cell walls consisting of hemicellulose, cellulose, lignin, and several other extractive ingredients that bind strongly to inhibit the digestibility of feed nutrients.

The highest organic matter degradation value was in E3P3 (Energy 14.5 MJ/kg + Protein 16.5%) of 70.01%. This is presumably due to the influence of the energy and protein levels given, the higher the energy and protein levels can increase the digestibility of organic matter. According to Noorsati *et al* [14] degradation of organic matter is influenced by silica and lignin contained in cell walls which together form complex compounds with hemicellulose and cellulose. This compound is difficult to penetrate by enzymes from rumen microbes so that it will reduce digestibility and cell contents including organic matter. According to Dewi *et al* [15] the OMD value was higher than the DMD value, this was because the dry matter still contained of ash, while the organic matter no longer contained of ash. Ash in the feed causes slow feed degradation therefore organic matter without ash is easier to digest by rumen microbes. If in the analysis there is a decrease in OMD, it is suspected that the microbes in the rumen have had sufficient nutrition, so they are not able to utilize feed which has an impact on decreasing the performance of rumen microbes. Crude protein content is very influential on digestibility.

4 Conclusion

The results showed that a complete feed with energy of 14.5 MJ/kg and protein 16.5% was the best treatment with a nutrient content value of DM 92.51%, OM 90.33%, Ash 9.67% CP 16.57, CF 19.29%, EE 1.77% and NFE 53.70% and the TDN content is 67.14%. The content of DMD is 66.14% and OMD is 70.01%, so it can be concluded that feed with 14.5 MJ/kg of energy and 16.5% protein can meet the needs of livestock.

References

1. R. Rukmana, *Budi Daya Rumput Unggul* (Penerbit Kanisius, Yogyakarta, 2005)
2. A. Koloud, C.L. Kaunang., A.V. tuturoong, M.R. Waani, Zootech **40**, 428-438 (2020)
3. T. Naibaho, Despal, Permana, I.G, Buletin Makanan Ternak **104**, 12-20 (2007)
4. D.S. Widodo, *Pengaruh lama fermentasi dan penambahan inokulum Lactobacillus plantarum dan Lactobacillus fermentum terhadap kualitas silase tebon jagung* (Zea mays, Malang, Indonesia, 2014)
5. D. N. Adli, Liverstock Research Rural Development **33**, 10 (2021)
6. O. Sjofjan, D. N. Adli, M.H. Natsir, Y. F. Nuningtyas, T. S. Wardani, I. Sholichatunnisa, S. N. Ulpah, O. Firmansyah, Jurnal Ilmu Ternak dan Veteriner **26**, 39-48 (2021)
7. A. Budiman, T. Dhalika, B. Ayuningsih. Jurnal Ilmu Ternak **6**, 132-135 (2006)
8. Riswandi, Muhakka, M. Lehan, Jurnal Peternakan Sriwijaya **4**, 35-46 (2015)
9. K.J. Harvatine, Allen MS, Journal of Dairy Science **88**, 4018-4027 (2005)
10. I.T. Sutowo, Adelina, D. Febrina, Jurnal Peternakan **12**, 41-47 (2016)
11. M. Mastopan, Tafsin, N. D. Hanafi, Jurnal Peternakan Integratif **3**, 37-45 (2014)
12. A. Indah, I. G. Permana, Despal, Sains Peternakan **18**, 38-43 (2020)
13. Kusmartono, Marjuki, Jurnal Ilmu-Ilmu Peternakan **24**, 8-9 (2014)
14. M. Noorsati, L. K. Nuswantara, A. Subrata, Animal Agriculture Journal **1**, 143-158 (2012)
15. O. Dewi, Suryani, I. M. Mudita, E-journal Peternakan Tropika **8**, 587-605 (2020)

Ruminal profile of completed feed as influenced by myristic and tannins addition

Siti Chuzaemi* and *Iftitah Nuzulia* Qur'any

Faculty of Animal Science, Universitas Brawijaya, Malang 65145, Indonesia

Abstract. The study aimed to determine the effect of adding tannin and myristic acid in complete feed on nutrient content and the Profile of volatile fatty acids (VFA) in vitro. The materials used for this research were corn straw, condensed tannins, myristic acid, coffee waste, rice bran, tapioca by-product, soybean meal, copra meal, and palm kernel meal. The method was used in this experiment was laboratory experiment with four treatments and three replications. The treatment consisted of T_1 complete feed (40% corn straw + 60% concentrate (be based DM)), T_2 (complete feed + condensed tannins 3%/kg DM and myristic acid 2%/kg DM), T_3 (complete feed + condensed tannins 3%/kg DM and myristic acid 3%/kg DM), and T_4 (complete feed + condensed tannins 3% /kg DM and myristic acid 4%/kg DM). Every treatment was repeated three times and analysed statistically using Randomized Block Design (RBD). The adding of 4% myristic acid to the complete feed (T4) gives the best results, it increased nutrition and propionic acid, but decreased acetic acid, butyric acid, total VFA and C2/C3 ratio.

1 Introduction

One of the factors determining the success of a livestock business is feed. Animal feed is all feed ingredients that can be given and are beneficial for livestock and do not harm the body of animals. Each feed given must contain complete nutrition with a balanced composition so that the feeding is efficient according to the needs of the livestock [1]. However, ruminant feed is not available throughout the year, especially during the dry season. It is challenging to find forage which is the main feed for ruminants. Therefore, sufficient feed resources need to be sought to supplement the value of low forage quality and reduce dependence on grass. Sources of feed should be easy to obtain, available in large quantities at a relatively low cost, such as the utilization of corn straw waste. Corn straw consists of stalks (stems) and leaves. The percentages of each waste were 50% stalks, 20% leaves, 20% corn cobs, and 10% corn cobs [2]. In Indonesia, the amount of corn straw is 7,015,950 tons of dry matter and 25,056,965.21 tons of fresh material [3].

Corn straw is classified as a quality forage and low digestibility with a CP content of 6.37% and CF 27.61% [4]. [5] stated that corn and agro-industrial wastes have the potential

* Corresponding author: schuzaemi@ub.ac.id

as ruminant animal feed. [6] stated that corn straw as ruminant feed and a substitute for fibre sources must be balanced with concentrate to meet the animal's need. Concentrate as an additional feed in a complete feed is a feed ingredient that can improve the nutritional harmony of the whole feed and is mixed as a supplement (complementary). The concentrate has a CF content below 18% and is easy to digest. Concentrates are made from a mixture of several feed ingredients, namely sources of energy (grains), protein sources (types of cake and nuts), vitamins, and minerals [7].

Complete feed contains nutrients for livestock at a certain physiological level that is formed and given as the only feed that can meet the basic needs of production without additional substances, except water [8]. Complete feed technology is one of the efforts to increase the utilization of waste, either from agriculture or plantations, and the addition of non-conventional feed ingredients with physical treatment, supplementation, and considering the nutritional needs of livestock, both crude fiber, protein, energy, and others [9].

[10] stated that 15% of soybean meal protected with 1% tannin in complete feed contributed to the proportion of propionate and increased blood glucose in thin-tailed sheep but had not been able to reduce methane gas production. Adding more than 5% tannin in the diet will decrease both in vitro and in vivo digestibility and feed palatability. This will interfere with the use of nutrient digestibility and livestock productivity [11]. [11] stated that the addition of 2.5%, 5.0%, and 7.5% oil in animal feed was able to reduce the number of protozoa, reduce methane production without a negative effect on NH3 levels, VFA levels, activity CMC-ase, and microbial protein levels.

Volatile Fatty Acid (VFA) is one of the products of carbohydrate fermentation in the rumen, which is the primary energy source for ruminants and can contribute 55-60% of energy needs. The concentration of VFA in rumen fluid can be used to measure feed ferment ability and is closely related to rumen microbial activity [11]. Based on the description above, it is necessary to research the effect of adding condensed tannins and myristic acid to complete feed based on corn straw on nutrient content and concentration of VFA profile in rumen fluid in vitro.

2 Materials and methods

Making complete feed with each treatment in this study was carried out at the Laboratory of Animal Nutrition and Feed, Faculty of Animal Science, Brawijaya University, Malang. Analysis of the nutrient content of feed ingredients was carried out at the Nutrition and Animal Feed Laboratory, Faculty of Animal Science, Brawijaya University, Malang, and VFA concentrations were analysed at the Centre for Food and Nutrition Studies, Gadjah Mada University.

2.1 Materials

The main objective in analysing time series is to understand, interpret and evaluate changes in the phenomena in the hope of more correctly anticipating the course of future events. The materials used in this study were corn straw, concentrates consisting of coffee husks, rice bran, cassava, palm oil meal, copra meal, soybean meal, molasses, urea, salt, feed premix, condensed tannins, myristic acid, chemicals for proximate analysis and measuring in vitro gas and rumen fluid production.

2.2 Method

The research method used in this study was a laboratory experiment with four treatments and three replicate groups. Grouping based on the time of taking different cow rumen fluid. Complete feed was composed of forage and concentrate in a ratio of 40:60 (based on the percentage of DM), the concentration was composed of iso-protein with a CP content of 14% with the addition of condensed tannins (3%/kg DM) and myristic acid (2, 3, 4% /kg DM), as follows:

T_1: complete feed (60% concentrate + 40% corn straw).

T_2: complete feed (60% concentrate + 40% corn straw) + condensed tannins 3%/kg DM+ myristic acid 2%/kg DM.

T_3: complete feed (60% concentrate + 40% corn straw) + condensed tannins 3%/kg DM + myristic acid 3%/kg DM.

T_4: complete feed (60% concentrate + 40% corn straw) + condensed tannins 3%/kg DM + myristic acid 4%/kg DM.

Variables observed included nutrient content [12] and concentration of Profile Volatile Fatty Acids (VFA) [13].

2.3 Data analysis

The research data were analysed using a Randomized Block Design (RBD) and continued with Duncan's Multiple Range Test (DMRT).

3 Results and discussion

3.1 Proximate profile

Complete feed is a complete ration that has been formulated so that it contains all the nutrients according to the needs of livestock and is given as the only feed. [14] explained that complete feed for ruminants is a mixture of concentrate and forage feed ingredients. Feeding of complete feeds form to the ruminant animals must pay attention to rumen condition especially rumen pH. The results of the analysis of the nutrient content of the feed ingredients that make up the treatment feed, namely corn straw, concentrate, condensed tannins, and myristic acid, are presented in Table 1.

Corn straw has a CP content of 5.13%, CF is 36.43%, and EE is 0.63% (Table 1). The analysis results are higher than the analysis conducted by [15], namely the CP of 4.46%; EE is 0.85%, and CF is 33.12%. [16] stated that corn straw as forage is classified as low-quality forage with low digestibility because the CP content is 6.37% and CF is 27.61%. [16] revealed that forage feeding on fattening cattle would not significantly affect high body weight gain in a short time. The weight gain of cows is higher with a relatively short fattening time if the cows are given a ration consisting of concentrate and forage.

The concentrate used in this study had a CP content of 21.43%. This protein content is excellent to meet the nutritional needs of the diet. Concentrate function on complete feed corn straw-based with condensed tannins and myristic acid can improve the nutritional quality of diet. This is comparable to the statement by [17] that the provision of concentrate feed, which has a higher nutritional value than forage, is intended to provide opportunities for livestock to maximize growth or production. This is clarified by [16] statement that high CP content in concentrate can increase rumen microbial growth, resulting in increased activity in digesting feed ingredients. Giving concentrate can help rumen microbes to digest CF faster and increase levels of propionate, which is useful in the formation of meat.

Mimosa leaves (*Mimosa pudica* L.) used as tannins in this study had a CP content of 0.59%; CF is 0.47%, and EE is 0.85%. This is following [18] statement that shy daughter contains 8.37% CP; EE 1.43%; CF 3.3% and contains flavonoid compounds, alkaloids, phenols, glycosides, tannins, saponins, terpenoids, and coumarins. [19] added that providing tannins in feed ingredients can have a defaunation effect. The presence of protein-tannin complexes can pressure the rumen protozoa population, thus causing an indirect impact on the decline of protozoa.

Myristic acid used in this study has an EE content of 1.05%. [20] explained that myristic acid is one of the three most abundant saturated fatty acids in coconut oil, in addition to lauric acid around 44-52% and palmitic acid around 7.5-10, 5%.

Table 1. Nutrient content of diet treatments' ingredients

Feedstuff	**Nutrient value (%)**						
	DM	**Ash***	**OM***	**CP***	**CF***	**EE***	**Tannin**
Corn straw	94.46	10.17	89.83	5.13	36.43	0.63	-
Concentrate	90.18	11.42	88.58	21.43	18.74	7.12	-
Condensed tannin	90.22	6.41	93.59	0.59	0.47	0.85	23
Myristic acid	90.41	0.03	99.97	-	-	1.05	-

Feed Analysis result from Laboratory of Animal Feed and Nutrient, Faculty of Animal Husbandry, University of Brawijaya (2019).
*) Based on 100% Dry Matter

The primary saturated fatty acids in meat are myristic, palmitic, and stearic. The use of coconut oil in feed can reduce the population of protozoa in the rumen and reduce methane gas emissions per kg live weight of sheep. A decrease of total volatile fatty acids in the rumen and the concentration of acetic and butyric acids were also seen due to the administration of coconut oil. The results of the analysis of the nutrient content of the treatment feed consisting of corn straw, concentrate, condensed tannin, and myristic acid in each treatment are shown in Table 2. The results showed that the CP content in the complete feed was following the initial target of the formulation of 14%. The composition of the CP content of complete feeds is based on the CP needs of ruminants ranging from 9-15%, and especially the CP needs of beef cattle ranging from 13-16% [18].

The CP content in each treatment ranged from 13.05 to 16.18% (Table 2). The CP content in the T1 treatment had the highest value from each treatment of 16.18% because there was no addition of condensed tannins and myristic acid in this treatment. The CP content in treatments T2, T3, and T4 increased in each treatment. This is presumably due to the addition of protein sources derived from concentrates and condensed tannins, and myristic acid. The CP content in each treatment was close to the formulation target of 14%. [19] stated that the high levels of CP in complete feeds are thought to come from additional protein sources from legumes with high protein content. [21] stated that feed with high protein content could be used as a nitrogen source for microbial growth in the rumen of ruminants. The benefits of tannins in feed include preventing bloating in cattle and sheep. Condensed tannins also protect the protein from rumen microbial degradation (increase bypass protein), so the small intestine can directly absorb that protein.

The content of CF in each treatment ranged from 22.87-24.00% (Table 2). The CF content in each treatment decreased due to condensed tannin and myristic acid with a CF content of

3.30%. [22] explained that CF is the primary energy source for ruminants. This is supported by [23] that ruminants can digest CF in DIET well, and an average of 70-80% of energy needs come from fiber. High CF tends to reduce protein digestibility. If an increase follows the increase in protein in the diet in CF, there will be a slight change in protein digestibility. Still, if the CF is reduced and protein is increased, protein digestibility will increase. The EE content in each treatment ranged from 2.94-6.19% (Table 2). The EE content in each treatment decreased due to condensed tannins and myristic acid in the complete feed. The EE content in the T3 treatment had the highest value of 6.19%; the high EE content in ruminant feed ingredients could interfere with the fermentation process of feed ingredients in the livestock rumen. The EE content in this study was higher than the research by [24] that the EE content of the feed consumed was relatively high, around 3.78%, but was still within the safe limit of the EE content for consumption ruminants (below 5%). [25] stated that fat content in the feed that is too high (above 5% of the total diet) would negatively affect the digestibility of CF in the rumen. It can affect the ability of livestock to utilize the nutrients of the feed consumed.

Table 2. Nutrient content of diet treatments

Treatments	DM (%)	OM*(%)	Ash*(%)	CP*(%)	CF*(%)	EE*(%)
T_1	93.44	90.09	9.91	16.18	24.00	2.94
T_2	93.66	90.56	9.44	13.73	23.31	5.33
T_3	93.51	90.69	9.31	13.05	23.26	6.19
T_4	93.46	90.56	9.44	14.12	22.87	3.35

Feed Analysis result from Laboratory of Animal Feed and Nutrient, Faculty of Animal Science, University of Brawijaya (2019).
*) Based on 100% Dry Matter

3.2 VFA Profile

Volatile Fatty Acid is one of the end products of carbohydrate fermentation from feed given in the rumen of livestock. Volatile Fatty Acid also affects the digestibility of feed in the rumen because VFA results from the breakdown of carbohydrates in the rumen. [26] explained that VFA is the main product of carbohydrate digestion in the rumen and the final product of the fermentation of organic matter, which is used as the primary energy source for ruminants from the rumen. [21] added that the production of VFA in rumen fluid can be used to measure feed fermentability and microbial activity and characterize the amount of VFA produced. In otherwise, the provision of feed for animals is a major contributor to land and water use greenhouse gas emission [19, 21]. The most common compositions in rumen fluid are acetic, propionic, and butyric acids, while small amounts of VFA are formic, isobutyric, valeric, isovaleric and caproic acids [27]. The results of the VFA concentration can be seen in Table 3.

The addition of condensed tannins and myristic acid in complete feed on the concentration of acetic acid (C2) had a very significant effect and decreased ($P<0.01$), and propionic acid (C3) had a very significant effect and increased ($P<0.05$), while the addition of condensed tannins and myristic acid to complete feed had no effect on butyric acid (C4) ($P>0.05$).

The lowest concentration of acetic acid occurred in treatment T_4 of 14.76 mMol with the addition of 3 grams of condensed tannin and 4% myristic acid/kg DM. [28] explained that myristic acid supplemented in a complete diet of 5% could reduce 36% of CH_4 gas production without affecting feed consumption and milk production. Condensed tannins can reduce nutrient availability and enzyme activity and affect fiber degradation). The concentration of acetic acid in this study was 14.76±3.30-28.21±1.36 mMol. If the acetate produced is low, then the production of CH_4 gas has been successfully suppressed. This indicates that the T_4 treatment suppressed the CH_4 gas, which produced lower acetic acid than the other treatments.

The highest propionic acid concentration occurred in treatment T4 of 13.31 mMol with the addition of 3 grams of condensed tannin and 4% myristic acid/kg DM. The concentration of propionic acid in this study ranged from 7.30±0.64-13.31±1.88 mMol. [29] revealed that the total concentration of VFA decreased due to defaunation, but there was an increase in propionate concentration. [11] showed that the average propionic acid content of coconut oil was higher than that of sunflower seed oil and palm oil, each of which had a propionic acid content.

Table 3. The average of VFA concentration from rumen of in vitro gas production with 48 hours incubated

Treatments	**VFA (m Mol)**				
	Acetic Acid (C2)	**Propionic Acid (C3)**	**Butyric Acid (C4)**	**VFA Total**	**C2/C3 Ratio**
T_1	28.21 ± 1.36^{c}	7.30 ± 0.64^{a}	6.44 ± 1.80	41.95 ± 1.94^{b}	3.88 ± 0.29^{b}
T_2	21.17 ± 0.99^{b}	9.78 ± 1.92^{ab}	4.44 ± 0.93	35.39 ± 3.69^{a}	2.21 ± 0.32^{a}
T_3	19.89 ± 0.47^{b}	10.74 ± 1.08^{ab}	4.22 ± 1.89	34.86 ± 2.49^{a}	1.87 ± 0.22^{a}
T_4	14.76 ± 3.30^{a}	13.31 ± 1.88^{a}	3.99 ± 1.84	30.98 ± 5.08^{a}	1.14 ± 0.38^{a}

Notes: Different superscripts in the column mean value of acetic acid, total VFA, and C2/C3 ratio of 48 hours incubation showed a very significant difference ($P<0.01$). Different superscripts in the column mean value of propionic acid incubation 48 hours showed significant differences ($P<0.05$).

In this study, the concentration of butyric acid had no significant effect on the addition of condensed tannins and myristic acid in complete feed ($P>0.05$). The proportion of butyric acid at T_2 was 4.44 mMol; T_3 was 4.22 mMol, and T_4 was 3.99 mMol. The concentration of butyric acid in this study ranged from 3.99±1.84-6.44±1.80 mMol. [27] explained that the formation of C_4 has faster absorption properties than C2 and C3, so the proportion of C4 in VFA is also low. [30] added that if starch increases or propionate and butyrate increase, the pH will decrease to 4.5-5. At low pH conditions, it will inhibit the growth of cellulolytic bacteria, inhibiting the digestion of forage. [31] stated that propionate and acetate have slower absorption properties than butyrate. The formation of butyric acid is closely related to the formation of acetic acid so that when the concentration of acetic acid increases, the concentration of butyric acid increases. Some of the butyric acid formed is used as a precursor for milk fatty acids into the blood and is used as an energy source for body tissues.

The total VFA concentration in this study had a very significant effect and decreased ($P<0.01$) in each treatment. The total VFA concentration ranged from 30.98±5.08 to 41.95±1.94 mMol. Volatile Fatty Acid produced in each treatment was lower than the control, so all treatments could suppress CH_4 gas. The T_4 treatment had the lowest total VFA value compared to the other treatments. [32] explained that the total VFA concentration (mmol) in the rumen fluid resulted from complete feed degradation with UDP supplementation of 0, 2.5, 5.0, 7.5, and 10.0%, respectively was 100.82; 94.12; 87.02; 106.32, and 88.33 mmol. The statistical analysis results showed that the difference in the level of UDP supplementation in complete feeds did not affect the concentration of total volatile fatty acids. The concentration of VFA in this study was not affected by the level of

UDP supplementation given in 1% formaldehyde-protected soybean meal. One of the causes of the low concentration of total VFA can be influenced by the amount of non-structural carbohydrates composed of complete feeds. This is supported by [33] stating that the decrease in total VFA concentration was due to VFA absorption and VFA as a carbon framework for rumen microbial protein synthesis.

The ratio of acetic acid and propionic acid (ratio C2/C3) aims to measure the efficiency of ruminant energy use and the quality of the product produced. The C2/C3 ratio results in the study had a very significant effect and decreased between treatments ($P<0.01$). The results of this study obtained the results of C2/C3, namely T_2 (2.21 mMol), T_3 (1.87 mMol), and T_4 (1.21 mMol). The low ratio of C2/C3 in this study is thought to be due to the high efficiency of livestock energy use, so the quality of the resulting product is more optimal. If the proportion of propionic acid increases, the balance of acetic acid will decrease. This causes the C2/C3 ratio to decline, and the low C2/C3 ratio is good for beef cattle.

4 Conclusion

The results of the study can be concluded, namely treatment T_4 [complete feed with the addition of the proportion of condensed tannins (3%/kg DM) and myristic acid (4% /kg DM)] is the best treatment by increasing the nutrient content and acid concentration. Propionic acid can reduce acetic acid, butyric acid concentration, total VFA concentration, and the ratio of C2/C3. Complete feed with the addition of condensed tannins (3%/kg DM) and myristic acid (4%/kg DM) resulted in the best nutrient content and concentration of VFA profile suitable for application to livestock. Further research needs to be done by applying complete feed based on corn straw with condensed tannins and myristic acid directly on livestock in vivo.

References

1. A.S. Pernata, Agromedia Pustaka, **146** (2010)
2. J. McCutheon, D. Samples, Ohio State University Extension, **10** (2002)
3. B. Nappu, T. Yuniarsih, Jurnal Serealia **329** (2013)
4. E. Mayasari, B. Ayuningsih dan R. Hidayat, Student Journal Fakultas Peternakan Universitas Padjajaran **4** (2015)
5. U. Umiyasih, U. E. Wina, Buletin Ilmu Peternakan Indonesia **18** (2008)
6. A. Parakkasi, Jakarta: Universitas Indonesia Press. (1999)
7. Y.A Hadiyanto, Surono, M. Christiyanto, Journal Animal Agriculture **1** (2012)
8. Abdurachman, S. Askar, Media Peternakan **7** (2001)
9. H. Hartadi, S. Reksohadiprodjo, A. D. Tillman. *Ilmu Makanan Ternak Dasar* (Gadjah Mada University Press, Yogyakarta, Indonesia, 2005)
10. S. Nuraliah, A. Purnomoadi, L. K. Nuswantara, Jurnal Veteriner **16** (2015)
11. P. Sitoresmi, L. M. Yusiati, H. Hartadi, Buletin Peternakan **33** (2009)
12. AOAC. Official Methods of Analysis of the Association of Official Analytical Chemists. Published by the Association of Official Analytical Chemist. Marlyand (2005)
13. Z. Bachruddin, Fakultas Peternakan. Universitas Gajah Mada. Yogyakarta (1996)
14. R.S. Budiono, R.S. Wahyuni, R. Bijanti. Jurnal Veteriner **1** (2003).
15. Y.N. Anggraeny, U. Umiyasih, D. Pamungkas, Jurnal Veteriner **147** (2005).
16. E. Mayasari, B. Ayuningsih, R. Hidayat, Student E-Journal **4** (2015).
17. N. Laryska, T. Nurhajati, Agroveteriner**1** (2013).
18. J.A. Momot, K. Maaruf, M. R. Waani, Ch. J. Pontoh. Jurnal Zootek **34** (2014).
19. R. Dianita, Jurnal Pastura **2** (2012).
20. G. Sajati, W. H. E. Prasetyo, Surono. Animal Agricultural Journal **1** (2012).

21. D. N. Adli, Livest. Res. Rural. Dev, **33:94** (2021)
22. E. Gustiani, K. Permadi, Jurnal Peternakan Indonesia **17** (2015)
23. O. Sjofjan, D. N. Adli, M.H. Natsir, Y. F. Nuningtyas, T. S. Wardani, I. Sholichatunnisa, S. N. Ulpah, O. Firmansyah, Jurnal Ilmu Ternak dan Veteriner **26** (2021)
24. H. Suprapto F. M. Suhartati, T. Widiyastuti. Jurnal Ilmiah Peternakan **1** (2013)
25. S. Sitompul, Martini, Jurnal Nasional Tenaga Fungsional **1** (2005)
26. B. Haryanto,Wartazoa **22** (2012)
27. E. Wina, I. W. R., Susana, Wartazoa **22** (2013).
28. N. Siska, T. Widiyastuti, T. R. Sutardi, Jurnal Ilmiah Peternakan **1** (2013).
29. D. Pamungkas, Y. N. Anggraeny, Kusmartono, N. H. Krishna, Jurnal Peternakan dan Veteriner **197** (2008).
30. Mcdonald, P., Edwards, R.A., Greenhalgh, J.F.D. and Morgan, C.A. *Animal Nutrition* (Benjamin Cummings, 2002)
31. N. Odongo, M. M. Or-Rashid, E. Kebreab, J. France, B. Mcbride, J. Dairy Sci, **90** (2007)
32. H. Wei-lian, W. Yue-ming, L. Jian-xin, G. Yan-qiu, Y. Jun-an. J. Zhejiang Univ. Sci, **6** (2005)
33. B. Suhartanto, R. Utomo, Kustantinah, I. G. S. Budisatria, L. M. Yusiati, B. P. Widyobroto, Buletin Peternakan **38** (2014)

Volatile fatty acids concentration and efficiency of microbial protein synthesis of concentrate cassava peel diet with different levels of protein sources

Arya Dwi Saputra[1], *Kusmartono* Kusmartono[2*], *Mashudi* Mashudi[2], *and Poespitasari Hazanah* Ndaru[2]

[1]Postgraduade Programe, Faculty of Animal Science, Universitas Brawijaya, Malang 65145, Indonesia
[2]Faculty of Animal Science, Brawijaya University, Malang 65145, Indonesia

Abstract. This study was conducted to evaluate the effects of different levels of protein sources in a concentrate cassava peel diet on volatile fatty acid (VFA) concentration and efficiency of microbial protein synthesis (EMPS) by *in vitro* method. The five diet treatments in this study consisted of cassava meal (CM), cassava peel (CP), cassava leaves (CL) and moringa leaves (ML) with the proportion applied were T1=CM70% + CP20% + CL5% + ML5%; T2=CM60% + CP20% + CL10% + ML10%; T3=CM50% + CP20% + CL15% + ML15%; T4=CM40% + CP20% + CL20% + ML20%; T5=CM30% + CP20% + CL25% + ML25%. The experiment used a randomized block design with five treatments and three replications. The results showed that increasing the level of protein sources in the ration tended to increase the value of total VFA, acetate, propionate, CO_2, and CH_4 in which T3 had the highest values. However, the effect of treatments were significant in EMPS value ($P<0.01$). It was concluded that increasing levels of cassava and moringa leaves in the concentrate cassava peel diet tended to increase the value of total VFA, acetate, propionate, butyrate, CO_2, and CH_4 which led to a significant increase in EMPS values.

1 Introduction

Carbohydrates are fermented by rumen microbes to produce volatile fatty acids (VFA) which consist of 60-70% acetic acid, 15-20% propionic acid, and 10-15% butyric acid [1]. The concentration of VFA is influenced by the level of feed ingredient fermentability in the rumen [2]. Apart from their function as a source of energy for the host, VFAs also serve as a source of carbon skeletons as well as ATP for microbial protein synthesis [3]. The microbial protein produced in the rumen and flows into the abomasum provides the majority of protein for further digestion by HCl-pepsin and trypsin to form amino acids in the small intestine for absorption (50-80%). Bach *et al.* [4] concluded that rumen ammonia is the main precursor,

* Corresponding author: kusmartono_ansi@ub.ac.id

and its efficiency of conversion into microbial protein is determined by the type and amount of feed consumed by ruminants. Ramaiyulis *et al.* [5] stated that rumen microbial growth, largely determined by the availability of essential nutrients such as soluble carbohydrates and protein.

Cassava (*Manihot utilissima* Crantz) is a major plant in Indoneisa with 19.3 million tons production in 2018 [6]. This perennial plant contains high energy in the root [7] and a high protein in the leaves [8]. Cassava meal which formed from the drying and grinding process of cassava tubers contains 87.3% dry matter (DM), 2.13% crude protein (CP), and 3.2% crude fiber (CF) [9]. Lukuyu *et al.* [10] mentioned that cassava meal provides 96.2% organic matter (OM) and non-fiber carbohydrates (NFC) 82%. Cassava peels contained 7.1% OM, 7.1% CP, 2% extract ether (EE), and 66.4% NFC. Hang *et al.* [11] mentioned that cassava leaves are rich in protein with 29.4% DM and 22.1% CP.

Retnaningrum *et al.* [12] reported that the use of cassava meal as an energy source to supplement maize stover basal diet had the highest digestibility and daily weight gain of crossbred Limousine. [13] mentioned that cassava peel can replace corn as an energy source in ration rated from ruminal parameters on small ruminants. Mulyati *et al.* [14] stated that cassava and moringa leaves potentially as a supplement on low-quality roughages to support rumen microbial growth and activity due to high protein content. This study aimed to evaluate the effects of different levels of protein sources in a concentrate cassava peel diet on VFA concentration and efficiency of microbial protein synthesis (EMPS) by in vitro method.

2 Materials and methods

2.1 Location and time

This research was conducted in the Nutrition and Animal Feed Laboratory Faculty of Animal Science Brawijaya University in November.

2.2 Materials

Materials used were set of tools for volatile fatty acids (VFA) and efficiency of microbial protein synthesis (EMPS) analysis and feedstuffs consisted of cassava meal (CM), cassava peel (CP), cassava leaves (CL) obtained from Ngajum, Malang, and moringa leaves (ML) took from Kambingan Village, Tumpang, Malang.

2.3 Method

This study used a Randomized Block Design (RBD) with five treatments and three replication. The treatments applied were T1=CM70% + CP20% + CL5% + ML5%; T2=CM60% + CP20% + CL10% + ML10%; T3=CM50% + CP20% + CL15% + ML15%; T4=CM40% + CP20% + CL20% + ML20%; T5=CM30% + CP20% + CL25% + ML25%. Table 1 showed the chemical composition of feedstuffs and treatments.

The *in vitro* method used Makkar *et al.* [15] protocol, in which after the incubation, the residue were centrifuged to separate sediment and supernatant parts. The separated sediment was dried and weighed for measuring fermented organic matter (FOM) value. The separated supernatant was used to determine the VFA total, acetate (C_2), propionate (C_3), and butyrate (C_4) by gas chromatographic techniques. The result of VFA concentration also used to determine CO_2 and CH_4 by Van Soest [16] equation. For determining of EMPS value, the residues were refluxed using NDS solution and calculated with Bach *et al.*[4] equation.

Table 1. Chemical composition of feedstuffs and treatments

Feed	Chemical Composition (%)							
	DM	Ash*	CP*	EE*	CF*	NFE*	ADF*	NDF*
Cassava Meal (CM)	84.41	2.63	2.86	0.84	2.90	90.77	2.18	8.96
Cassava Peel (CP)	24.32	7.51	9.18	1.72	9.88	71.72	9.36	28.36
Cassava Leaves (CL)	25.73	9.87	25.48	9.72	18.46	36.47	47.15	46.50
Moringa Leaves (ML)	24.23	12.97	27.38	6.78	10.35	42.51	13.39	17.51
T1	90.62	4.26	9.10	1.22	4.76	80.66	7.34	16.03
T2	91.47	5.03	11.50	1.69	5.66	76.12	8.50	17.50
T3	91.87	5.98	12.56	2.35	6.82	72.29	10.19	19.03
T4	92.29	6.94	14.46	3.16	7.56	67.88	11.68	22.57
T5	90.88	7.87	19.28	4.65	8.57	59.62	13.09	33.14

*) Based on 100% of DM [17]

2.4 Statistical analysis

The data obtained were analyzed by analysis of variance (ANOVA) and followed by Duncan's Multiple Range Test if the treatments gave a significant effect on the variables measured according to Sudarwati *et al.* [18].

3 Results and discussion

Table 2 presents the result of VFA total, acetate, propionate, butyrate, CO_2, CH_4, and EMPS by treatments. The results showed the treatment did not significantly affect the value of all parameters (P>0.05), except EMPS which showed a significant by treatments (P<0.01). Treatment T3 achieved the highest value of total VFA (148.6 mMol/l), whilst the lowest total VFA value was recorded in T1 (97.4 mMol/l). Similar trends were observed in acetate, propionate, butyrate, CO_2, and CH_4 values. Increased proportions of cassava and moringa leaves in the ration has led to EMPS value being highest for treatment T5 (27.28 g N/kg FOM), whilst T2 had the lowest value of EMPS (12.10 g N/kg FOM). This evidence indicates the potential of cassava and moringa leaves as protein sources to supplement cassava peel and cassava meal as carbohydrate sources.

The VFA total in this study was high because the concentration of VFA was generally from 70-150 mMol [19] and this may be due to a high fermentation rate of cassava meal and peels in the rumen. The high starch diet in ration increases acetate and propionate value [20] and increases the amylolytic microbial population [19]. Suharlina *et al.* and Gumilar *et al.* [21, 22] stated that fibrous feed becomes acetic in the rumen which probably caused acetate and propionate to be higher. In contrast, T3 has a higher concentration of VFA than T4 and T5 which have higher fiber. It also means that the high protein content in the ration without being followed by an adequate energy source cannot be utilized properly by microbes. Retnaningrum *et al.* [12] reported that the VFA increased with the addition of cassava meal in the ration, but at the 70% proportion of cassava meal, drastically decreased total VFA, acetate, propionate, and butyrate due to inadequate proportion of protein source. Santos *et al.* [13] reported that cassava peel tended to decrease propionate and increase the acetate concentration in replacement of corn.

Tabel 2. Total VFA, acetate, propionate, butyrate, CO_2, CH_4 and EMPS of treatments

Parameters	Treatments					*SEM*
	T1	T2	T3	T4	T5	
Total VFA (mMol/l)	97.40	133.10	148.63	136.13	137.60	8.298
Acetate (mMol/l)	61.67	86.43	95.63	87.40	85.30	5.538
Propionate (mMol/l)	21.63	28.0	30.53	29.90	30.67	2.074
Butyrate (mMol/l)	14.10	18.67	22.47	18.83	21.63	1.516
CO_2 (Mol)	58.86	58.82	60.21	57.90	60.0	0.688
CH_4 (Mol)	32.63	34.19	34.77	33.74	33.25	0.841
EMPS* (g N/kg FOM)	12.71^{a}	12.10^{a}	17.37^{ab}	20.11^{ab}	27.28^{b}	1.910

Different superscript in the same row showed the significant effect at P<0.01 (*)
EMPS= Efficiency of Microbial Protein Synthesis
FOM= Fermented Organic Matter

Apart from VFA levels, CH_4 can be assessed for feed efficiency parameters. In this study, the CH4 trend was similar to the VFA value which the highest value was T3 (34.77 Mol) and the lowest in T1 (32.63 Mol). McDonald *et al.* [1] mentioned that the organic matter degradation by rumen microbes will produce VFA, CO_2, CH_4, and other gasses. The higher VFA formed higher the gasses produced. Muchlas *et al.* [23] stated that the higher CH_4 production illustrates more energy wasted, which indicates the more inefficient the feed was.

Karsi and Russel [24] mentioned that EMPS values generally range from 10-70 g N/kg FOM and the optimal value for EMPS above 30-40g N/kg FOM, thus the EMPS value in this study is low. It may be caused by the low fiber diet and a high concentrate that caused the low rumen pH as stated by Muchlas *et al.* [23]. The low EMPS in this study may be causing a high value of NH_3, in which according to Bach *et al.* [4], the available energy and amino acid (AA) will be used directly for microbial protein synthesis, while limited energy will cause amino acid (AA) to be deaminated and the carbon skeleton fermented into VFA. The excess of AA will be removed from the cytoplasm into ammonia because some rumen microbes lack of mechanism for emitting AA.

4 Conclusion

In conclusion, increasing levels of cassava and moringa leaves in the concentrate cassava peel diet tended to increase the value of total VFA, acetate, propionate, butyrate, CO_2, and CH_4 which led to a significant increase on EMPS values.

The authors wish to thank Rector of Brawijaya University for financial support through Professor research grant.

References

1. P. Mc Donald, R.A. Edwards, J.F.D. Greenhalgh. C.A. Morgan, L.A. Sinclair, and R.G. Wilkinson, *Animal Nutrition 7th Edition* (Benjamin Cummings, 2011)
2. C.J.B. Sendow, C.T. Noviandi, and R. Utomo, Buletin Anim. Sci. **41** (2017)
3. E.R. Orskov, *Protein Nutrition in Ruminants, 2nd Edition* (Academic Press, 1992)
4. A. Bach, S. Calsamiglia, and M.D. Stern, J. Dairy Sci. **88**, 9-21 (2005)
5. Ramaiyulis, R.W.S. Ningrat, M. Zain, and L. Warly, Pak. J. Nutr. **18** (2019)
6. A. Amir, B.P. Purwanto, Nahrowi, A. Atabany, Salundik, and A. Yani, LRRD **33** (2021)
7. P. Sumadong, A. Cherdtong, S. So, and M. Wanapat, BMC. Vet. Res. **17** (2021)
8. S. Suharti, H. Oktafiani, A. Sudarman, M. Baik, and K. G. Wiryawan, AOAS. **66** (2021)
9. S. Retnaningrum, Kusmartono, Mashudi, K.J. Harper, and D. Poppi, IOP Conf. Ser. Earth Environ. Sci. **478** (2020)
10. B. Lukuyu, I. Okike, A. Duncan, M. Beveridge, and M. Blummel, *Use of Cassava in Livestock and Aquaculture Feeding Programs* (ILRI Editorial and Publishing Services, Addis Ababa, Ethiopia, 2014)
11. L.T.T. Hang, T.R. Preston, N. Xuan Ba, and D. Van Dung, LRRD. **31** (2019)
12. S. Retnaningrum, Kusmartono, Mashudi, K.J. Harper, and D.P. Poppi, Animal, **15**, 1-7 (2021)
13. V.L.F. Santos, M.A. Ferreira, M.C.B. Siqueira, T.T.B. Melo, J.L. Silva, I.B. Andrade, A.A. Soares, and C.T.F. Costa, Small Rumin. Res. **113**, 88-92 (2015)
14. Mulyati, Kusmartono, Hartutik, and Rusdi, LRRD, **27** (2015)
15. H.P.S. Makkar, M. Blümmel, and K. Becker, British J. Nutri. **73**, 897-913 (1995)
16. J. P. Van Soest, *Nutrition Ecology of Ruminant, 2nd Edition* (Cornell University Press, Ithaca, 1994)
17. A.D. Saputra, Kusmartono, Mashudi, P.H. Ndaru, IOP Conf. Ser. Earth Environ. Sci. **888** (2021)
18. H. Sudarwati, M.H. Natsir, and V.M.A. Nurgiartiningsih, *Statistika dan Rancangan Percobaan dalam Bidang Peternakan* (Universitas Brawijaya Press, Malang, Indonesia, 2019)
19. A.D. Putridinanti, C.T. Noviandi, Gunawan, A. Agus, K. Harper, and D. Poppi, IOP Conf. Ser. Earth Environ. Sci. **387** (2019)
20. B. Hatew, S.C. Podesta, H. Van Laar, W. F. Pellikaan, J.L. Ellis, J. Dijkstra, and A. Bannink, J. Dairy. Sci., **98**, 486-499 (2015)
21. Suharlina, D.A. Astuti, Nahrowi, A. Jayanegara, and L. Abdullah, JITAA, **41** (2016)
22. D.A.K.W. Gumilar, E. Rianto, and M. Arifin, IOP Conf. Ser. Earth Environ. Sci. **119** (2018)
23. M. Muchlas, Kusmartono, and Marjuki, Jurnal Ilmu-Ilmu Peternakan, **24**, 8-19 (2014)
24. M.A. Karsli and J.R. Russel, J. Vet. Anim. Sci. **25**, 681-686 (2001)
25. J. Achmadi, E. Pangestu, Surahmanto, A. Subrata, and M. A. Harahap, IOP Conf. Ser. Earth Environ. Sci. **518** (2020)

Morphometric and gross anatomy studies of gonads in day-old Arabic chick (*Gallus turcicus*) lead to gonadal asymmetry

Iswati Iswati[1,*] and *Gatot* Ciptadi[2]

[1]Politeknik Pembangunan Pertanian Malang, 65200, Indonesia
[2]Faculty of Animal Science, Universitas Brawijaya, Malang, 65145, Indonesia

Abstract. The differentiation of avian gonads occurred since the embryonic stage, resulted in asymmetric morphology. This study aimed to analyse the morphometric and gross anatomy of gonads in Day old Arabic chick. This study utilized 116 Day old Arabic chick. After necropsy, they were divided into 61 male and 55 female. The variables studies were length, width, volume, location, colour, and shape of the gonads. Data analysis used descriptive analysis and independent T test. The results showed a significant difference ($P<0.01$) between the right and left ovary lengths 1.67±0.08 mm and 3.69±0.05, respectively. A significant difference ($P<0.01$) between the width ovaries 0.64±0.03 mm (right ovary) and 1.18±0.03 mm (left ovary), respectively. It was found a significant difference ($P<0.05$) between the length of right and left testis 2.47±0.08 mm and 3.29±0.07mm, and between volume of testis 1.46±0.12 mm^3 (right testis) and 2.03±0.13 mm^3 (left testis). However, the right and left testes width was not significant ($P>0.01$). The testes colour was creamy white, sometimes pink, and grey to black, the colour of ovaries was creamy yellow and pink. Morphometric and gonadal morphology of Day old Arabic chick showed gonadal asymmetry, and the left side is larger than the right side.

1 Introduction

Arabic chicken is one of the native chickens, layer-type chicken with superior egg production [1]. Its production potential has attracted farmers, so it has grown widely in several areas in Indonesia. In addition to having high egg production, Arabic chickens can adapt to various environmental conditions. Arabic chicken also does not have the brooding trait to extend their egg-laying period [2, 3].

Freshly hatched Arabic chickens are generally monomorphic, difficult to determine between males and females because they have almost the same phenotypic characteristics. Recently, research has been developed about sexing Day-old-chick in Arabic chickens to distinguish sex quickly and accurately. Research on Day-old Arabic chick sexing requires

* Corresponding author: iswati@polbangtanmalang.ac.id

proof of sex as soon as possible. One of the proofs is through gonad identification (post-mortem) in day-old chicks because they are relatively more available than molecular sexing [4]. Gonads are reproductive organs, ovaries to produce ova in females, and testes to produce spermatozoa in males [5]. Because the size of the gonads of freshly hatched chickens is relatively small, it is necessary to observe their morphology to determine the testes or ovaries.

The avian gonads have location and morphology differences from mammals; birds' testes are in the abdominal cavity, normally, the right ovary regresses. The left ovaries have been identified as functional gonad, and asymmetric development of the right and left ovaries has been reported since the embryonic phase [6]. The male, both gonads develop into testes [7]. Testicular asymmetry is also common in poultry, especially in passerines, and the degree of asymmetry decreases with the increasing age of chickens. The right and left testes are functional, although the left is larger in most species [8]. The size of the testes is often associated with their reproductive ability [9]. Testicular morphometric studies have been carried out on adult Nigerian local chickens under physiological conditions to predict the reproductive ability of the testes [10]. Left-Right (L:R) asymmetry could be detected in both sexes in chickens [11].

Based on the description, it is necessary to reveal the shape and size of the ovary and testes of Day-old Arabic chick. This study aimed to analyse the morphometric and gross anatomy of the testes and ovaries of Day-old Arabic chick. Descriptions and information about their morphological characters are essential to confirm gonadal observations for purposes of proving to sex, to predict male quality through post-hatching development, and detect the gonadal abnormalities in Day-old chick (DOC).

2 Materials and methods

2.1 Day-old-chick Arabic chicken

This study used 116 Day-old Arabic chick, resulting in mating between hen and Arabic gold rooster in previous studies [4]. Day-old chick obtained from the chicken breeding unit in Politeknik Pembangunan Pertanian Malang. The research was conducted at the Poultry Unit and the Reproduction Laboratory of the Politeknik Pembangunan Pertanian Malang.

The chicks have been estimated for their sex by feather sexing, then were euthanized by cervical dislocation. The necropsy procedures incise the ventral abdomen until the peritoneal cavity opens and removes the viscera organs, so the gonads are visible. The testes and ovaries were not removed from the abdominal cavity and were not weighed because relatively very small. The chicken was fixed in a dorsal lying position with the gonads still attached to the abdominal cavity, observed with direct eyes and then clarified under a trinocular stereo microscope connected to a computer to identify the gonads is easier.

2.2 Research variable

The gonads were examined for colour, shape, and location. Ovarian morphometrics includes the length and width of the right and left ovaries, while the testes include length, width, and volume. Data collection on the shape, colour, and location of the gonads was carried out under a stereo microscope with 10x magnification. The stereo microscope used is the Euromax SB 1903-P series produced by Holland with a DC 5000P camera. Measurement of the length and width of the gonads using the micrometre software from the camera. Quantitative data on testicular length and width were tabulated to calculate testicular volume.

Length of testes and ovaries was measured from the anterior end to the posterior end of the gonad, and gonad width was measured at the broadest part of the gonad. Testis's shape is ellipsoid [9], so the testicular volume was calculated using formula = $(4/3) \times \pi \times (a)^2 (b)$. Where (a) is: half of the width and (b) is half of the length [12] or with formula (simplified) namely testicular volume = $0.523 \times (\text{Width})^2(\text{Length})$ [8].

2.3 Statistical analyses

The results of the gross anatomy identification of testes and ovaries were shown with explanation descriptions. The morphometric data were analysed using descriptive analysis. Differences in size between the right and left gonads were analysed by independent T-test at significance $P<0.01$, using SPSS version 16. Data was presented in mean ±SEM.

3 Results and discussion

3.1 Gonads overview of day-old Arabic chick

Freshly hatched chicks were identified to be 55 females and 61 males after necropsy. The testes and ovaries were located in the abdominal cavity and could be seen when the viscera organs were removed. Results of observations to the ovaries and testes morphology are shown in Figure 1. Postmortem examination of chick testes and ovaries showed that the gonads consist of two right and left at above the surface of the kidney. The gonads were connected to the backbone on the surface of the two kidneys.

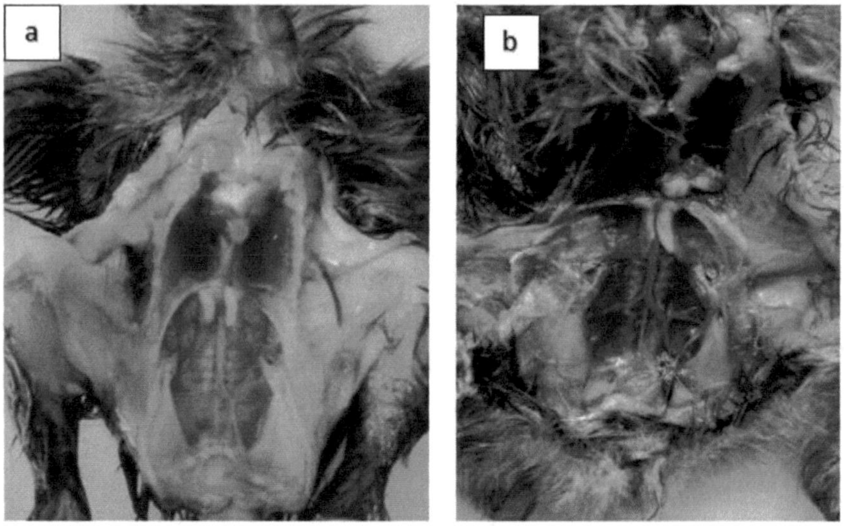

Fig. 1. Showing Gonads of Day-old Arabic chick a). Male chick ; b) Female chick.

3.2 Morphometric and Gross Anatomy of Day-old Arabic Chick Ovaries

Gross anatomy observation of the gonads showed that the shape of the ovary is irregular, and the left ovary has a more defined shape than the right ovary. The ovaries are close to the adrenal glands, where the right side is caudolateral of the adrenal gland, while the left side covers the ventral adrenal gland. The ovaries are located in the ventromedial mesonephros although the mesonephros is sometimes not visible due to regression.

In general, the ovaries of Day-old Arabic chick have an elongated shape with slightly curved laterally or towards the kidneys. The left ovary shows a curvature towards the right kidney with various shapes and uneven edges. The right ovary is smaller with an elongated and irregular shape. Almost all of the left adrenal glands are covered by the ovaries, and the right adrenal gland is more visible. The colour of the ovaries is creamy yellow, transparent pink, and sometimes grey. The surface of the left ovary appears smoother, slightly compact. This ovarian character is still similar to the morphology of the ovaries of 1-week old chickens, as described by [6]. The gross anatomy of the ovary of Day-old Arabic chick was shown in Figure 2.

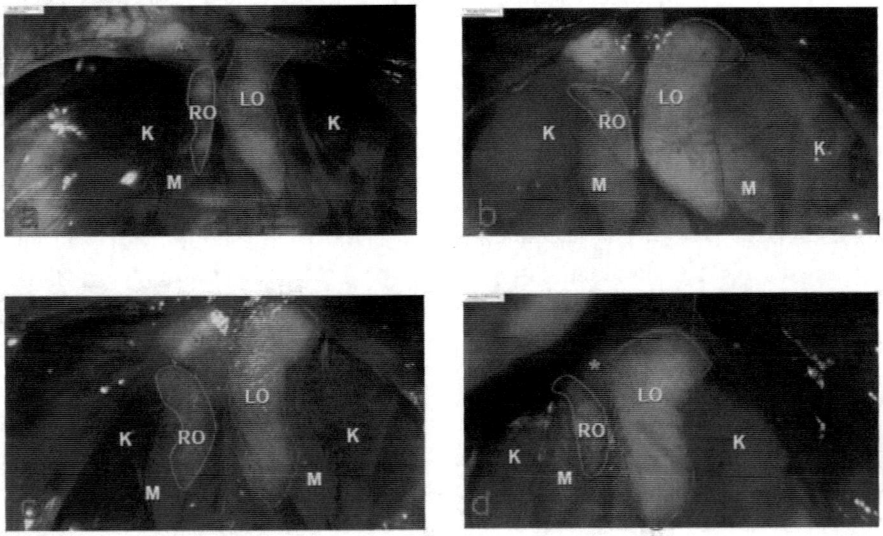

Fig. 2. Gross anatomy of ovaries Day-old Arabic chick. RO=Right ovary; LO=Left ovary; K=kidney; M=mesonephros; Blue (*) = Adrenal Gland. The images were taken using a Euromax SB 1903-P stereo microscope and a DC 5000P camera, magnification: 10x, scale:0.655 mm.

Figure 2, part (a) the shape of the right ovary: small elongated, the right mesonephros appears small, and the left mesonephros is not visible; (b). the left ovary is much larger, the right and left mesonephros are still visible clearly; (c) the right ovary is like a bean arch, the right, and left mesonephros are still visual even though they are smaller; (d). The right ovary is small, like a bean arch and the adrenal gland and the right mesonephros are not visible. The difference between the right and left ovaries: the left ovary is more elongated than the right ovary, anterior end of the left ovary is wider than the posterior end. The left ovary appears thinner than the right ovary.

Table 1. Ovaries Morphometric of the Day-old Arabic chick (Average ± SEM) and T-test result.

Characteristics	Right ovary	Left ovary
Length of ovary (mm)	1.67±0.08	3.69±0.05
Width of ovary (mm)	0.64±0.03	1.18±0.03
Longest ovary size (mm)	3.34	4.55
Shortest ovary size (mm)	0.66	2.23
Difference in length (left-right) of the ovary (mm)	0.97±0.09	
Difference in width (left-right) of the ovary (mm)	0.58±0.06	
The T-test between right and left ovary length	*P-value* <0.0001**	
The T-test between the width of the right and left ovaries	*P-value* <0.0001**	

Note: **: indicates a significant difference (P<0.01)

Table 1 shows that the length and width of the ovaries vary widely. T-test analysis showed that the size difference between the right and left ovaries was very significant (P<0.01). The size difference of the chick ovary results from the asymmetry process that has occurred since the embryonic phase. In domestic birds, the left and right ovaries begin to grow around 72 hours of initial incubation [13] and morphological differentiation of gonads occurred on day 6.5 of incubation [5]. In the early stages before sexual differentiation, the ovaries of chicken embryos develop symmetrically without showing significant left-right asymmetry. As the embryo develops, the right ovary in the medulla develops without a cortex. This cortical development is considered essential for developing the left ovary [13]. There is a size difference between the right and left ovaries, resulting in ovarian asymmetry on the 20th day of incubation [6] [14]. From this study, ovary asymmetry in Arabic chicken was identified after hatching. The hen has only one ovary and oviduct on the left side [5].

3.3 Morphometric and Gross Anatomy of Day-old Arabic chick testes

Gross anatomy observations showed the testes are located in the ventromedial mesonephros. The mesonephros has regressed from the embryonic phase at hatching. It appears as a small part at the bottom of the gonads. The invasion of primordial germ cells forms embryonic gonads into the coelomic epithelium that partially covers the mesonephros [15]. The cranial portion of the testes is located close to the ventral border of the lungs. In contrast, the caudal part is located in the cranial-ventral kidney.

The colour of the testes in Arabic chicken was creamy white, slightly pink, grey, and some were black in the left or right testicles, or even black in both right and left testes. The testicles of certain breeds of fowl may be mixed with black colour and white patches or maybe entirely greyish black due to the presence of the pigment melanin in the melanoblasts, in the connective tissue of the testes [14]. The gonads of sexually immature or resting birds (in seasonal birds) have yellowish-white to black or grey colouration, becoming moderately black in regressing testes and containing melanoblasts [14]. Testicular colour variability can be caused by sexual activity in testes and chicken's age. Bright yellow in juveniles and grey-white in adults of the quails and ostrich. The bright yellow colour in juveniles is due to the accumulation of lipids in the interstitial cells and the grey-white colour in adults due to the spread of interstitial cells by enlarging seminiferous tubules after sexual maturity [20]. The results of this study, most of the testicle's colour in DOC Arab chickens is yellow-white (creamy), the findings of the other colours besides bright yellow, still have to be confirmed histologically to support the determination of its quality.

The description of the location, shape, and colour of the Day-old Arabic chick testes is presented in Figure 3.

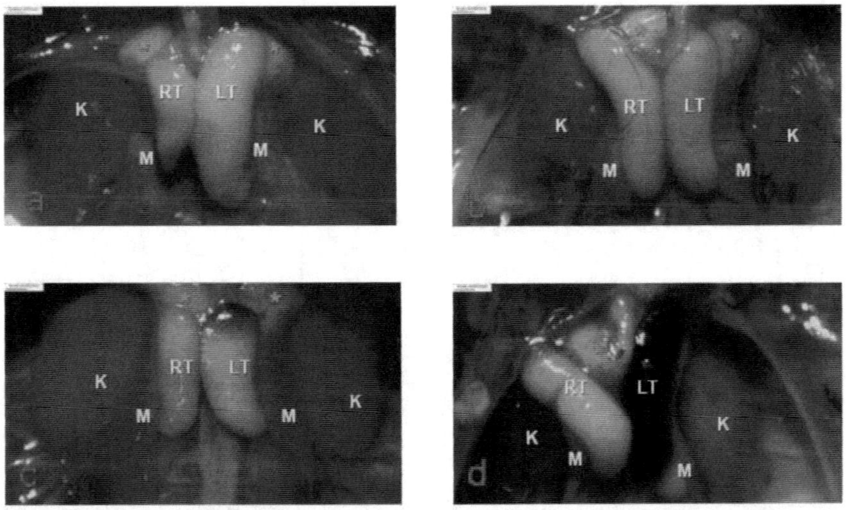

Fig. 3. Gross anatomy of Day-old Arabic chick testes. R= Right testes; LT= Left testes; K=kidney; M=mesonephros; Blue (*): Adrenal Gland. The images were taken using a Euromax SB 1903-P stereo microscope and a DC 5000P camera, magnification: 10x, scale:0.655 mm.

The testes of birds in this study are bean-shaped and elongated ovals, generally belonging to the category of an ellipsoid or prolate spheroid shape [8]. The form of the testes tends to be straight, lengthwise cranial and caudal, or slightly curved towards the kidney. Figure 3 shows that (a) the left testis is larger than the right testis, the right and left sides show clear mesonephros; (b) the size of the right testis is larger than the left testis; (c). The length of the right testis is almost the same as that of the left. However, the left testis is bigger and grey with a wider anterior end and black colour (d). the left testis is longer than the right testis, most of the left testis is darker in colour, and the anterior end of the right testis is grey. The results on Day-old Arabic chick, the number of detected black colour on the testes was 22.95%. Melanoblasts cause dark colour in the interstitium, so the testes are melanistic. Bicoloured testes are not considered gonads abnormal. The testes with bicolour do not show any histologic abnormalities in African flamingos [16]. In this study, the position of the right and left testes tends to be symmetrical, but there are differences in size. The morphometrics of Day-old Arabic chick is presented in Table 2.

Table 2 showed the size (length, width and volume) of left testis is larger than right testis. The results of the independent T-test analysis showed that the length and volume of the testes were significantly different between the right and left testes ($P<0.01$). In contrast, the testes' width was not significantly different between the right and left ($P>0.01$). This study showed that the number of Day-old Arabic chick with the left testis was larger than the right, namely 72.13%, and the number of Day-old Arabic chick with the right testis was larger than the left testis only 27.86%. It has been described in previous study that the left testis larger than the right in birds commonly, but there are exceptions in some species where there is no difference in size, and the right testicle is larger than the left testis [14]. The left testicle is larger than the right testis in Gallus domesticus is 63% in birds and 57%.

The right testis is larger than the left testis, only 26.5%, while both testes are the same in 8.2% of birds [8].

Table 2. Testes Morphometric of Day-old Arabic chick (Average ± SEM)

Characteristics	**Right testis**	**Left testis**
Length of the testis (mm)	2.47±0.08	3.29±0.07
Width of testis (mm)	0.98±0.03	1.04±0.02
The volume of the testis (mm³)	1.46±0.12	2.03±0.13
Difference in length (left-right) of the ovar (mm)	0.82±0.08	
Difference in width (left-right) of the ovar (mm)	0.05±0.03	
Volume total of testis (mm³)	3.49±0.22	
T-test between right and left testis length	P-value 0.001 **	
T-test between right and left testis width	P-value 0.194*	
T -test between the volume of right and left testis	P-value 0.002**	

Note: * = not significant difference; **= a significant difference (P<0.05).

Testicular asymmetry may be caused by irregular migration of the Primordial Germ Cells (PGC) to the left and right sides during embryonic development. This testicular asymmetry has been reported that the right gonadal cortex loses its chemotactic attraction to PGCs. Thus, it leads to the left gonadal cortex on day 3th of embryogenesis [14]. Similar results were reported by [11] and [18] that the number of PGCs in the left testis is more than in the right testis, and in the chicken embryos on the 6th day of incubation, the size of the left testis relatively larger. Despite the lack of a relationship between male quality and the degree of testicular asymmetry, but several studies have found that testicular asymmetry increases with age [19]. The results of this study can be used as a basis for Day-old chick testicular morphometrics to be compared with the age afterward in the same type of chicken.

4 Conclusion

Studies of Gonad morphometrics in Day-old Arabic chick have shown gonadal asymmetry, left ovary are larger than the right ovary. The size and volume of the left testis are larger than the right testis. Gross anatomy and evidence of gonadal asymmetric in Arabic chicken can be used as reference when observing the gonads and as information on the gonad's development at hatching. Further research is needed to reveal the histological and molecular asymmetry, and the observation of gonadal size at a later age, which can be associated with the quality of the chicken.

Thank you to the Head of the Agricultural Human Resources Extension and Development Agency, Ministry of Agriculture as the scholarship provider. The Director of Politeknik Pembangunan Pertanian Malang has been permitted to use research facilities, especially the Poultry Livestock Installation and Reproduction Laboratory.

References

1. R. Hartawan, N.L.P.I. Dharmayanti, HAYATI J. Biosci. **23**, 160–167 (2016).
2. T. Sartika, K. Diwyanto, S. Iskandar, *Mengenal plasma nutfah ayam Indonesia dan*

pemanfaatannya, (Balai Peneliti Ternak, Sukabumi, Indonesia, 2007).

3. G. Indra, Achmanu;, A. Nurgiartiningsih, Junal Ternak Tropika. **14,** 8–14 (2013).
4. Iswati, M.H. Natsir, G. Ciptadi, T. Susilawati, Journal of Animal & Plant Sciences, **44**, 7708–7716 (2020).
5. S. Guioli, S. Nandi, D. Zhao, J. Burgess-Shannon, R. Lovell-Badge, M. Clinton, Sex. Dev. **8**, 227–242 (2014).
6. J.D.L. Mfoundou, Y.J. Guo, M.M. Liu, X.R. Ran, D.H. Fu, Z.Q. Yan, M.N. Li, X.R. Wang, Poult. Sci. **100**, 101191 (2021).
7. R. Tóth, B. Lázár, Á. Südy, A. Nagy, A. Kidane, M. Anand, E. Gócza, N. Biotechnol. **33**, S212 (2016).
8. G.R. Graves, The Auk. **121**, 473 (2004).
9. N.A. Bachmid, F.Y. Purba, A.M.S. Apada, D.K. Sari, I*OP Conf. Ser. Earth Environ. Sci*. **343** (2019).
10. A.K. Okpe Godwin Chidozie, Nwatu Ugochukwu, Anim. Res. Int. **7,** 1163–1168 (2010)
11. S. Intarapat, O. Satayalai, Anat. Res. Int. **2014**, 1–9 (2014).
12. A. Dawson, J. Avian Biol. **34**, 19–123 (2003).
13. Y. Ishimaru, T. Komatsu, M. Kasahara, Y. Katoh-Fukui, H. Ogawa, Y. Toyama, M. Maekawa, K. Toshimori, R.A.S. Chandraratna, K.I. Morohashi, H. Yoshioka, Development. 135 (2008) 677–685.
14. B.G.M. Jamieson, *Reproductive Biology and Phylogeny of Birds* (CRC Press, Boca Raton, Florida, United States, 2011).
15. C.G. Scanes, *Sturkie ' s Avian Physiology* (Academic Press, San Diego, United States, 2015).
16. L. Crosta, H. Gerlach, M. Bürkle, L. Timossi, Exot. Anim. Pract. **6,** 57–83 (2003).
17. S. Intarapat, C.D. Stern, J. Poult. Sci. **51,** 352–358 (2014).
18. T.K. TriBudi, T. Kostaman, Majalah Ilmiah Peternakan, **23,** 107–112 (2020).
19. T.R. Birkhead, K.L. Buchanan, T.J. Devoogd, E.J. Pellatt, T. Székely, C.K. Catchpole, Song, Anim. Behav. **53** 965–971 (1997).
20. P. Dharani, S. Ushakumary, V. Sundaram, C. Joseph, G. Ramesh. Int. J. Morphol. **36**, 909-914 (2018).

The relationship of estrus character before double dose artificial insemination in Limousin Crossbred cow

Putri Utami[1], *Muhammad Zainul* Hanif[1], *Aulia Puspita Anugra* Yekti[2], *Joko* Riyanto[3], *Chusnul* Hanim[4], and *Trinil* Susilawati[2*]

[1]Post Graduate Programe Faculty of Animal Science, Universitas Brawijaya, Malang, Indonesia
[2] Faculty of Animal Science, Universitas Brawijaya, Malang, Indonesia
[3] Faculty of Agriculture, University of Sebelas Maret, Surakarta, Indonesia
[4] Faculty of Animal Science, University of Gadjah Mada, Yogyakarta, Indonesia

Abstract. Ongole cattle in smallholder farms on Java, Indonesia, have been crossed with Limousin, which is not controlled repeatedly. The character of estrus affects the success of Artificial Insemination. This study aimed to determine the relationship between estrus characters before double dose artificial insemination using frozen semen sexing. The research was conducted in Senggreng Village, Sumber Pucung District, Malang Regency. The material was 45 head of Limousin crossbred cow inseminated using sexing semen produced by the Singosari National Artificial Insemination Center (SNAIC). The sexed semen processed with the Percoll Density Gradient Centrifugation (PDGC) method. The average body value of cows between 3-5 (scale 1-9) with an average age of 1.8-7 years, normal reproduction, clear signs of estrus. The reproductive organs were examined by rectal palpation to confirm that the cows are not pregnant. Parameters of estrus character were vulvar color, cervical mucus, vaginal temperature, pH, vulvar swelling, and heat detector value. The result showed that the estrus characters, including vulva color, vulva swelling, cervical mucus, vaginal temperature, cervical mucus pH, and heat detector score in the first and second artificial insemination have a low relationship. The regression models produced can be used but are less precise to estimate the relationship between estrus characters because the coefficient of determination shows a low influence.

1 Introduction

Beef cattle on smallholder farms have been crossed with limousine or Simental cattle repeatedly with a low success rate of pregnancy and high service per conception, which is more than two times insemination to conceive successfully [1]. Artificial insemination (AI) is one of the reproductive technologies widely applied in smallholder farms to increase

* Corresponding author: tsusilawati@ub.ac.id

cattle productivity by utilizing superior male spermatozoa, thereby increasing the number of offspring quickly [2].

The timing of AI is one of the factors that influence the success of pregnancy in livestock. The appearance of signs of estrus becomes a guideline for artificial insemination. Human resources, especially breeders and inseminators, need to detect estrus in one period of the estrus cycle so that they can determine the best time for AI [3]. Detection and observation of estrus character are one essential factor in determining the success of pregnancy. Cows in estrus will show characters such as a reddened, swollen vulva, increased body temperature, slimy (mucus or mucus coming from the cervix is transparent or clear in color). When a male ride it, it will remain silent [4]. The accuracy of detection is related to the time of artificial insemination. Lack of farmers' knowledge in detecting estrus can affect the delay in insemination time, resulting in pregnancy failure. The success of mating pregnancy can be evaluated through the reproductive efficiency of female cattle. To assess the success of pregnancy, it can be known through several parameters, namely the conception rate (CR) and pregnancy rate (PR) [5]. Based on the description above, it is essential to know the relationship between estrus characters, which can later be used to determine the time of Artificial Insemination.

2 Materials and methods

The study was conducted from March 5th 2021, to June 20th 2021. The post-thawing spermatozoa quality test was carried out at the Reproductive Laboratory of the Faculty of Animal Husbandry, Universitas Brawijaya, Malang and the AI in Senggreng Village, Sumber Pucung District, Malang Regency, East Java. The material used in this study was 45 female Limousin Crossbred cow which were artificially inseminated using sexed semen produced by Singosari National Artificial Insemination Center (SNAIC). The material were selected based on the average body value between 3-5 (scale 1-9) with age 1.8-7 years old had a health condition, showed signs of heat, and their reproductive organs were examined by rectal palpation so that the acceptors were confirmed to be in unpregnant.

The method used in this study is an experimental method with a field experiment. The sampling technique used the purposive sampling method. The artificial insemination technique is done with 4+ deposition (cornua uteri). Insemination used a double dose at the 2nd (first AI) and 8th hours (second AI) after estrous, and injection of "Rheinbio" as BioATP+. Observations of estrus characters included vulvar color, vulvar swelling, cervical mucus, vulvar temperature, cervical mucus pH, and heat detector values. The observation process of the estrus character was carried out before the first and second artificial insemination.

The data obtained from the study results were the estrus character of livestock before the first and second artificial insemination. Data were analyzed using the simple linear regression correlation analysis method using Microsoft Excel software. Correlation analysis was used to find the direction and strength of the relationship between two variables. In contrast, regression analysis predicts how far the value changes between variables when it changes up or down. Simple linear regression analysis was used to find a regression line between two variables, namely X and Y, described in coordinate points [6].

Table 1. Description of Estrus Characteristic

Parameters	Description
Vulva color (reddening vulva)	The appearance of vulva are assessed by a modified score consisting of: evenly red with a score of 3, uneven red (pink) with a score of 2, and pale with a score of 1 [7].
Swelling vulva	Swelling of the vulva was assessed by a modified score consisting of: very swollen with a score of 3, slightly swollen with a score of 2, and not swollen with a score of 1 [7].
Cervical mucus	Cervical mucus is assessed by a modified score consisting of: There is a lot of mucus with a score of 3, there is a little mucus with a score of 2, and the mucus is dry or none with a score of 1 [8].
Vaginal temperature	Vaginal temperature before AI was measured using a thermometer [7]
pH	Measurement of the pH of cervical mucus is done using litmus paper which will be expressed in the range 1-12 [7]
Heat detectors	Heat detectors the measurement results of the heat detector value will appear in the form of numbers 0-60Ω [9].

3 Results and discussion

3.1. Relationship Between Cervical Mucus and Heat Detector Value Before The First And Second AI

Table 2. Percentage of cattle on each character of criteria servical mucus and heat detector value before first AI

Condition servical mucus before first AI	**Head (%)**	**Heat detector value in first AI**	**Head (%)**
Abundant	3 (6,67)	20-29	45 (100)
Moderate	34 (75,56)	30-40	0 (0)
Low/Absent	8 (17,78)		
Total	45 (100)	Total	45 (100)

Table 3. Percentage of cattle on each character of criteria servical mucus and heat detector value before second AI

Condition servical mucus before second AI	**Head (%)**	**Heat detector value in second AI**	**Head (%)**
Abundant	18 (40,00)	25-29	43 (95,56)
Moderate	24 (53,33)	30-40	2 (4,44)
Low/Absent	3 (6,67)		
Total	45 (100)	Total	45 (100)

Based on the analysis of variance results, it can be concluded that the linear regression equation of the relationship between cervical mucus and value of heat detectors in the first and second AI is not significant (P>0.05). The regression analysis results showed that the correlation coefficient (r) of cervical mucus and the value of cow heat detector in the first and second artificial insemination were -0.158 and -0.087, which had a low level of relationship. The direction of the resulting relationship is negative. The coefficient of determination was 2.508%, meaning that the heat detector value was determined by cervical mucus in the first artificial insemination and the second by 0.763%. In comparison, 97.492% and 99.237% were determined by other factors. The regression analysis results using the significance level (0.05) indicate that this regression model can be used. Still, it is not appropriate to estimate the impacts of the relationship between cervical mucus and heat detector because the results of the coefficient of determination show a common effect. The estradiol from mature de Graff follicles in estrus conditions would cause changes in the maximum tubular reproductive tract. The condition of the uterus becomes tense, and cervical mucus secretion increases. High levels of the hormone estrogen affect vaginal vasodilation so that the volume of cervical mucus secreted by goblet cells will increase. The appearance of cervical mucus changes during estrus. These changes include a transparent appearance at the beginning of estrus, gradually cloudy in the middle of estrus, and clear and shiny at the end of estrus [10]. Cervical mucus has relatively high conductivity, so that the volume of cervical mucus has an increasing effect on the lower electrical resistance produced [11]. The lower electrical resistance means that the value that appears on the screen of the heat detector shows a small number ranging from 30-40 [9]. The numbers displayed on the screen of the heat detector indicate that the volume of cervical mucus produced is relatively large. The hormone estrogen level in the blood increases, and the cattle are in the estrus phase.

3.2. Relationship Between Cervical Mucus and Vaginal Temperature Before The First And Second AI

Table 4. Percentage of cattle on each character criteria of servical mucus and vaginal temperature before first AI

Condition servical mucus before first AI	Head (%)	Vaginal temperature before first AI	Head (%)
Abundant	3 (6,67)	37-38,5	33 (73,33)
Moderate	34 (75,56)	>38,5	12 (26,67)
Low/Absent	8 (17,78)		
Total	45 (100)		45 (100)

Table 5. Percentage of cattle on each character criteria of servical mucus and vaginal temperature before second AI

Condition servical mucus before second AI	Head (%)	Vaginal temperature before second AI	Head (%)
Abundant	18 (40)	37-38,5	25 (55,56)
Moderate	24 (53,33)	>38,5	20 (44,44)
Low/Absent	3 (6,67)		
Total	45 (100)		45 (100)

Based on the analysis of variance results, it can be concluded that the linear regression equation of the relationship between cervical mucus and vaginal temperature in the first and second artificial insemination was not significant ($P>0.05$). The analysis results showed that the level of closeness between cervical mucus and vaginal temperature in the first and second AI was -0.164 and 0.045. The direction of the relationship produced by cervical mucus in the first AI was negative, and in the second AI was positive. The regression analysis results using the significance level (0.05) indicate that this regression model can be used. Still, it is not appropriate for estimating the relationship between cervical mucus and vaginal temperature because the coefficient of determination shows a common effect. The coefficient of determination was 2.703%, meaning that the vaginal temperature was determined by cervical mucus in the first AI 2.703% and the second at 0.205%, while other factors decided 97.297% and 99.795%.

Environmental factors are one of the factors that influence changes in vaginal temperature. Temperature changes could be caused by increased livestock activity during estrus because GnRH and estradiol influence the central regulation of body temperature. In addition, housing management can also have an effect so that the cattle's space is limited and results in a significant increase in body temperature during estrus. However, infection or unhealthy livestock conditions, especially after calving, can affect the temperature observed around estrus. An increase in temperature in livestock under unsanitary conditions can be mistaken for an increase in estrus. High levels of the hormone estrogen during estrus cause the blood vessel network to multiply by developing the reproductive tract. At the same time, estrogen increases blood flow to the reproductive tract, resulting in a high vulvar temperature [12]. Cow's vagina in each estrus phase showed a difference. The diestrus phase is 37.40°C, and the proestrus phase is 37.86°C; in the estrus phase, the highest temperature is 38.39°C, and the metestrus phase is 38.05°C. At the time of estrus, more increased blood flow flows into the reproductive tract, which causes the temperature to increase. The highest average vaginal temperature was obtained at the time of estrus. Changes in vaginal temperature occur due to the vascularization process involving the hormone estrogen. The blood vessels become full, accompanied by thermal flowing along with the increase in vaginal temperature. In addition, the consistency of cervical mucus varies in each cycle, and the cervix provides the energy (glucose) needed for the movement of spermatozoa after they are separated from the semen. Cervical mucus during estrus, besides being a medium for spermatozoa to reach the site of fertilization, also functions as a selector for spermatozoa [13].

3.3. Relationship Between The Color Of The Vulva and Ph of Cervical Mucus Before The First and Second AI

Table 6. Percentage of cattle on each character criteria of the vulva color and pH cervical mucus before first AI

The color of the vulva before first AI	Head (%)	pH of cervical mucus before first AI	Head (%)
Evenly red	15 (33,33)	6-7	20 (44,44)
Uneven red (pink)	19 (42,22)	7,1-8	25 (55,56)
Pale	11 (24,44)		
Total	45 (100)	Total	45 (100)

Table 7. Percentage of cattle on each character criteria of the vulva color and pH cervical mucus before second AI

The color of the vulva before second AI	Head (%)	pH of cervical mucus before second AI	Head (%)
Evenly red	19 (42,22)	6-7	20 (44,44)
Uneven red (pink)	12 (26,67)	7,1-8	25 (55,56)
Pale	14 (31,11)		
Total	45 (100)	Total	45 (100)

Based on the analysis of variance results, it can be concluded that the linear regression equation of the relationship between the color of the vulva and the pH of cervical mucus in the first and second AI was not significantly different (P>0.05). The regression analysis results showed that the correlation coefficient (r) between the color of the vulva and the pH of cow's cervical mucus in the first and second inseminations was 0.023 and 0.100. The direction of the resulting relationship between the two in the first and second AI was positive. The coefficient of determination showed that the pH value of cervical mucus was determined by the color of the vulva in the first and second AI was 0.053% and 1.002%, while other factors decided 99.947% and 98.998%.

The degree of acidity (pH) is an essential parameter of cervical mucus in sperm transfer into the reproductive tract of female cattle, and pH affects the viability of sperm in the cervix. During estrus, the cervix secretes mucus which has a pH of 6.6-7.5 in cattle [14]. The difference in the pH value of cervical mucus in each animal is influenced by biophysical and biochemical conditions of the mucus produced by the cervix, controlled by hormones that play a role in the estrus cycle [15]. The higher the estrogen hormone in the blood during estrus, the volume of cervical mucus will increase with an alkaline pH. The hormone estrogen affects the increase in mucoprotein, sodium chloride, and cervical mucus levels, which change the pH of cervical mucus during estrus [16]. The alkaline pH of cervical mucus affects spermatozoa's motility, viability, and penetration power [17]. That each estrus cycle stage produces a different pH value. The presence of estrogen indicates that the ovarian cycle is functioning and the discharge of cervical mucus. Estrogen levels are closely related to cervical mucus, where the pH value of cervical mucus at the peak of estrus is more alkaline. This is because increased levels of the hormone estrogen affect sodium chloride levels that increase. The pH value of cervical mucus is closely related to the transport medium of spermatozoa. The pH value range of 7.0-8.5 is the optimal condition for the viability and motility of spermatozoa, and if the pH value is below 6, it will decrease the motility of spermatozoa [18].

3.4. Relationship Between The Color of The Vulva and Heat Detector Value Before The First and Second AI

Table 8. Percentage of cattle on each character criteria of the vulva color and heat detector value before first AI

The color of the vulva before first AI	Head (%)	Head detector value before first AI	Head (%)
Evenly red	15 (33,33)	20-29	45 (100)
Uneven red (pink)	19 (42,22)	30-40	0 (0)
Pale	11 (24,44)		
Total	45 (100)	Total	45 (100)

Table 9. Percentage of cattle on each character criteria of the vulva color and heat detector value before second AI

The color of the vulva before second AI	Head (%)	Head detector value before second AI	Head (%)
Evenly red	19 (42,22)	25-29	43 (95,56)
Uneven red (pink)	12 (26,67)	30-40	2 (4,44)
Pale	14 (31,11)		
Total	45 (100)	Total	45 (100)

Based on the analysis of varian result, the relationship between the color of the vulva and the heat detector in the first AI was not significant, and the second AI was significantly different ($P<0.05$). The regression analysis results showed that the correlation coefficient (r) between the color of the vulva and the heat detector of cows in the first and second AI was -0.174 and -0.311. The direction of the resulting relationship between the two characters in the first and second AI was negative. The coefficient of determination value was 3.002% and 9.672%, which means that the heat detector value was determined by the color of the vulva 3.002% and 9.672%, while other factors determined 96.998% and 90.328%.

Another factor that affects the value of the heat detector is the concentration of mucus. Changes in the heat detector value or low electric current resistance in vaginal mucus in estrus cattle conditions are caused by high hormone estrogen levels, resulting in vasodilation in the vagina and an increase in the number and volume of ions secreted by the vagina. These ions are electrolytes and have reasonably high conductivity, so that the more ions, the lower the electrical resistance produced [11]. Cervical mucus consists of protein, fat, cholesterol, sorbitol, reducing sugar, sodium, potassium, calcium, organic phosphorus, and chlorine [19]. The appearance of clear mucus in cattle under estrus is directly proportional to the low resistance. The low value of the resulting inhibition may be due to an increase in estrogen, which causes vasodilation in the vagina [17]. Cervical mucus can prevent the entry of microbes into the female reproductive tract and facilitate the transportation of spermatozoa. Cervical mucus changes from a small volume to become cloudy and thick in mid or late estrus. Variations in color, consistency, and volume of cervical mucus indicate the physiological properties of cervical mucus change during the estrus cycle [8].

3.5. Relationship Between The Color of The Vulva and Swelling Vulva Before The First and Second AI

Table 10. Percentage of cattle on each character criteria of the vulva color and swelling vulva before first AI

The color of the vulva before first AI	Head (%)	Swelling vulva before first AI	Head (%)
Evenly red	15 (33,33)	Very swollen	26 (57,78)
Uneven red (pink)	19 (42,22)	Slightly swollen	19 (42,22)
Pale	11 (24,44)	Not swollen	0 (0)
Total	45 (100)	Total	45 (100)

Table 11. Percentage of cattle on each character criteria of the vulva color and swelling vulva before second AI

The color of the vulva before second AI	Head (%)	Swelling vulva before second AI	Head (%)
Evenly red	19 (42,22)	Very swollen	26 (57,78)
Uneven red (pink)	12 (26,67)	Slightly swollen	19 (42,22)
Pale	14 (31,11)	Not swollen	0 (0)
Total	45 (100)	Total	45 (100)

Based on the analysis of variance results, it can be concluded that the linear regression equation of the relationship between vulvar color and vulvar swelling in the first and second AI was not significant ($P>0.05$). The regression analysis results showed that the correlation coefficient (r) between vulvar color and vulvar swelling of cows in the first AI was 0.101 and the second AI was 0.271. The direction of the resulting relationship between the two characters is positive. The coefficient of determination values were 1.013% and 7.335%, meaning that the value of vulvar swelling was determined by the color of the vulva 1.013% and 7.335%, while other factors determined 98.987% and 92.665%. The regression analysis results using the significance level (0.05) indicate that this regression model can be used but is not appropriate to estimate the impacts of the relationship between vulvar color and vulvar swelling because the coefficient of determination shows a common effect.

Elevated levels of the hormone estrogen in the blood have been linked to changes in color and swelling of the vulva. The increase in the level of the estrogen hormone tends to occur when the cattle are in estrus. The hormone estrogen will stimulate the thickening of the vaginal wall and increase vascularization so that the vulva will swell and change color to reddish [20]. The condition of the hormone estrogen continues to increase when cattle estrus stimulates blood flow to the reproductive tract and related genital organs. This condition causes the blood vessel network to multiply according to the development of the reproductive tract. At the same time, estrogen increases blood flow to the reproductive tract [21]. The presence of males as stimulators also influences the sexual behavior of female cows. The company of male cattle can provide a biostimulant effect on female cows, which affects sexual maturity and can accelerate the emergence of estrus postpartum [22].

4 Conclusion

The estrus characters, including vulvar color, vulvar swelling, cervical mucus, vaginal temperature, cervical mucus pH, and heat detector in the first and second artificial insemination, have no relationship. The regression models produced can be used but are less precise to estimate the relationship between estrus characters because the coefficient of determination shows a low influence.

Thank you to the Ministry of Research and Technology through the Indonesian Collaborative Research Program (PPKI) between University of Brawijaya, University of Gadjah Mada, and University of Sebelas Maret for providing funding for this research.

References

1. K. Kastalani, H. Torang, A. Kurniawan, Jurnal Ilmu Hewani Tropika (Journal of Tropical Animal Science), **8** (2020)
2. T. Susilawati, S. Rahayu, S. Udrayana, H. Sudarwati, E. Nugroho, American-Eurasian Journal of Sustainable Agriculture, **8** (2014)
3. J. Labetubun, I.P. Siwa, F. Reressy, Agrinimal, **8** (2020)
4. A. Rachmawati, Ismaya, B.P. Widyobroto, S. Bintara, T. Susilawati, Jurnal Ilmu-Ilmu Peternakan, **28** (2018)
5. T. Susilawati, L. Wahyudi, N. Isnaini and A.P.A. Yekti, The 6th International Seminar on Tropicel Production Integrated Approach in Developing Sustainable Tropical Animal Production **20** (2015)
6. H. Sudarwati, M.H. Natsir, V.M.A. Nurgiartiningsih. 2019. Statistika dan Rancangan Percobaan Penerapan Dalam Bidang Peternakan. Malang: UB Press.
7. D. Saputra, S. Sumartono, dan N. Humaidah, Dinamika Rekasatwa. **2**(2018)
8. A. Murtaza, M.I.U.R. Khan, M. Abbas, N. Hameed, W. Ahmad, I. Mohsin and M.Z. Tahir, Animal reproduction science, **213** (2020)
9. F.A. Annashru, M. N. Ihsan, A. P. A. Yekti and T. Susilawati, Jurnal Ilmu-Ilmu Peternakan. **27** (2017)
10. M.N. Ihsan, *Ilmu Reproduksi Ternak Dasar*, Malang : UB Press. ISBN. 978-6028960-00-7 (2020)
11. K.K.Verma, S. Prasad, A. Kumaresan, T.K. Mohanty, S.S. Layek, T.K. Patbandha and S. Chand, Veterinary World, **7**(2014)
12. F. Randi, M. McDonald, P. Duffy, A.K. Kelly and P. Lonergan, Theriogenology, **110** (2018)
13. A. Furqon, Dinamika Rekasatwa, **1** (2016)
14. T. Tsiligianni, G.S. Amiridis, E. Dovolou, I. Menegatos, S. Chadio, D. Rizos and A.

Gutierrez-Adan, The Canadian Journal of Veterinary Research, **75** (2011)
15. W.A. Prasdini, S. Rahayu and M.S.Djati, International Journal of ChemTech Research, **7** (2015)
16. A. Rizki, P. Srianto, E. Suprihati, T. Sardjito, I. Ismudiono and M.A. Al Arif, Ovozoa, **8** (2019)
17. S.S. Layek, T.K. Mohanty, A. Kumaresan, K. Behera and S. Chand, Livestock Science, **152** (2013)
18. P.I. Sitaresmi, B.P. Widyobrot, S. Bintara and D.T. Widayati, IOP Conference Series: Earth and Environmental Science **387** (2019)
19. T.H. Tsiligiann, A. Karagiannidis, P. Brikas and P.H. Saratsis, Theriogenology, 55 (2001)
20. R.D. Frandson, W. L. Wilke and A. D. Fails, *Anatomy And Physiology Of Farm Animal.7th Edition*, Lippincottwilliams And Wilkins. Philadelphia (2009)
21. S.C. Scolari, S.G. Clark, R.V. Knox and M.A. Tamassia, Journal Of Swine Health And Production, **19** (2011)
22. E. Baliarti, M.H. Ali, T.S.M. Widi, D.E. Yulianto, B.A. Atmoko, H. Maulana, J. Effendhy, P.W. Prihandhini and D. Pamungkas, IOP Conference Series: Earth and Environmental Science, **465** (2011)

Role of high quality feed supplement additional to the success of artificial insemination using sexing semen on Limousin Crossbred Cow

Muhammad Zainul Hanif[1], *Putri* Utami[1], *Rizki* Prafitri[2], *Chusnul* Hanim[3], and *Trinil* Susilawati[2,*]

[1] Post Graduate, Faculty of Animal Science, Universitas Brawijaya, Malang 65145, Indonesia
[2] Faculty of Animal Science, Universitas Brawijaya, Malang 65145, Indonesia
[3] Faculty of Animal Science, Gadjah Mada University, Yogyakarta 55281, Indonesia

Abstract. The success of artificial insemination must be supported by a balance of nutrients for livestock. AI combined with sexing produces offspring according to wishes of the breeder. This study aims to determine the increase in concentration on the success of Artificial Insemination using frozen semen sexing. This research was conducted at the people's beef cattle farm, Sumber Pucung District, Malang Regency. The limousine crosser used was 30 cows. Cattle aged 1.8 – 5 years with BCS 3-5 (score 1-9) Treatment T0 : The usual feed given by farmers (Control). T1: Control+HQFS. Parameters NRR-1, NRR-2, Conception Rate (CR), Pregnancy Rate (PR). The additional feeding of concentrate did not significantly affect the reproductive parameters. The results showed that at T0 the results of NRR-1 and NRR-2 were 80% and 73.33%, respectively. while in T1 it was 73.3% and 66.67%. The results showed that the CR in both treatments was 26% and the results showed that the PR for T0 was 26% and for T1 was 33%. Insufficient feed consumption for all treatments based on NRC requirement data, T0 and T1 dry matter consumption respectively 6.6 and 7.3 Kg/head/day, Protein consumption 0.36 and 0.3 Kg/head/day and TDN consumption of 3.08 and 3.20. kg/head/day.

1 Introduction

The increase in meat consumption in Indonesia continues to increase, to fulfill the needs meat in Indonesia the government must import meat from abroad to fulfill the needs in Indonesia, not only that, a program to increase the amount of livestock in Indonesia must be implemented one of which is the use of productive females with artificial insemination [1]. The artificial insemination program not only increases the livestock population but also helps breeders in obtaining superior offspring, besides that it can help breeders in determining the sex of the offspring, the technology used in determining the sex of the offspring is sexing Spermatozoa [2]. Sexing Spermatozoa is a reproductive technology by separates X and Y spermatozoa [3]. Wahyudi *et al* [4] sexing treatment of spermatozoa

* Corresponding author: tsusilawati@ub.ac.id

resulted in a decrease in the quality of spermatozoa seen from membrane damage and decreased motility of spermatozoa. This is what makes the basis for artificial insemination using frozen double doses of semen sexing or two straws for one time AI with more and more doses inseminated will increase the number of spermatozoa so that it is expected to increase the chances of higher pregnancy success. The results of Yekti *et al* [5] showed that AI using double dose sexing semen had a value of CR = 68.75% and the results of research by [3] have a value of CR = 55.56%.

High-Quality Feed Supplement (HQFS) is a high-quality concentrated supplement that is a combination of protein bypass with high energy source material in the form of total digestible nutrients (TDN) and a mixture of minerals and macro minerals to meet the nutritional needs of livestock. HQFS also contains vitamins and probiotics as well as essential oils. HQFS feed is formulated from corn grain, corn gluten meal (CGM), pollard, soy beans Meal(SBM), and wheat flour which is first processed in compact pellet form. The finished pellets were then mixed with an Agromix amplifier (Mineral mix Ca 26.45%, P 0.62%, K 0.22%, Mg 0.26%, Na 4.70%, S 681.7 ppm, Fe 1,44%, Zn 300.2 ppm, Cu 229.7 ppm, Se 0.541 ppm, and Co 7.7 ppm, vitamins A, D, E, and K, probiotics (Lactobacillus acidophilus, Enterococcus faecium, Saccharomyces cerevisiae) and essential oils processed from walnuts, olives, soybeans, lavender, eucalyptus, coconut, orange, peppermint and sesame seeds), HQFS consists of 81.9% CM, 90.3% OM, 19.56% CP, 10.15% CF, 0.66 % CF and 75% TDN Astuti *et al* [6]. The higher the nutritional content of the feed provided increases the success rate of artificial insemination with semen sexing so that it will produce male offspring with AI success parameters including Non-Return Rate (NRR), Conception Rate (CR), and Pregnancy Rate (PR).

2 Materials and methods

This research was conducted on March 15th – June 20th, 2021, The Nutritional Content Test of High-Quality Feed Supplement was carried out at the Nutrition and Animal Feed Laboratory of Brawijaya University, and the Implementation of Artificial Insemination (AI) was carried out in Senggreng District, Malang Regency, East Java. The research material used was 30 Limousin Crossing Cow, 15 cows in each treatment with The concentrated feed (HQFS) from Widodo Makmur Cianjur.

The method used in this research was the experimental method with field experiments. The sampling technique was using the Purposive Sampling method with the criteria of cow aged 1.8-5 years based on permanent incisors, normal reproduction, BCS 3-5 with a scale of 1-9, not yet AI done again after giving birth. The sample of this study used 30 Limousine crossed cows that had not been AI after giving birth, with different feeding treatments for each treatment, namely:

T0: Control
T1: Control + Concentrated Feed (HQFS)
*Control : Feed From farmers

Table 1. The result nutrition of proximate in High quality feed supplement.

Dry Matter %	Organic Matter %	Crude Protein %	Crude Fiber (%)	Extract ether (%)	TDN (%)
86.63	93.52	10.99	10.11	2.00	61.37

The stages of the research include 1) Selection of the cow as acceptors 2) feeding according to the treatment 3) on day 14 of treatment the feed is synchronized using PGF 2 alpha with the brand Lutelyse if it shows estrus, we will do AI 4) Thawing straws with tap water temperature for 30 seconds 5) AI is carried out based on observations of signs of estrus that appear in livestock and is supported by using a heat detector to ensure that cow is ready for AI. AI was carried out using frozen semen, Sexing in double doses, at the 2nd and 8th hours, semen deposition in the 4+ position (deep insemination) According to Susilawati [1] Semen deposition in the 4+ has a higher success rate of pregnancy when compared to position 4, and then injected with bio-ATP with the brand "Rheinbio" which was carried out after the AI was carried out. 6) After one cycle day 19 – 22, NRR-1 is observed, if it shows signs of estrus in AI return, using the same AI method and on day 38 – 42, NRR-2 is observed, then after more than 30 days, pregnancy examination is carried out using Ultrasound.

2.1 Measurement of body weight (BW)

Measurement of body weight of livestock using the method of estimation of body weight by measuring the chest size (CS) using the Dalton tape which is a chest size measuring device that already has an estimated body weight in kilograms. Measurement of body weight to determine the condition of the livestock body and the nutritional needs of livestock. The formula for calculating body weight estimation is as follows:

$$CW = \frac{(CS+22)^2}{100}$$

Description:
CW: Cow Weight
CS: Chest Size
Periembawe *et al* [7]

1. Body Condition Score (BCS):

Observation of BCS in livestock is an indicator to determine the nutritional status of livestock. Observations were made by observing 6 observation points and doing palpation, namely at the base of the tail (Tail head), the pelvis (Rump), pelvis (Hip Bone), sitting bones (Pin Bone), spine (Back Bone) and ribs (Ribs) by categorizing thin for BCS (1-4), ideal (5-7) and fat (8-9) [8]

2. Non-Return Rate (NRR) including NRR1 and NRR2

The NRR value was obtained by observing the acceptors on days 18 – 21 and days 40 – 42 after AI was implemented. NRR is the percentage of cattle that do not show signs of estrus after AI and can be assumed to be pregnant cattle [9].

$$NRR = \frac{number\ of\ AI\ cattle - number\ of\ returning\ AI\ cattle}{number\ of\ AI\ cattle} \times 100\%$$

3. Conception Rate (CR)

Conception Rate (CR) is the percentage of pregnant cattle in AI in the first AI and is also called the conception rate [10] with the following formula:

$$CR=\frac{number\ of\ pregnant\ cattle}{number\ of\ AI\ cattle}\text{ x }100\%$$

4. Pregnancy Rate

Pregnancy rate is the percentage of pregnant cow based on rectal palpation compared to the number of cow in AI [11]

$$\text{Pregnancy Rate}=\frac{number\ of\ pregnant\ cattle}{total\ of\ female\ cattle\ breed}\text{ x }100\%$$

2.2 Intermediate Variable:

Feed Intake

The feed used by breeders as acceptors is weighed in the morning and evening and weighing the rest of the provision in the morning before giving in the morning by the breeder, weighing the provision and the rest of the feed aims to determine the amount of feed consumption in livestock for 1 day in acceptor livestock.

$\sum$Feed Consumption = $\sum$ Feeding -$\sum$ Letover Feed

Dry Matter Intake(DM)

Calculation of crude materials consumption with the following formula:

$$\text{Dry Matter Intake} = \frac{dry\ matter}{100}\text{x feed consumption}$$

Calculating protein consumption by means of the percentage of protein content multiplied by the consumption of the CM feed, with the following formula:

$$\text{Crude protein consumption}=\frac{crude\ protein}{100}\text{x Dry matter consumption}$$

Periembawe *et al* [7]

Data obtained from the results of research on the success of AI include No-Return Rate (NRR), Pregnancy Rate, Conception Rate (CR), DM, CP, And TDN consumtion analyzed using descriptive, Acceptor performance research includes body weight (BW), Body Condition Score (BCS) was analyzed using T-test.

3 Results and discussion

3.1 Cow performance

Table 2. Cow performance

Treatment	Average BCS	Body weight
T0	3.87 ± 0.83	324.20 ± 33.91

T1	3.67 ± 0.62	341.00 ± 53.93

In this research, each treatment consisted of 15 cows, based on table 2. in treatment T0 and T1 had an average BCS = 3.87 ± 0.83 and 3.67 ± 0.62, there was no significant difference ($P > 0.05$). While treatments T0 and T1 had a bodyweight of 324.20 ± 33.91 Kg and 341.00 ± 53.93 Kg. Between treatments, no significant difference was shown ($P > 0.05$)

Body Condition Score (BCS) in a cow is closely related to nutritional adequacy status in livestock and will affect the condition of the reproductive organs according to Michael *et al.* [12] The BCS of a cow during childbirth is an important factor to determine the timing of estrus again after further calving. Sarah *et al.*[13] cows with a BCS of 4 – 6 have a better pregnancy rate than those with a lower BCS, This is related to the adequacy of nutrition in cows to produce hormones, reproduction, and meet the production needs of livestock.

Bodyweight in livestock is crucial to note because it illustrates the contribution of the quantity and quality of feed consumed by the cow, according to Sarah *et al.* [14] the increase in body weight has a relationship with livestock fertility, bodyweight at the beginning of the mating period immensely affects physiological and anatomical development of the reproductive organs to achieve reproductive success in livestock.

3.2 Feed Intake

Based on the average dry matter consumption at T0 and T1 were 6.6 and 7.3 kg/head/day, while the dry materials requirements based on NRC were 8.0 and 8.5 kg/head/day. Based on these data, it shows that the average dry materials consumed are lower or insufficient than they need based on the NRC at T0 and T1. The condition of insufficient dry ingredients consumed by livestock is influenced by one of them by the condition of feed that is around the research location which is mostly only straw.Survival of livestock, especially in the provision of quality forage for livestock. Hersom *et al.*[15] said that dry ingredients consumption in livestock is influenced by cow body weight, the quality of available forage (energy and protein concentration), and BCS, a cow with large BCS and body weight have the potential to consume greater forage than a cow with low body weight.

Based on the average crude protein consumption at T0 and T1 were 0.36 and 0.3 kg/head/day, while the crude protein requirements based on NRC were 0.89 and 0.9 kg/head/day. Based on these data, it shows that the average crude protein consumed is lower or insufficient than they need based on the NRC at T0 and T1. This is conveyed by Funston [16] that the intake of livestock feed will decrease when the content of feed consumed is low, the intake of forage will lower which is related to decreased microbial activity in the rumen, whereas if the crude protein is high, the intake of forage consumed will be higher. Funston [17] stated that excessive and deficient feeding has an impact on livestock reproduction if overfeeding leads to low pregnancy success and protein deficiency leads to weak estrus, low conception, and premature birth.

Based on the average consumption of Total Digestible Nutrient at T0 and T1 were 3.08 and 3.20 kg/head/day, while the need for Total Digestible Nutrient based on NRC were 5.2 and 5.5 kg/head/day. Based on these data, it shows that the average Total Digestible Nutrient consumed is lower or insufficient than they need based on the NRC at T0 and T1. According to Indah *et al.* [18], the TDN content will increase if the CP and CF in the feed ratio increase, according to Ayuningsih *et al.* [19] cows experiencing protein and energy deficiencies will have an impact on reproductive failure and loss of body weight. Lack of energy consumption can cause livestock to experience delayed ovulation according to Negussie [20] which states that livestock that experiences slow uterine involution and has low fertility rates are caused by the availability of energy in the livestock's body that is low or insufficient for the process of forming reproductive hormones. High energy availability is caused by high consumption of TDN by livestock so that livestock will use sufficient energy for productivity, basic life, and reproduction which will activate the LH hormone which functions for the follicle formation process [21]

3.3 The successful pregnancy results from artificial insemination using double doses

Table 3. Results of Observing the Success of IB Non-Return Rate 1, Non-Return Rate 2, Conception Rate (CR) and Pregnancy Rate (PR)

Treatment	NRR-1 total (%)	NRR-2 total (%)	Conception Rate Head (%)	Pregnancy Rate Head (%)
T0	80	73,33	(4) 26	(4) 26
T1	73,33	66,67	(4) 26	(5) 33

The results NRR1 and NRR2 at T0 observations are greater than T1. Based on Table 1 above, there was a decrease in the value from NRR 1 to NRR2 in treatment T0 and T1, namely T0 decreased from 80% to 73.33% and T1 from 73.33 to 66.67%. The decrease in the value of NRR 1 to NRR 2 may be due to the occurrence of an unknown heat by the breeder, the occurrence of silent heat (or calm heat) experienced by livestock according to Salim *et al.* [22] stated that the low NRR value might have been caused by lack of observations made by breeders so that breeders are late in providing information to inseminator officers for AI, in addition to that, an embryonic death can cause a decrease in the NRR value as well. Non-Return Rate (NRR) is one of the assessments of the success of AI by observing cows that are not in heat at NRR-1 19-21 days post-IB and NRR-2 39-41 post-AI according to Yus *et al.* [23] stating that NRR as Observations to assess the success of AI are often carried out as indicators of livestock fertility, age factors, environmental factors, and cage management affect livestock fertility.

Maintenance management is one of the provisions of feed by breeders. Providing adequate nutrition for female livestock is extremely essential in supporting the reproductive performance of female livestock. An adequate-protein consumption in female livestock will affect the performance of FSH and LH hormones which will begin to show signs of heat until pregnancy occurs according to Martin *et al.*[24] stated that feeding with high protein causes higher pregnancy rates compared to livestock that consumes low protein. The age factor of livestock affects the readiness of livestock reproductive organs for implantation, Saili *et al.* [25] the older the age, the higher chance of decrease in the ability to reproduce, in cows aged 3 – 4 years have a higher NRR value of cow aged 5-6 years, namely 100% and 67%.

The results of observations on the CR and PR values are very low, namely the CR T0 and T1 values are 26%, and the PR values at T0 are 26% and T1 are 33%, according to Ihsan *et al.* [26] stating that the condition of livestock in Indonesia with crosses that have Many have done and the natural conditions of Indonesia, which is a subtropical area, are said to have good CR when it reaches 45% 50%. The low value of CR is influenced by many factors, one of the factors in this study is the use of sexing cement which is produced from a long process, this is conveyed by Yekti *et al.* [27]. In males, sexing semen has lower quality than non-sexing semen because during the sexing semen process it goes through a long series and damages spermatozoa physically and physiologically Bonneya *et al.* [28] the centrifugation process reduces the motility of spermatozoa and damages the acrosome and membranes. Pregnancy examination was carried out by rectal palpation and using ultrasonography which was carried out when 2 months of age after AI was reached, according to Szelenyi *et al.*[29] stated that ultrasound examination of pregnancy is better done at gestational age above 28 days this is done to facilitate identification of the fetal heart). The low pregnancy rate is also caused by AI acceptors found in underdeveloped follicles, and ovarian hypofunction, the condition of these reproductive organs is influenced by reproductive hormones, one of which is the quality of feed provided by breeders. Meteer *et al.* [30] Feeding inadequate quantity and quality leads to reduced hormone and reproductive function in livestock, livestock that lacks feed experienced a decrease in body weight accompanied by a decrease in glucose, resulting in the minimization release of GnRH from the hypothalamus leading to a reduction in LH concentration needed for ovulation and fat where both is a precursor of the hormone progesterone and estrogen that helps to maintain pregnancy as well as signs of estrus.

4 Conclusion

The results show that at T0 of NRR-1 and NRR-2 are 80% and 73.33% respectively. while in T1 it is 73.3% and 66.67%. It is proven that the CR in both treatments was 26% and the outcome of the PR for T0 is 26% and for T1 is 33%. Feed consumption for all treatments is not sufficient based on NRC needs data, dry matter consumption T0 and T1 were 6.6 and 7.3 Kg/head/Day, protein consumption is 0.36 and 0.3 Kg/head/day and TDN consumption is 3.08 and 3.20. Kg/head/day. With the additional feed, as much as 1kg of concentrate has not affected reproductive parameters exceedingly, therefore no research is needed on the provision of concentrate on the success of the pregnancy.

We thank the University of Gadjah Mada and the University of Sebelas Maret for joining the Indonesian collaborative research program (Kemenristek)

References

1. T. Susilawati, Jurnal Ternak Tropika (Journal of Tropical Animal Production), **12** (2011)
2. T. Susilawati, *Sexing Spermatozoa*, 2014.
3. M. T. Fernanda, T. Susilawati, N. Isnaini, Jurnal Ilmu Ilmu Peternakan (Indonesian Journal of Animal Science), **24** (2014)
4. L. Wahyudi, T. Susilawati, N. Isnaini, Jurnal Ternak Tropika (Journal of Tropical Animal Production), **15** (2014)
5. A. P. A. Yekti, E. A. Octaviani., Kuswati., T. Susilawati, Jurnal Ternak Tropika (Journal of Tropical Animal Production), **20** (2019)
6. A. Astuti, A. Agus, D. S. Priyono, S. Budhi, Buletin Peternakan, **33** (2009)
7. D. K. A. Periembawe, R. Sutrisna, Liman, Jurnal Ilmu Peternakan Terpadu, **4** (2016)
8. E. E. Dan, F. B. Milyssa, B. H. John., E. D. Richard, Virginia Coorperative Extension, (2009)
9. M. R. Jainudeen, E. S. E. Hafez, *Cattle and Buffalo in Reproduction* in *Farm Animals* ed By E. S. E. Hafez and B. Hafez, (2020)
10. H. Ayres, R. M. Ferreira, J. R. S. T. Junior, C. G. B. Demetrio, M. F. S. Fihlo, L. U. Gimenes, L. Penteado, M. J. D. Occhio, P. S. Baruselli, Theriogenology, **82** (2014)
11. D. P. Berry, R. D. Evans, S. Mc. Parland, Theriogenoloy, **75** (2011)
12. J. D. Michael, S. B. Pietro, G. Campanile, Theriogenology, **125** (2019)
13. E. D. Sarah, F. E. Michelle, K. A. Lisa, B. E. Joshua, N. W. Bailey, W. D. Paul, P. R. Soren, H. B. Fernando, Journal of Animal Science and Biotechnology, **10** (2019)
14. E. M. Sarah, H. B. Fernando, Journal of animal science and biotechnology, **11** (2020)

15. Hersom, J. Matthew, *Basic Nutrient Requirements Of Beef Cows*, EDIS : 1-12 (2007)
16. Funston, Rick, The State Of Beef Conference, (2014[a])
17. Funston, Rick, *The AABP Proceedings*, **47** (2014[b])
18. S. A. Indah, L. G. Permana, Despal, Sains Peternakan, **18** (2020)
19. B. Ayuningsih, I. Hernaman, D. Ramdani, Siswoyo, Jurnal Ilmiah Peternakan Terpadu, **6** (2018)
20. T. Nigussie, J. Dairy Res. Tech., **1** (2018)
21. E. Mulyanti, F. K. Keraf, Jurnal Sain Peternakan Indonesia, **16** (2019)
22. A. Salim, A. P. A. Yekti, K. Kuswati, T. Susilawati, Ternak Tropika (Journal of Tropical Animal Production), **19** (2018).
23. E. Yus, F. J. Dieguez, Pesq Agropec Brasilia, **46** (2011)
24. J. L. Martin, K. A. Vonnahme, D. C. Adams, G. P. Lardy, R. N. Funston, J. animal science, **85** (2014)
25. T. Saili, L. Ode Baa, L. O. A. Sani, S. Rahadi, I. W. Sura, F. Lopulalan, Jurnal Ilmu Ternak, **16** (2016)
26. M. N. Ihsan, S. Wahjuningsih, Jurnal Ternak Tropika, **12** (2011)
27. A. P. A. Yekti, T. U. Kurniaesa, N. Isnaini, Kuswati, T. Susilawati, J. Ilmu Peternakan, **28** (2018)
28. Boneya, Galma, International Journal of Advanced Research in Biological Sciences, **8** (2021)
29. Z. Szelényi, O. Szenci, L. Kovács, Garcia-Ispierto, Animals, **11** (2021)
30. W. C. Meteer, D. W. Shike, F. C. D. Cardoso, Acta Scientiae, **43** (2015)

Effect of addition level of *Pandanus amaryllifolius* Roxb. on Broiler performances at starter period

Eko Widodo[1,*], *Jein Rini* Leke[2], *Khusnul Teguh* Pangestu[3], *Aulia* Hidayatul[4], *Dhimas Rodho* Purnomo[4] and *Muhamad* Farhan[4]

[1] Faculty of Animal Science, University of Brawijaya, Jl. Veteran, Malang 65145 – Indonesia
[2] Faculty of Animal Science, University of Sam Ratulangi, Bahu, Manado 95115 – Indonesia
[3] Magister Student at Faculty of Animal Science, University of Brawijaya, Jl. Veteran, Malang 65145 – Indonesia
[4] Undergraduate Student at Faculty of Animal Science, University of Brawijaya, Jl. Veteran, Malang 65145 – Indonesia

Abstract. *Pandanus amaryllifolius* Roxb. leaves are commonly used as food additives in many countries. Beside to improve taste, it is recognized to contain various compound groups namely alkaloid, saponins, flavonoids and tannins that have a role in antibacterial activity. However, the current research aim was to examine effect of using *Pandanus amaryllifolius* Roxb. leaves as feed additive in broilers. The method used was experiment, employing 6 treatments included P0: control feed; P0: control feed added with 0.1% zinc bacitracin; P1: control feed added with 0.5% *Pandanus amaryllifolius* Roxb. leaves powder; P2: control feed added with 1.0% *Pandanus amaryllifolius* Roxb. leaves powder; P3: control feed added with 1.5% *Pandanus amaryllifolius* Roxb. leaves powder; P4: control feed added with 2.0% *Pandanus amaryllifolius* Roxb. leaves powder. The variables measured were feed consumption, body weight gain and feed conversion ratio. The results indicated that in all variables no significant results were reported. Those indicated that addition of *Pandanus amaryllifolius* Roxb. leaves powder in broiler feed did not affect growth performances of broiler. It is concluded that the use of *Pandanus amaryllifolius* Roxb. leaves powder as feed additive did not change growth of broiler, probably due partly to its antibacterial effect.

1 Introduction

In the era of non-antibiotic removal in the poultry feed has encourage the elaborative use of antimicrobial agents from many different plants to maintain good productivity of animal. *Pandanus amaryllifolius* Roxb. leaves are commonly used as food additives in many Asian countries. It is commonly found in the coastal area. Beside to improve taste, it is recognized to contain various active compound groups namely alkaloid, saponins, flavonoids and tannins that have a role in antibacterial activity.

* Corresponding author: eko.widodo@ub.ac.id

The use of natural antimicrobial products to replace the use of antibiotics has been a current trend to fine safer product for the consumers. The use of *Pandanus amaryllifolius* which contain active substance such as alkaloid might be effective antimicrobial agent. Research in Indonesia reported by [1] reported *Pandanus amaryllifolius* showed antibacterial effect towards some dandruff bacteria. Other Research in Indonesia reported by [2] reported *Pandanus amaryllifolius* showed antibacterial effect against *Salmonella thypii*. In addition, research by [3] reported *Pandanus amaryllifolius* showed antibacterial effect towards *Candida albicans*. But, there has been limited report with regard to the use of *Pandanus amaryllifolius* meal as feed additive for poultry. The only report is research carried out by [4] in Thailand. The result indicated that under open house system the use of 5g/day improved 11% of egg production of laying chicken, but no improvement was obtained for poultry house equipped with evaporative cooling system. However, the current research aim was to examine effect of using *Pandanus amaryllifolius* Roxb. leaves as feed additive in broilers.

2 Materials and methods

2.1 Materials

Materials used were 270 one-day broiler chicks, concentrate feed and other feedstuffs were purchased from local poultry shop in Malang. The chicks were distributed to 6 different treatments and each treatment had 5 replications, of 9 chicks each. They were fed and provided water drink *ad libitum*. Raising period of the chicks was until 3 weeks old.

2.2 Method

The method used was experiment. The treatments used included

- P0 : basal feed without antibiotic
- P0+ : basal feed added 0.1% antibiotic
- P1 : basal feed + pandanus fluor 0.5%
- P2 : basal feed + pandanus fluor 1.0%
- P3 : basal feed + pandanus fluor 1.5%
- P4 : basal feed + pandanus fluor 2.0%

Each bird was fed ad libitum. Water was provided *ad libitum*. The basal feed used was a commercial concentrate feed produced by PT New Hope, Sidoarjo. The starter feed were adjusted by mixing concentrate feed with fish meal, soybean meal, moringa leaf meal, mineral and palm oil, to suit the nutrient requirement set by [5]. The variables included feed consumption, body weight gain and feed conversion ratio.

2.3 Statistical Analysis

Data were subjected to one-way analysis of variance. If significant effect exists, then continued by least significant different test.

3 Results and discussion

The results indicated that feeding treatments significantly ($P<0.05$) influenced feed consumption, body weight gain dan feed conversion ratio. On the basis on feed consumption, it showed that feeding 2% *Pandanus amaryllifolius* (P4) had similar level with antibiotic given group (P0+) and significantly higher that other treatments. Consequently, body weight gain of P0+ group was significantly ($P<0.05$) higher than other treatment. The summary results of research in indicated in Table 1.

Table 1. Effect of gradual inclusion of *Pandanus amaryllifolius* in feed on Broiler Starter Performance

Treatment	Feed consumption (g/bird)	Body weight gain (g/day)	Feed conversion Ratio
P0	$1034^a \pm 17$	$585^a \pm 15$	$1.77^a \pm 0.06$
P0+	$1077^b \pm 26$	$607^b \pm 36$	$1.78^a \pm 0.12$
P1	$1050^a \pm 38$	$557^a \pm 39$	$1.87^b \pm 0.07$
P2	$1044^a \pm 45$	$549^a \pm 35$	$1.91^b \pm 0.12$
P3	$1034^a \pm 29$	$565^a \pm 22$	$1.83^b \pm 0.10$
P4	$1072^b \pm 14$	$570^a \pm 42$	$1.89^b \pm 0.12$

The active substance especially flavonoid of *Pandanus amaryllifolius* leaf is a bit low, which was 1.46 g/100g much lower as compared to neem leaf (2.44 g/100g) or bitter lef (2.58 g/100g [6]. Presumbaly, the effect of active substance in *Pandanus amaryllifolius* leaf might be too low to influnce physiological changes of broiler chicks. The highest level of *Pandanus amaryllifolius* (P4) failed to attain high body weight gain might be attributable to the higher fiber content of the leaf of *Pandanus amaryllifolius*. The current reserach analyzed protein and crude fiber contents of *Pandanus amaryllifolius* which was 9.315 and 41.71%, respectively. However, [7] reported their findings that type of fiber influence the digestibility of nutrient of poultry, of which dietary lignin was more effective in increasing protein digestibility and therefore significantly increased weight gain than cellulose and hemicellulose. When the results were compared to control groups (P0 and P0+), broiler weight gain and feed conversion were impaired by any levels of inclusion of *Pandanus amaryllifolius* in the feed. Since there is no report yet on the fiber composition of the leaf of *Pandanus amaryllifolius*, it is speculated that the leaf contains more cellulose dan hemicellulose. This fiber content might also cause to lower protein digestion and as a result increase the feed conversation ratio. However, as the poultry grow older it could improve the ability to digest nutrients [8], or the use of enzyme might further improve performance of broiler [9]. Therefore, further direction of research must consider the use of older birds and or in combination with enzymes.

4 Conclusion

From the point of production, *Pandanus amaryllifolius* inclusion might impair production performances of broiler.
From the point of antibiotic removal, *Pandanus amaryllifolius* might not be able to effectively act as antimicrobial agent.

References

1. A. Ambarwati, T. A. Sujono, R. Sintowati, *The 3rd University Research Colloquium* 2016: 222-228 (2016).
2. E. Suwandi, S. Sugito, Jurnal Laboratorium Khatulistiwa, **1** (2017).
3. P. H. R. Elifas, Y. Nugraha, S. Y. Astuti, *Proceeding Book*, 656 (2018).
4. S. Sanguanphan, Agriculture and Natural Resources, **44** (2010).
5. National Research Council. *Nutrient requirements of poultry*, (1994)

6. Ofongo, R. T., Ohimain, E. I., Iyayi, E. A. European Journal of Medicinal Plants, **32** (2021).
7. M. Adibmoradi, B. Navidshad, M. Faseleh Jahromi, Italian Journal of Animal Science, **15** (2016).
8. A. B. Batal, C. M. Parsons, Poultry Science, **81** (2002).
9. S. Eidivandi, A. Mahdavi, F. Alemi, S. A. Hosseini, A. Mostafa Tehrani,V. Ghasemloo, Veterinary Researches & Biological Products, **24** (2011).

Biological test of compost made of rabbit faeces and the addition of banana peels on the vegetative growth of *Morus alba* and *Arachis pintoi*

Nur Cholis[1,2,*] and *Ita Wahju* Nursita[1,2]

[1]Department of Animal Production, Faculty of Animal Science, Universitas Brawijaya, Malang 65145, Indonesia
[2] Research Group of Animal Waste and Compost Technology, Faculty of Animal Science, Universitas Brawijaya, Malang 65145, Indonesia

Abstract. The purpose of this research was to determine vegetative growth performance of *Morus alba* and *Arachis pintoi* with different percentages of compost mixture (rabbit feces and *kepok* banana (*Musa paradisiaca* Linn) peels) and soil as media. The materials used was 72 *Morus alba* stacks, rabbit faeces, *kepok* banana peels, and *Azotobacter* culture as decomposer. This research was conducted by Completely Randomized Design with 4 treatments and 6 replications. The treatments were T_0 (100% soil), T_1 (20% of compost and 80% of soil), T_2 (40% of compost and 60% of soil), and T_3 (60% of compost and 40% of soil). The data was analysed using analysis of variance, followed by Duncan's Multiple Range Test (DMRT) if the results are significantly different. The treatment gave higher leaves number, roots weight, and roots length ($P<0.05$) in *Morus alba*. Same also on plant length, branches number and wet weight ($P<0.05$) of *Arachis pintoi*. The conclusion of this research is that the application of organic compost based on rabbit manure and banana peel with *Azotobacter* microbial culture decomposer can improve the growth performance of *Morus alba* and *Arachis pintoi*. The best level was T_1 with 20% of compost and 80% of soil.

1 Introduction

The number of rabbits in East Java Province in 2019 reached 370,107 heads. This figure has increased by 1.56% from 2018 which could reach 367,983 heads. This makes East Java Province ranked 2nd with the largest number of rabbit populations in Indonesia [1]. Public awareness with a healthy lifestyle has made rabbit meat popular because of high protein but low cholesterol meat. The protein and fat content of rabbit meat is still better than chicken meat (20.8% vs 18.2%). Rabbit fat content is 10.2% while chicken fat is 25%. This makes rabbit farmers compete to increase rabbit meat production. The increase in the number of

* Corresponding author: ncholis@ub.ac.id

rabbits was followed by an increase in the feces produced. Adult rabbits weighing 1 kg produce 28 grams of feces per day [2].

Rabbit feces contain better nitrogen, phosphorus, and potassium than other livestock such as cattle, sheep, buffalo, horses, and chickens [3]. Tumimbang *et al.* [4] stated that rabbit manure has a total N value of 2%, P_2O_5 1.5%, K_2O <1%, and organic C- 10%.

Bananas are one of the most abundant fruits, especially in the tropics. Based on data by Abdullah *et al* [5], the *Kepok Tanjung* banana has a yield potential of 20-30 tons/ha/year. Banana is one of the fruits that people are interested in because it has good benefits, especially for health. However, the consumption of bananas will leave waste in the form of banana peels. The weight of the banana peel waste is quite a lot, which is about ⅓ of an unpeeled banana [6].

The selection of *Kepok* (*Musa paradisiaca* Linn) banana peels was because the higher nitrogen content than that of *Ambon* bananas and *Raja* bananas. According to Sriharti and Takiyah [7] nitrogen in *Kepok* banana peels is 1.34% while *Ambon* bananas 0.21% and *Raja* bananas 0.07%. Rabbit faeces and *Kepok* banana peels are wastes that still contain good nutrients for plants after going through the composting process.

Pintoi beans (*Arachis pintoi*) are one of the most palatable food crops for livestock, have high nutritional value and have several functions, namely as feed for both ruminants and non-ruminants, increasing soil fertility and preventing erosion [8]. The nutritional content of *Arachis pintoi* is protein 17.06%; crude fibre 17.96%; 3.54% fat; BETN 53.43% and ash 8.01%.

Mulberry (*Morus alba*) is an animal feed plant that is generally used as feed for silkworms. *Morus alba* has the characteristics of pinnate leaves with jagged edges. The development of mulberry planting patterns needs to be improved because mulberry leaves are the only food for silkworms to be able to produce cocoons which will become silk fibers. However, big and small ruminants also like mulberry leaves.

This study aims to determine the ability of the composted media from rabbit faces with the addition of *Kepok* banana peel waste in the vegetative growth of *Morus alba* and *Arachis pintoi* plants, and to determine the best proportion in their use.

2 Materials and methods

The materials used in this study were rabbit faeces *Kepok* banana peel waste (*M. paradisiaca* Linn), *Azotobacter* culture decomposer, river sediment soil, *Morus alba* and *Arachis pintoi* plants.

About 97.2 kg of rabbit faeces were obtained from rabbit breeders in Arjosari. *Kepok* banana peel waste as much as 32.4 kg was obtained from the *Kepok* banana chips factory which is located at Jln. Mentaraman Gang VII Talok Turen Malang 65175. Banana peel waste which had previously been dried under the sun and then chopped. Microbial culture was purchased online and used as much as 525 mL.

The composting equipment consists of a large straw chopper, 150 L plastic bucket, portable hanging scales 0-10kg/5g, 10-50kg/10g, litmus pH indicator on a scale of 1-14, trash bag plastic sacks, mercury thermometer. Planting media equipment consists of polybags measuring 35 cm x 30 cm, measuring 0-2 meters, digital scales 0-5 kg scale, folio paper and scissors

The equipment for making the decomposer solution consists of a 26 L plastic bucket, 1 L measuring cup and a stirrer. The decomposer solution was prepared by mixing 525 mL of *Azotobacter* culture (trademark MA110) with 2% sugar and 26.25 L of water.

Mulberry seedlings are stem cuttings with a diameter of 1-1.5 cm and a minimum of 5 buds and have been grown for 1 month before being planted in the treatment planting medium. *Arachis pintoi* plants with an age of approximately 30 days were uniformed by trimming so that they had a length of around 8 cm. Compost composition is 75% rabbit feces and 25% banana *Kepok* waste [9].

The study was conducted using a Completely Randomized Design/CRD experiment consisting of 4 treatments and 6 replications. The treatments were T_0 (100% soil), T_1 (20% compost + 80% soil), T_2 (40% compost + 60% soil), and T_3 (60% compost + 40% soil). The study began with composting rabbit faces and *Kepok* banana peel waste for 8 days with the help of the *Azotobacter* decomposer. The composted fertilizer was then applied to the *Morus alba* plant to observe its vegetative growth for 29 days. Observations were made on the leaves number, root weight, and plant root length.

The composted fertilizer was then applied to *Arachis pintoi* plants to observe their vegetative growth for 29 days. Observations were made on plant height, branches number, and plant fresh weight. Planting the tested plant seeds on the planting media begins with opening the polybag of plant seeds using a sharp knife so that the lumps of soil in the polybag are not damaged. After the seeds are placed in the holes in the planting medium, the surrounding soil is loosened and compacted. Plant seeds are selected that are uniform.

The planting environment is shaded using para-net or black nets to provide shade for plants of about 40%. Watering is done twice a day, giving water as much as 250ml/polybag/day. Watering is done 1 week after planting, if there are plants that die, they are replaced with new plants with the same age and height as before, the goal is that the plants can grow uniformly. Weeding is done when there are weeds that interfere with plants. Weeding is done once a week or according to weed development. The data obtained were then analysed by Analysis of Variance and continued with Duncan's Multiple Distance Test to determine the best treatment.

3 Results and discussion

3.1 Physical properties of compost

Composting was carried out for 8 days. Table 1 below informs the results of the color, texture, and odor of the compost during the study. In general, the mature compost has physical characteristics such as blackish brown color and smells close to the smell of earth. This is because the material it contains already resembles soil material.

Table 1. Compost color, texture, and smell during the study

Days to	Color	Smell	Texture
1.	Light brown	Banana peel	Mushy
2.	Light brown	Sour and banana peel	Mushy
3.	Light brown	Sour and banana peel	Mushy
4.	Dark brown	Sour and banana peel	Mushy
5.	Dark brown	Sour and banana peel	Mushy
6.	Dark brown	Sour and banana peel	Mushy
7.	Dark brown	Smell of earth	Mushy
8.	Dark brown	Smell of earth	Crumb

Blackish brown color formed by the influence of stable organic matter. The final form of composting does not resemble the original form because it has been destroyed due to natural decomposition by microorganisms that live in the compost. The texture at the beginning of composting tends to be soft until the last day the compost texture when you hold it no longer sticks to your hands (crumbs). This is in accordance with Ayunin *et al.* [10] that the mature compost is blackish brown in color, smells like soil and has a crumb texture. The size of the compost pile becomes smaller than at the beginning of composting which indicates that there is degradation activity by various bacteria in the compost [11].

3.1.1 Compost temperature

Figure 1 shows the development of the compost temperature during the process. On the first day of composting (October 22, 2020) the temperature was 28°C, on the second day the temperature rose to 30.67°C, then on the third day to day 8 (October 8, 2020) the temperature gradually dropped slowly.

Fig. 1. Graph of temperature development compost

The increase in temperature occurs because the decomposition activity carried out by bacteria will produce heat and the temperature will rise until it reaches the maximum temperature, after that the temperature will decrease like the initial temperature [12]. It is supported by Irawan [13] that in the early stages of composting there are mesophilic bacteria. These bacteria live at a temperature of 10–40°C. The decrease in temperature during the composting process is in line with the decrease in the number of bacteria due to reduced nutrients and the compost returns to its initial temperature [14]. *Azotobacter* is a mesophilic bacterium that can live at a temperature of 20-30°C.

3.1.2 Compost pH

The results of the observation of pH during the composting process can be seen in Figure 2. During the observation of fluctuating pH, the first day the pH showed acid, then on the second day the pH began to rise to a peak point on the third day of 7.5, then gradually decreased to 6 and increased until the pH neutral. According to Supadma and Arthagama [15] at the beginning of composting, the pH tends to be acidic due to the formation of organic acids, then the pH increases due to protein decomposition and the release of ammonia. The rise and fall of pH is also a sign of bacterial activity in decomposing organic matter [16].

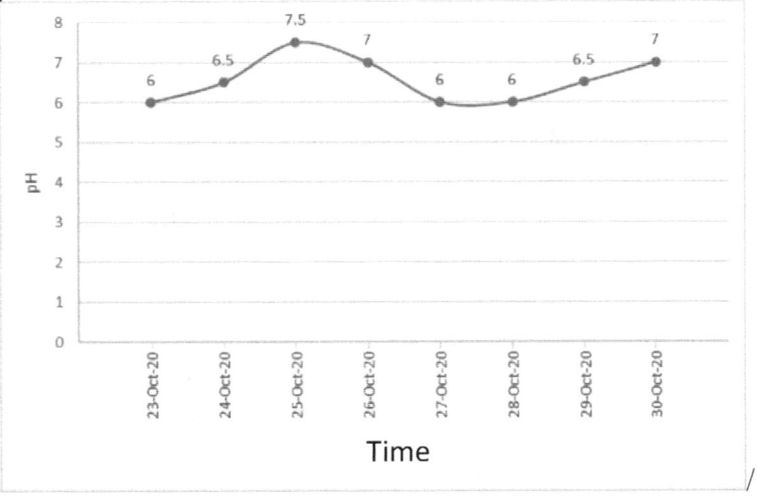

Fig. 2. Compost pH development

3.2 Compost nutrient content

The nutrient content and chemical properties of the compost are known after the samples were tested in the UPT Lab for the Development of Agribusiness for Food Crops and Horticulture in Bedali – Lawang as shown in Table 2.

T_1 has the highest water content, namely 21%, followed by T_2 16%, then T_3 14% and pure compost 17%. This happens because of the influence of plant soils that have high water content.

Table 2. Compost nutrient content

Element	SNI*	Compost	T_0	T_1	T_2	T_3
% Water	Max 50	17	21	21	16	14
C-org	9,8	24.6	1.5	9	9	6
C/N	Min 10%	23.21	-	10	10	7-79
% N	Min 0.4	1.06	-	0.9	0.9	0.77
% P_2O_5	0.1	1.05	-	0.72	0.8	0.55
% K_2O	0.2	3.22	-	0.95	1.5	2.05
OM	27-58	42.39	2.6	15.5	15.5	10.3

According to Mayasari *et al.* [17], the water content of the soil is 43.55%. Each soil has a different water content depending on the condition of the plant soil. MA-11 is composed of *Rhizobium* sp. bacteria combined with bacteria taken from the rumen of cows, namely cellulolytic bacteria, proteolytic bacteria, and amylolytic bacteria which are tasked with breaking down cellulose so that it is easily consumed by *Rhizobium* sp. bacteria which are active in binding free nitrogen.

Analysis of P content in compost has exceeded SNI 19-7030-2004, compost only (T_0) 1.05; T_1 0.72; T_2 0.8, and T_3 0.55 where the standard is 0.1. Ratna *et al.* [18] said that in the composting process some phosphorus is absorbed by microorganisms to form proteins in the body. The more microorganisms, the faster the compost will ripen.

K content in compost is quite high, pure compost only (T_0) has a level of 3.22, far above SNI 19-7030-2004 with a minimum K content of only 0.2. This is influenced by the addition of *Kepok* banana peels in the composted material. Akbari *et al.* [19] said that based on laboratory tests at the University of North Sumatra, it was known that the K_2O element content in the kepok banana peel was 1.478%.

3.3 Compost biological test

3.3.1 Morus alba

Table 3 presents the results of the biological test of rabbit faeces compost with the addition of *Kepok* banana peels on *Morus alba* vegetative growth.

Table 3. Leaves number (sheets), root weight (g), and root length (cm) of *Morus alba*

Variable	T_0	T_1	T_2	T_3
Leaves number	13.7±1.25^{a}	27.7±5.01^{bc}	24.5±8.78^{bc}	19.7±9.07^{ab}
Root weight	4.3±1.25^{a}	16.0±3.37^{c}	12.9±2.33^{bc}	10.3±4.36^{b}
Root length	9.9±1.53^{a}	23.1±3.49^{bc}	18.6±7.39^{bc}	14.2±1.98^{ab}

Leaf growth indicates the presence of cell metabolism in plants. The results showed that the T_1 (20% compost + 80% soil) gave the highest average leaves number ($P<0.05$) compared to other treatments. Based on Duncan's multiple-distance follow-up test, it was found that the T_1 and T_2 treatments gave the highest effect and did not differ from the T_3 treatment compost.

Leaf growth is affected by some factors, including the level of nitrogen absorbed by plant roots. Adi *et al.* [20] that nitrogen is an important element for plants in vegetative growth. Nitrogen plays a role in stimulating overall plant growth, especially in stems, branches, and leaves and the formation of green leaves (chlorophyll). Furthermore, Augustien and Suhardjono [21] explained that leaf formation by plants is strongly influenced by the availability of nitrogen and phosphorus nutrients in the medium and those available to plants. These two elements play a role in the formation of new cells and are the main components of organic compounds in plants such as amino acids, nucleic acids, chlorophyll, ADP, and ATP.

The application of treated compost fertilizer at the level of 20% (T_1) was considered to have the best effect on the growth of the leaves number of the mulberry plant (*Morus alba*). The nitrogen (N) content in the P1 treatment fertilizer was 0.9%. Nitrogen levels in compost can help plants in the formation of leaves. Nitrogen makes plant parts green because it contains chlorophyll which plays a role in photosynthesis. This element is also useful for accelerating plant growth in height, increasing the tillers number, influencing leaf width and length, and making them large, increasing protein and fat levels of plants.

In addition to nitrogen, phosphorus (P) also plays a role in plant growth, including leaf growth. In Liferdi's research [22] stated that phosphorus had a significant effect on plant height, branch length, branches number, and leaves number. The leaves number increased as the phosphorus dose increased. This is because phosphate is needed by plants for the formation of cells in plant tissues.

In addition, the element potassium (K) also has a role in the growth and leaves number. Potassium can maintain cell turgor pressure and water content in plants, increase plant resistance to disease and drought, and improve yields and crop quality [23]. This is in accordance with Uke *et al.* [24] who stated that the addition of potassium has a very significant effect in increasing the leaves number. Potassium is also a source of strength for plants in the face of drought and disease. Plants that are deficient in potassium have symptoms of shrivelled and dry leaves.

Based on the analysis of variance that has been carried out, it is known that giving compost treatment T_1 (20% compost + 80% soil) gave the highest average root weight compared to other treatments ($P<0.01$). Furthermore, with Duncan's follow-up test, it can be concluded that the treatment of T_1 compost gave the highest root weight and did not differ from that of the T_2 treatment compost.

Root growth is an important factor in plant growth. Roots play a role in absorbing elements in the planting medium so that plants can use them for growth. Sutarta *et al.* [25] explain that in the mechanism of nutrient absorption by plants, roots are organs that function in absorbing nutrients from the soil, so roots are an important factor in determining plant productivity. Root growth potential is necessary.

The mulberry plant (*Morus alba*) has a taproot system so that its growth requires phosphorus to strengthen its roots. Phosphorus content in compost has a very significant effect on the average weight of plant roots. This is in line with Adi *et al.* [20] which states that the element phosphorus (P) helps in root growth and strengthens roots. Phosphorus is needed in large quantities so that it is a macro element needed by plants, especially in the vegetative phase of plants. Phosphorus plays a role in increasing root development and as a source of energy by forming ATP [26].

Other elements such as nitrogen and potassium also contribute to the root weight of *Morus alba*. In the research of Sunarto *et al.* [27], it was stated that the administration of KNO_3 influenced one of the variables in increasing root weight. However, it is inversely proportional to the increase in nitrogen which reduces root weight. So that the element of potassium also contributes to increasing the weight of plant roots because potassium functions in improving the root system [28].

Roots are an important part of plants because they have a role in finding and absorbing nutrients and nutrients contained in the soil to be used during plant growth. So that the soil used as a planting medium must contain the nutrients that plants need. One of the ways to increase nutrients and soil quality is the addition of organic fertilizer or compost. The ability of plant cuttings to form roots is influenced by natural growth regulators in plants (phytohormones), as well as growth regulators given exogenously, to produce a good root system [29].

Root growth is in line with the leaves number that grow. In the leaves, photosynthesis occurs which will produce a certain amount of energy to carry outgrowth for plants, one of which is root growth. In this process, potassium (K) is needed which plays a role in the process of photosynthesis and metabolism as well as nutrient transport in plants. Potassium contained in pure compost because of research is 3.22%. This value is in accordance with the SNI Compost standard which states that the potassium content in compost is at least 0.2%. Potassium levels in T_1, T_2, and T_3 treated compost also showed values that were in accordance with SNI standards with values of 0.95%, 1.49%, and 2.05%, respectively. Satriawi *et al.* [30] explained that potassium plays a role in the movement of photosynthate out of the leaves to the roots and can increase the supply of energy for root growth. Therefore, the more nutrients that are absorbed by the plant, the more photosynthate will accumulate in the roots for growth [31]. Plants that have an adequate supply of potassium, the photosynthetic and metabolic processes will run well [20].

3.1.2 Arachis pintoi

Table 4 presents the results of the biological test of rabbit faeces compost with the addition of *Kepok* banana peels on *Arachis pintoi.*

Table 4. Plant length (cm), root weight (g), and fresh weight (g) of 58 days old *Arachis pintoi*

Variable	T_0	T_1	T_2	T_3
Plant length	14.6 ± 0.96^{a}	19.11 ± 2.51^{c}	17.9 ± 3.20^{bc}	16.7 ± 1.08^{ab}
Branches number	12.2 ± 3.3^{a}	19.8 ± 5.98^{bc}	12.9 ± 2.33^{bc}	14.6 ± 2.30^{ab}
Fresh weight	21.2 ± 1.73^{a}	29.8 ± 6.99^{c}	26.6 ± 1.65^{bc}	26.6 ± 1.65^{bc}

Arachis pintoi plants do not grow upright but propagate above the soil surface. Measurement of plant length is more accurate for this close relative of the peanut plant (*Arachis hypogea*) than plant height. Some of the advantages of this plant are drought resistance, can grow on various types of soil or low soil fertility, low pH, or aluminium-saturated soil [32] and shade with an intensity of up to 50% [33].

In Table 4 above, the planting medium without the addition of rabbit faeces compost and *Kepok* banana peel (T_0) had the lowest plant length ($P<0.05$). Plant media with treatment received additional supply of N, P, K from rabbit faeces and *kepok* banana peels as can be seen in Table 2 regarding the nutrient content of compost. Duncan test results showed that T_1 plants gave a significant difference with T_3 and T_0 and did not differ from T_2. Similarly, the effect of adding compost to the growing media on the branches number and fresh weight.

The branches number produced is due to the high levels of N, P, K elements found in rabbit faeces. According to R. Anwar and Djatmiko. [34] rabbit faeces contained 2.28% nitrogen, 2.31% P_2O_5 and 1.34% K_2O. Andri and Wawan [35] state that the increase in the branches number occurs due to the process of cell division and elongation as well as nutrients that can increase plant productivity, such as N, P and K elements. Plant height growth is a physiological process in which cells divide.

According to Astutik *et al* [36] that potassium has an important role in the photosynthesis of protein and cellulose formation, in addition to strengthening plant stems and accelerating the growth of existing stems. One of the functions of nitrogen is to form chlorophyll which functions in the process of photosynthesis. The higher the nitrogen application, the amount of chlorophyll formed will increase.

The increase in the amount of chlorophyll has an impact on the rate of photosynthesis which also increases so that plant growth is faster and maximum. According to Koryati [37] the results of photosynthesis are used for the growth of plant organs, where the larger the plant organs formed, the more water content that can be bound by plants. In addition, the higher the plant height and leaf area, the higher the fresh weight of the plant. This is supported by Prasetya *et al.* [38] that the fresh weight of plants is influenced by plant height and leaf area, where the higher and larger the leaf area, the fresh weight of the plant will be higher.

4 Conclusion

The application of organic compost based on rabbit manure and banana peel with *Azotobacter* microbial culture decomposer can improve the vegetative growth of *Morus alba* and *Arachis pintoi*. The best level was T_1 with 20% of compost and 80% of soil.

References

1. Direktorat Jenderal Peternakan dan Kesehatan Hewan, Populasi Kelinci 2014-2019 (2018)
2. S. M. Sholihah and M. A. Wahyunigrum, Jurnal Ilmiah Respati Pertanian, **2**, 650 – 658 (2016)
3. M. Nurrohman, A. Suryanto, Karuniawan Puji W., Jurnal Produksi Tanaman, **2**, 649-657 (2015)
4. M. Tumimbang, Z. E. Tamod, W. Kumolontang, Eugenia, **22**, 123-133 (2016)
5. M. A. Abdullah, J. Wardie and A. P. Berka, Journal Socio Economic Agricultural, **14**, 48 – 56 (2019)
6. I. S. Manis, I. Said, Jurnal Akademia Kim, **6**, 219-226 (2017)
7. Sriharti and Takiyah, Prosiding Seminar Nasional Teknoin, **1**, B72-B77 (2008)
8. N. G. K. Roni, N. N. Candraasih, M. Witariadi, N. W. Siti, Majalah Ilmiah Peternakan, **20**, 29-32 (2017)
9. H. N. A. Kurniawan, S. Kumalaningsih, A. Febrianto, Buletin Peternakan, **1**, 1-8 (2014)
10. R. W. Ayunin, W. D. Nugraha, G. Samudra, Jurnal Teknik Lingkungan, **5**, 1-10 (2016)
11. R. Kumalasari, E. Zulaika, Jurnal Sains dan Seni, **5**, 64-66 (2016)
12. F. P. Rahayu and E. Zulaika, Jurnal Sains dan Seni, **6** (2017)
13. B. Irawan, Metana, **10**, 18-24 (2014)
14. R. Bulan, T. Mandang, W. Hermawan, Desrial. Jurnal Rona Teknik Pertanian, **9**, 135-146 (2016)
15. A. A. N. Supadma, D. M. Arthagama, Jurnal Bumi Lestari, **8**, 113-121 (2008)
16. F. Firdaus, Kualitas pupuk kompos campuran kotoran ayam dan batang pisang menggunakan bioaktivator MOL tapai, Undergraduate Thesis, IPB University (2011)
17. U. Mayasari, B. S. Dewi, L. Afriani, S. P. Harianto, JITEK, **1**, 1-13 (2018)
18. D. A. P. Ratna, G. Samudro, S. Sumiyati, Jurnal Teknik Mesin, **6**, 1-5 (2017)
19. W. A. Akbari, Y. Fitriningsih, D. R. Jati, JITEK, **2**, 1-10 (2015)
20. I. P. T. S. Adi, M. S. Yuliartini, I. G. B. Udayana, Sustainable Environment Agricultural Science, **4**, 151-156 (2020)
21. N. Augustien, H. Suhardjono, Agritrop: Jurnal Ilmu-Ilmu Pertanian (Journal of Agricultural Science), **14**, 54-58 (2016)
22. L. Liferdi, Jurnal Hortikultura, **20**, 18-26 (2010)
23. N. Sumarni, R. Roslaini, R. S. Basuki, Y. Hilman, Jurnal Hortikultura, **22**, 233-241 (2013)
24. K. H. Y. Uke, H. Barus, I. S. Madauna, E-Jurnal Agrotekbis, **3**, 655-661 (2015)
25. E. S. Sutarta, Winarna, M. A. Yusuf, Jurnal Pertanian Tropik, **4**, 84-94 (2017)
26. Sudrajat, A. Dawis, A. Wachjar, Indonesian Journal of Agronomy, **42**, 222-227 (2014)
27. Sunarto, M. Ihsan, S. J. Rachmawatie, Agrodinamika, **12**, 74-79 (2018)
28. A. Murtilaksono, T. Nurmala, A. Suriadikumah, Student e-Jurnal, **3**, 1-11 (2014)
29. R. Prihatin, EKSAKTA: Berkah Ilmiah Bidang MIPA, **18**, 62-68 (2017)
30. W. Satriawi, E. W. Tini, A. Iqbal, Jurnal Penelitian Pertanian Terapan, **19**, 115-120 (2019)
31. D. K. Anam, Jurnal Ilmiah Pertanian, **15**, 31-36 (2019)

32. Maswar, Kacang hias (*Arachis pintoi*) pada usahatani lahan kering (2004)
33. A. Fanindi, S. Yuhaeni, E. Sutedi, Oyo, Seminar Nasional Teknologi Peternakan dan Veteriner, **1**, 701-706 (2009)
34. R. Anwar and Djatmiko, Jurnal Agroqua, **6**, 152-158
35. R. K. Andri, Wawan, JOM Faperta, **4**, 1-14 (2017)
36. D. Astutik, D. Suryaningdari, and U. Raranda, Jurnal Citra Widya Edukasi, **11**, 67-76 (2019)
37. T. Koryati, Jurnal Penelitian Bidang Ilmu Pertanian, **2**, 13-16 (2004)
38. B. Prasetya, S. Kurniawan, M. Febrianingsih, 2009, Jurnal Agritek, **17**, 1022-1029 (2009)

The effect of meniran extract (*Phyllanthus niruri L.*) addition on viscosity, foam overrun, foam stability and foam's microscopic of nano whey protein isolate

Winda Fransisca Saragih[1*], *Abdul* Manab[2], *Manik Eirry* Sawitri[2], *Premy Puspitawati* Rahayu[2], and *Ria Dewi* Andriani[2]

[1]Master Student of Animal Science, Universitas Brawijaya, Jl. Veteran, Malang 65145, East Java, Indonesia
[2]Lecturer of Animal Science, Universitas Brawijaya, Jl. Veteran, Malang 65145, East Java, Indonesia

Abstract. Meniran is an Indonesian herbal plant which has health benefits due to its bioactive compounds. It can be applied to food product, whey protein isolate, were easily absorbed for body. The purpose of this research was to determine the effect of meniran extract addition to nano whey protein isolate on viscosity, foam overrun, foam stability and foam's microscopic. Data collection method was laboratory experiment with 4 treatments of different meniran extract addition levels which consisted of P0 (without meniran extract), P1 (30 ug/mL), P2 (60 ug/mL) and P3 (90 ug/mL) which repeated 3 replications respectively. Experiment was designed by Completely Randomized Design (CRD). The data was analyzed using Analysis of Variance (ANOVA), if there were significantly differences, it would be continued by Duncan Multiple Range Test (DMRT). The results showed that meniran extract addition to nano whey protein isolate gave significant effect ($P<0.05$) on viscosity. However, there was no effect on foam overrun and stability. It is concluded that the addition of meniran extract P3 (90 ug/mL) to nano whey protein isolate resulted in viscosity and foam's microscopic (stable, uniform and thick).

1 Introduction

Proteins are heterogeneous macromolecules with various structural (primary, secondary, tertiary, quaternary). The primary structure of protein related to peptide bonds between amino acid components and sequence in their molecule. Proteins consist of five categories namely casein, whey, milk fat globule membrane, enzymes and other minor proteins [1]. The main components of milk protein are casein (80%) and whey (20%). Whey protein is a by-product of dairy processing or casein production. It consists of α-lactalbumin (α-LA, 20%), β-lactoglobulin (β-Lg, 50%), bovine serum albumin (BSA, 10%), immuno-globulins (10%) and peptones protease (<10%) [2].

* Corresponding author: windafsaragih@student.ub.ac.id

Whey proteins have high solubility, viscosity through water binding, gelling, emulsifying, fat binding, foam forming and able to improve the milk's color, taste and texture [3]. Whey protein component that has good viscosity and foam-forming properties is β-lactoglobulin [3], [4]. Foam formation is affected by materials adsorption on the air-water surface and the ability to decrease surface tension. Polyphenols decrease surface tension because of interaction hydrophobic groups of amino acids, therefore optimal foam formed. Another advantage of polyphenols addition is foam stabilization [5].

Meniran is an Indonesian herbal plant that has health benefits, such as antioxidant and immunomodulator. Bioactive compounds of meniran consist of alkaloid, flavonoid, phenolic, phyllanthin, saponin, steroid and tannin [6]. They can be obtained using microwave-assisted extraction (MAE) method by organic solvent and microwaves. The advantages of MAE are used solvent minimally, time efficiently and environmentally friendly. Besides health benefits, the application of meniran in food products can improve the functional properties of whey protein through covalent and non-covalent binding (hydrophobic interactions, van der Walls and hydrogen bonds). Based on the description above, it was carried out to evaluate the effect of meniran extract addition with different concentration treatments to determining functional properties on viscosity, foam overrun, foam stability and foam's microscopic of whey protein isolate.

2 Materials and methods

2.1 Materials

The materials used for this research were meniran powder that was obtained from Balai Materia Medica Batu (Malang, East Java), Whey Protein Isolate (WPI) 90 was obtained from local market. NaN_3 and aquadest were obtained from Panadia Laboratory (Malang, East Java). The equipment used were microwave at level medium high power (70°C) (Sharp Model R – 222Y(S)), hotplate stirrer (SBS), optical microscope (Olympus CX21FS1), viscometer (Brookfield DV3TLVCJ0), mini hand mixer (Seniora) and filter paper (Whatman No. 4).

2.2 Methods

The research methodology was laboratory experiment using a Completely Randomized Design (CRD) with four different treatments containing meniran extract (control, 30ug/mL, 60ug/mL and 90ug/mL) (w/v) which repeated three replications respectively.

2.2.1 Meniran Extraction And Evaporation

Meniran extraction was described by [7] using Microwave-Assisted Extraction (MAE) method. Meniran powder 3 g in 100 mL aquadest was macerated for 24 hours. After that, meniran solution was extracted with Microwave-Assisted Extraction method at 70°C for 10 minutes (one minute radiation, two minutes off and so on) and filtered by filter paper Whatman No.4. The extract (10 mL) was evaporated at 50°C for 6 minutes using microwave evaporator. The final form of the extract had dry and crumbly texture.

2.2.2 Interaction of Whey Protein Isolate (WPI) and Meniran Extract

Five grams of WPI were heated using a hotplate stirrer at 65-70°C for 40 minutes. After that, meniran extract was gently added according to treatments until reached 100 mL volume. The addition of meniran extract changed functional properties on viscosity, foam overrun, foam stability and foam's microscopic of whey protein isolate.

2.2.3 Viscosity

Viscosity was analysed according to [8] procedure with slight modification using Viscometer Brookfield. Sample was taken as much as 100 mL and analysed at 37°C and 20 rpm for 20 seconds using spindle No.40. The result was observed and recorded.

2.2.4 Foam Overrun and Stability

Foam overrun was determined using [9] method. Sample was taken as much as 20 mL and placed into measure glass 100 mL, homogenized using mini hand mixer for 3 minutes and waited for 30 seconds. Foam overrun result ready to be observed. After that, the foam was waited for 30 minutes to continue foam stability analysis. The results of foam overrun and stability was measured by formula:

$$\textit{Foam Overrun (\%)} = \frac{Vt}{V0} \times 100\ \%$$

Description: V0 = Total volume before stirring (mL)
Vt = Total volume after stirring (mL)

$$\textit{Foam Stability (\%)} = \frac{Vt}{V0} \times 100\ \%$$

Description: V0 = Total volume before stirring (mL)
Vt = Total volume after stirring and waiting 30 minutes (mL)

2.2.5 Foam's Microscopic

Foam's microscopic analysis was done based on [7] procedure using optic microscope. Sample (20 mL) was homogenized for 60 seconds, placed the foam on object glass and covered with cover glass. Optical microscopy was observed and photographed at 40x and 100x magnification.

2.3 Statistical Analysis

The data for viscosity, foam overrun and foam stability were analyzed by analysis of variance (ANOVA) was carried out using the Microsoft Excel Software. The data for foam's microscopic was analyzed descriptively and compared to data published in literature. The level of significance was determined at P<0.05.

3 Results and discussion

3.1 Viscosity

The addition of meniran extract (control, 30ug/mL, 60ug/mL and 90ug/mL) gave different effects (P<0.05) on viscosity (Table 1). Whey protein isolate without meniran extract (P0) has high viscosity at 10,42 cP because protein is a food ingredient that has good solubility and viscosity properties. High viscosity indicates the strength of water-binding. But, the research of meniran extract addition on whey protein isolate decreased solubility and viscosity because of complex form between proteins and polyphenols, mainly formed by hydrophobic interactions and hydrogen bonds. It has an impact a change conformation of them and predicted decrease in viscosity of whey protein-meniran solution. [10] explained high viscosity has a high protein content because water-soluble protein form globules, which is hydrophobic inside and hydrophilic outside with a larger water-binding capacity. The water-binding capacity can be affected by cohesion forces include density and distance intermolecular. The high cohesion of whey protein was comparable to high water-binding capacity; thus, the flow solution becomes low.

Heating process of whey protein solution at 65-70ºC causes change in protein structure to become open, that can interact with polyphenols from meniran through hydrogen bonds. The interaction between whey protein and polyphenols meniran reduces viscosity. It is assumed that more addition of meniran extract result more intensive changes, especially in conformation change of whey protein and polyphenols; decrease surface hydrophobicity due to interaction between whey protein and phenolic compounds. The interaction thought to reduces water holding capacity and hydrophobic interaction in whey protein molecules, resulting decrease in whey protein flowability. [11] explained that protein and polyphenols form a complex structure, generally formed by hydrophobic interaction and hydrogen bonds, thus change the conformation of protein-polyphenols and decrease surface hydrophobicity. [12] reported that temperature during the heating process of whey protein is very affecting to their structures, too high temperature can make β-lactoglobulin denaturation which can reduce viscosity.

Table 1. The average viscosity of whey protein added with meniran extract

Treatment (ug/mL)	**Average ± SD (cP)**
0	$10.42^{b} \pm 0.87$
30	$5.30^{a} \pm 1.57$
60	$7.50^{ab} \pm 2.75$
90	$4.21^{a} \pm 2.32$

Description: The addition of meniran extract with different concentrations to whey protein gave significant effect ($P<0.05$).

3.2 Foam Overrun

The results of foam overrun in Table 2 indicate that the addition of meniran extract (control, 30ug/mL, 60ug/mL and 90ug/mL) did not give significant effect. It could be due to low concentration of meniran extract. The more concentration of meniran extract to whey protein impact more decrease foam overrun. Sample P3 with meniran extract 90ug/mL had the lowest percentage of form overrun. [13] described meniran bioactive compounds will interact with amino acids through non-covalent bond interactions consisting of hydrophobic interactions, van der Waals, hydrogen bridges and ionic bonds. However, it contrasts to the statement given by [14] that interaction between β-lactoglobulin and polyphenols will form biopolymer film which reduces surface tension and increases foam overrun.

Whey protein was a surfactant because it was able to reduce surface tension and has foaming properties through intermolecular interaction due to the stirring process. Whey protein without meniran extract gave highest percentage of foam overrun at 325,83%, there is no interaction to food ingredients, thus protein structures were stable with hydrophobic groups lead to air-phase and hydrophilic to water-phase. [15], [16] reported whey protein contributes to foam formation by diffusing and concentrating on the air-liquid interface, which reduces surface tension.

Another determining factor of foam overrun analysis is processing and stirring time. Sample was placed as much as 20 mL to measure cup 100 mL and stirred for 3 minutes using a mini hand mixer. [17] explained that the optimal time for protein agitation was 3 minutes. When the stirring duration is continued until 8 minutes, the foam overrun will decrease drastically because of hand mixer agitation.

3.3 Foam Stability

Data of foam stability are presented in Table 2. Foam stability is foam analysis to maintain foam formed, it is continuing of foam overrun. Foam stability occurs when the air on molecular cavity is persistent and no bubble collapse. [16] described the interaction between amino acids and polyphenols were increasing foam stability due to surface hydrophobicity decrease.

WPI is a good foam-forming agent, however not always stable, this research was done to improve foam stability by the addition of meniran extract containing phenolic compounds. This research aimed to improve interfacial properties and foam stability. [18] reported the addition of polyphenols on WPI can increase foam stability 1.9-2.1 times compared to WPI only.

Temperature and humidity were external factor to stabilized foam. In this study, foam stability was analysed at 28°C of room temperature and 65% of humidity. According to [19], foam stability was identified by bubble collapse against the temperature and humidity. When dried environment, the bubble collapse rate is higher than wet environment.

Table 2. The average foam overrun and stability of whey protein added with meniran extract

Treatment (ug/mL)	Foam Overrun (%)	Foam Stability (%)
0	325.83 ± 1.44	233.33 ± 14.43
30	325.00 ± 0.00	225.83 ± 1.44
60	325.00 ± 0.00	225.00 ± 0.00
90	317.50 ± 12.99	234.17 ± 13.77

Description: The addition of meniran extract with different concentrations to whey protein did not give significant effect.

3.4 Foam's Microscopic

Foam's microscopic was observed to find out shape and distribution foam at 40x and 100x magnification using an optic microscope, which is presented in Figure 1. These observations showed different result, respectively. Sample without meniran extract (P0) had a larger shape and uneven distribution foams. Increasing concentrations of meniran extract result in uniform shape and spread foams equally. [14] reported that the addition of polyphenols on protein will result in hydrophobic interactions and elasticity surface film. The decrease of elasticity surface film throughout foaming showed stable and firm bubbles which characterized by decrease of foam's volume after stirring process.

Increasing polyphenols induce air pressure on the small bubbles (Fig. A2 with triangle line) will diffuse to the larger bubbles (Fig. C2 with ellipse line) and develop foams almost uniform (Fig. D2 with circle line). [20], [21] explained the increasing concentrations of polyphenols from tea leaves extract and carrageenan were able to change the protein structures and thick the film which led to stability of protein structures. In certain thicknesses of film layer, it collapse and merge bubbles because of molecules' surface adsorption.

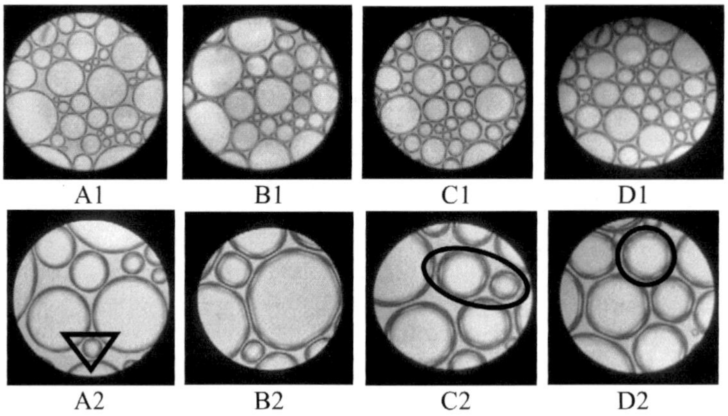

Figure 1. Optical Microscopy Whey Protein Isolate-Meniran

Description: A1 (control), B1 (30ug/mL), C1 (60 ug/mL), D1 (90ug/mL) observed at 40x magnification;
A2 (control), B2 (30ug/mL), C2 (60 ug/mL), D2 (90ug/mL) observed at 100x magnification

4 Conclusion

Based on the results showed that meniran addition can reduce viscosity because a change conformation of proteins and polyphenols' form and decrease surface hydrophobicity. The combination between whey protein isolate-meniran extract has improve the foams microscopic. The significant result was obtained on P3 (90ug/mL) that has viscosity at 4,21 cP, foam overrun at 317,50%, foam stability at 234,17% and foam's microscopic (stable, uniform and thick).

The acknowledgment of this study was supported by Penelitian Penerimaan Negara Bukan Pajak (PNBP) 2020, Universitas Brawijaya.

References

1. A. Manab, M. E. Sawitri, and K. U. A. Awwaly (UB Press, Malang, Indonesia, 2017).
2. C. Harna, C. M. Kusharto, and K. Roosita, Media Kesehat. Masy. Indones., **13/4**, 354-361 (2017).
3. R. F. Hutama and R. Andoyo, Prosiding Seminar Nasional Fakultas Pertanian UNS, 39–46 (2019).
4. R. I. Baeza, C. C. Sanchez, J. M. R. Patino, and A. M. R. Pilosof, E. Dickinson (Ed.), *Food Colloids: Interactions Microstructure and Processing.* (The Royal Society of Chemistry, Cambridge, UK, 2005).

5. S. D. Rodríguez, M. Von Staszewski, and A. M. R. Pilosof, Food Hydrocoll., **50**, 108–115 (2015).
6. H. Rivai, R. Septika, and A. Boestari, J. Farmasi Higea, **5(2)**, 15-23 (2013).
7. P. P. Rahayu, L. E. Radiati, and A. Manab, Curr. Res. Nutr. Food Sci. J., **3(3)**, 224–236 (2015).
8. M. Abughoush, M. Al-Mahasneh, M. Samhouri, M. Al-Holy, and T. Herald, Int. J. Food Eng., **4(7)**, 1–13 (2008).
9. M. Zheng, Z. B. Jia, and J. X. Jiang, Adv. J. Food Sci. Technol., **6(2)**, 238–240 (2014).
10. Z. Herceg and V. Lelas, J. Food Eng., **66(4)**, 433–438 (2005).
11. C. Li, T. Dai, J. Chen, X. Li, T. Li, C. Liu, D. J. McClements, Food Chem., **339**, 128145 (2021).
12. S. El-Shibiny, A. F. Farrag, G. El-Garawany, and F. M. Assem, Int. J. Dairy Sci., **2**, 196–206 (2007).
13. K. Nagy, M. Courtet-Compondu, G. Williamson, S. Rezzi, M. Kussmann, and A. Rytz, Food Chem., **132(3)**, 1333–1339 (2012).
14. M. Von Staszewski, V. M. P. Ruiz-Henestrosa, and A. M. R. Pilosof, Food Hydrocoll., **35**, 505–511 (2014).
15. X. Xiong, M. T. Ho, B. Bhandari, and N. Bansal, Int. Dairy J., **109**, 104758 (2020).
16. Y. Cao, Y. L. Xiong, Y. Cao, and A. D. True, Food Hydrocoll., **82**, 379–387 (2018).
17. C. A. Padiernos, S.-Y. Lim, B. G. Swanson, C. F. Ross, and S. Clark, J. Dairy Sci., **92**, 3049–3056 (2009).
18. J. T. Diaz, E. A. Foegeding, and M. A. Lila, Food Funct., **11(6)**, 5091–5104 (2020).
19. X. Li, S. I. Karakashev, G. M. Evans, and P. Stevenson, Langmuir, **28(9)**, 4060–4068 (2012).
20. C. D. Kanakis, I. Hasni, P. Bourassa, P. A. Tarantilis, M. G. Polissiou, and H. Tajmir-Riahi, Food Chem., **127(3)**, 1046–1055 (2011).
21. Z. Wang, S. Zhang, and B. Vardhanabhuti, J. Food Sci., **80(8)**, 1893–1902 (2015).

Effect of energy and dose of vitamin E selenium on improving the reproduction performance of Joper brood stock

Nining Haryuni[1], *Hartutik* Hartutik[2*], *Eko* Widodo[2], and *Sri* Wahjuningsih[2]

[1] Animal Science Doctoral Program, Faculty of Animal Science, University Brawijaya, Malang 65145, East Java, Indonesia
[2] Faculty of Animal Science, University Brawijaya, Malang 65145, East Java, Indonesia

Abstract. Production performance in Joper broodstock can be improved through increased feed energy and vitamin E-selenium supplementation. This study used 400 laying hens Isa Brown aged 35 weeks and 15 Sentul rooster aged 68 weeks. This study used a factorial completely randomized design with 2 factors, namely metabolic energy levels (2,700 and 2,800 kcal/kg) and vitamin E-selenium supplementation doses (0, 25, 50, 75 and 100 ppm) where each treatment used 10 laying hens and repeated 4 times. The dose of selenium is 0.001 ppm/mg vitamin E. The energy level has a very significant effect ($P < 0.01$) on (feed, protein and energy) intake and HDP. The dose level of vitamin E- selenium in feed had a very significant effect ($P < 0.01$) on HDP and significant ($P < 0.05$) on (feed, protein and energy) intake and FCR. The interaction between the two has a very significant effect ($P < 0.01$) on HDP. The average feed intake 113-115 g/day, protein intake 21.90-22.20 g/day and energy intake 308-317 kcal/bird. HDP 76.70-83.00%. The conclusion of this study is that feed with metabolic energy of 2800kcal/kg and vitamin E-selenium supplementation at a dose of 100 ppm can improve the productivity of Joper broodstock.

1 Introduction

The global economy that has occurred in recent years has led to changes in lifestyle. In Indonesia, this global economy has made fast food made from chicken more popular and the spread of restaurants is getting wider. The national consumption of poultry meat in 2017 was 2,305,000 tons [1]. Poultry meat supplies 69% of the total national demand for meat, consisting of 55% broilers, 9% native chickens, 4% layer hens and 1% ducks [2]. In Indonesia, the poultry business is able to provide employment for 12 million people so that it has an important role in the national economy. One of the government policies made to improve the economy and realize food self-sufficiency is the development of livestock with local genetic sources [3].

* Corresponding author: **hartutik@ub.ac.id**

Joper is a domestic chicken in Indonesia that produced from a cross between a laying chicken and a native rooster [4]. Joper demand has increased every year because it has a good immune system so that its maintenance system does not require antibiotics, the taste and texture of the meat is very distinctive [5]. In 2019, domestic chicken meat demand was only met by 30% [6]. The high demand for Joper meat is not balanced with Joper's population, so this business has great potential to be developed. The main problem with the limited supply of Joper is the lack of hatching eggs in the Joper breeding industry. According to [7] the broiler breeding industry is the main key in the successful development of the broiler industry. Reproductive factors have an important role in a broiler breeding industry. The reproductive performance of broodstock is a major factor in producing large amounts of DOC in a broiler breeding industry. The role of the ovaries is a major factor in reproduction and other supporting factors such as parent genetics, feed, age, environment and other factors. The same opinion was stated by [8] that reproductive performance determines the success of the broiler breeding industry and to maximize reproductive performance it is important to minimize the triggers of metabolic stress.

The limited supply of Joper DOC can be overcome by improving the reproductive performance of existing broodstock in the Joper breeding industry by supplementing with vitamin E-selenium and increasing feed energy. Vitamin E is an antioxidant that has an important role in regulating the metabolic system because it can protect cells from free radicals that damage cell work. Vitamin E works synergistically with selenium to protect tissue from oxidative damage and can be used to boost immunity. Optimum doses of vitamin E and selenium in feed can provide maximum protection during stress, maximum production and reproduction in hens [9].

Metabolic energy in the breeding industry will be utilized for maintenance requirements (basal metabolism, thermal regulation, immune response and activity), growth and egg production [10]. Energy balance plays an important role in the neuroendocrine system that affects the reproductive process [11]. Energy metabolism in laying hens produces glucose which is the main energy source for the brain [12]. The hypothalamus is a part of the brain that plays an important role in regulating the development of reproductive organs, follicular development, maturation and ovulation in laying hens [13]. Energy deficiency in industri broiler breeding causes a decrease in egg production due to disruption of hormone production by the hypothalamus. Energy will be stored in the form of body fat mass when maintenance requirements are fulfilled. Body fat mass has an important role in metabolism including as a precursor in the synthesis of reproductive hormones [14], to maximize the absorption of vitamin E-selenium and egg yolk lippoprotein synthesis [15]. Lippoprotein synthesis in broiler breeding is higher than that of laying hens for egg consumption purposes. The need for fat in broiler breeders is higher than that of laying hens because the lippoprotein in egg yolk is a source of nutrition for embryonic development. The combination of vitamin E-selenium supplementation and metabolic energy in feed is expected to improve the reproductive performance of Joper breedstock.

2 Materials and methods

2.1 Experiemental design and feed

This research is a biological experiment of feed treatment on Joper broodstock for 60 days. This study used a factorial completely randomized design method with 2 treatment factors. The treatment factors used in this study were 2 levels of metabolic energy (V) (2700 and 2800kcal/kg) and 5 doses of vitamin E-selenium (E) (0, 25, 50, 75 and 100 ppm). The repetition of each treatment in this study was 4 times and in each treatment 10 laying hens were used. The number of chickens used in this study were 400 laying hen strain Isa Brown aged 35 weeks and 15 Sentul rooster aged 68 weeks. The Breedstock Jopers were placed in individual cages with a cage length of 50x40x30cm and a front cage height of 37cm. Sentul rooster is placed in individual cages and separate from laying hens. Sentul rooster cage size is 70x50x100cm. The dose of selenium is 0.001 ppm/mg vitamin E. The frequency of feeding in this study was 2 times a day with 40% given at 6 am and 60% at 2 pm and drinking water was given ad libitum.

2.2 Semen collection and artificial insemination

Semen was collected using the massage method. Before doing the massage, the abdominal area to the cloaca was cleaned using a tissue that had been sprayed with 70% alcohol. The massage is done by pressing gently from the abdomen to the cloaca and is stopped when the male begins to lift his tail and then the semen is ready to be collected using a scale tube. Semen collection was carried out half an hour before the artificial insemination activity. Laying hens were inseminated every 4 days with a dose of 0.20 ml of semen per chicken. Total insemination for 2 months during the study was 15 times. Artificial insemination in this study used the method. The intravaginal method is the insemination method by inserting semen into the chicken's vagina with a depth of 2-4 cm. Artificial insemination (AI) was applied at 4 pm to avoid damage to the eggs in the uterus.

2.3 Parameters measure

Feed treatment with a combination of metabolic energy and dose of vitamin E-selenium supplementation was given to Joper broodstock 2 times a day and inseminated every 4 days and then its productivity is observed. The productivity parameters observed in this study were feed intake, protein intake, energy intake, hen day production (HDP), egg weight, feed conversion ratio (FCR) and mortality. Feed intake was measured by subtracting the feed given to laying hens with the rest of the feed being weighed the next day. Protein and energy intake is calculated by multiplying feed intake by the protein or energy content in the feed. Feed intake, protein intake and energy intake Feed intake, protein intake and energy intake are recorded daily and averaged for weekly data. HDP, egg weight and FCR were measured by collecting daily data and then weekly averaged. FCR was measured by calculating the total daily feed consumption divided by the total weight of eggs produced each day and then calculating the average for each week. The health of Joper brooders was observed every day and recorded if any died to determine mortality and used for FCR calculations.

2.3 Data analysis

All data obtained during the study were collected and analyzed using analysis of variance (ANOVA). Analysis of variance used a Factorial Completely Randomized Design (CRD) with 2 treatment factors and was repeated 4 times. Statistical analysis is continued with Duncan's test if the results obtained provide significant or very significant differences in influence.

Table 1. Composition and nutrient content of experimental diets

Ingredients	Treatment	
	E1	E2
Corn (%)	48.90	51.30
Soy bean meal, 45% CP (%)	21.60	22.10
Rice bran (%)	12.20	8.40
Meat bone meal (%)	8.00	8.00
Stone grit (%)	5.00	5.00
Stone powder (%)	3.30	3.10
Fat powder (%)	0.00	1.11
Complete premix (%)	0.50	0.50
Monocalcium phospate (%)	0.30	0.30
Salt (%)	0.10	0.10
Sodium bicarbonat (%)	0.07	0.07
Calculated Composition		
Metabolic energy (kcal/kg)	2,701	2,800
Crude protein (%)	19.00	19.00
Crude fat (%)	4.40	5.06
Crude fiber (%)	3.60	3.24
Lysine (%)	0.96	0.96
Metionin (%)	0.50	0.50
Methionine + cystine (%)	0.87	0.87
Threonine (%)	0.75	0.75
Triptopane (%)	0.22	0.22
Calsium (%)	3.99	3.91
Total phosphorus (%)	0.83	0.78
Phospor avail (%)	0.50	0.49
Sodium (%)	0.13	0.13

Chloride (%)	0.15	0.15

Calculation of feed composition using Brill Formulation software

3 Results and discussion

Based on research on improving the performance of Joper broodstock production, data on productivity of Joper brooders with increased metabolic energy, dose of vitamin E-selenium supplementation, and interaction between metabolic energy level and dose of vitamin E-selenium supplementation were shown in Table 2.

Different superscripts in the same column showed that the energy level had a very significant effect ($P < 0.01$) on the performance of laying hens; dose of vitamin E-selenium gave a very significant effect ($P < 0.01$) on HDP and significantly ($P < 0.05$) on feed intake, protein intake, energy intake, FCR; the interaction between energy level and dose of vitamin E-selenium gave a very significant effect ($P < 0.01$) on HDP.

3.1 Feed intake

The results of this study showed that there was no interaction between metabolic energy levels and the dose of vitamin E-selenium supplementation in the feed on feed consumption. Feed consumption was influenced by a single factor of metabolic energy level and vitamin E-selenium supplementation. Statistical analysis showed that a single factor of energy level had a very significant effect ($P < 0.01$) on feed consumption, while a single factor of vitamin E-selenium supplementation had a significant effect ($P < 0.05$) on feed consumption. The feed intake in this study ranged from 113 - 115 g/bird according to the Isa Brown Commercial Management Guide (2014) 113 g/bird at 35 weeks of age.

Energy is one of the limiting factors for feed intake. The central nervous system (CNS) in poultry has an important role in regulating energy utilization and reproduction. The hypothalamus is a part of the CNS that has a major role in regulating feed intake and energy expenditure [11]. Homeostatic regulation of energy in the hypothalamus is in the form of regulating hunger. Hunger is an integration of metabolic signals that indicate energy status when the body needs energy intake and is responded to in the form of an appropriate appetite [16]. Appetite encourages the behavior of laying hens to increase feed intake so that higher feed energy will reduce appetite because energy requirements have been fulfilled. [17] and [18] reported a decrease of 3.46 - 18.50% of feed intake when the feed was increased by about 100 - 310 kcal/kg of metabolic energy.

Vitamin E-selenium has a very important role in the Joper breeder industry. Vitamin E plays a role in fat metabolism which is used as a precursor for reproductive hormones, functions as an enzyme to increase metabolism and works synergistically with selenium to protect oxidative damage to cells, especially in the process of embryonic development. Increasing the concentration of vitamin E-selenium in the feed will be transferred to the egg yolk and will increase hatchability [19]. Supplementation of vitamin E-selenium in the diet of Joper breeders has an impact on optimal lipid metabolism and a healthier digestive system so that it can increase feed efficiency and decrease feed intake.

Table 2. Effect of metabolic energy level on productivity of Joper broodstock

Treatment	Variable						
	Feed Intake (g/bird)	Protein Intake (g/bird)	Energy Intake (kcal/bird)	HDP (%)	Egg Weight (g)	FCR	Mortality (%)
Metabolic energy level							
E1	114.00^{b}	22.20^{b}	309.00^{a}	78.60^{a}	62.20^{a}	2.42^{b}	0.01
E2	113.00^{a}	21.90^{a}	318.00^{b}	80.70^{b}	64.00^{b}	2.28^{a}	0.00
Suplementasion vitamin E-selenium							
V0	115.00^{b}	22.12^{b}	314.00^{b}	76.70^{a}	62.80	2.41^{b}	0.02
V1	114.00^{a}	22.10^{a}	313.00^{a}	79.20^{ab}	63.20	2.38^{a}	0.02
V2	114.00^{a}	22.10^{a}	313.00^{a}	79.00^{a}	63.10	2.39^{ab}	0.00
V3	114.00^{a}	22.11^{ab}	313.00^{a}	80.60^{b}	63.20	2.30^{a}	0.00
V4	114.00^{a}	22.10^{a}	313.00^{a}	82.70^{c}	63.30	2.28^{a}	0.00
Interaction of vitamin E-selenium supplementation and metabolic energy							
V1V0	114.50	22.20	309.10	77.40^{a}	61.70	2.46	0.04
V1V1	114.30	22.20	308.50	77.60^{a}	62.30	2.45	0.04
E1V2	114.20	22.20	308.20	77.30^{a}	62.30	2.48	0.00
E1V3	114.30	22.20	308.60	78.30^{a}	62.40	2.38	0.00
E1V4	114.30	22.20	308.60	82.60^{ab}	62.50	2.34	0.00
E2V0	113.40	22.00	317.60	76.00^{a}	63.80	2.35	0.00
E2V1	113.40	22.00	317.60	80.70^{a}	64.20	2.31	0.00
E2V2	113.40	22.00	317.40	80.70^{a}	63.90	2.30	0.00
E2V3	113.40	22.00	317.60	83.00^{b}	64.10	2.22	0.00
E2V4	113.40	22.00	317.50	82.90^{b}	64.10	2.22	0.00

3.2 Protein intake

Protein intake is influenced by the energy and protein content in feed. Protein intake is calculated by multiplying the amount of feed intake by the protein content of the feed [20]. Protein consumption in Joper broodstock serves to meet the protein needs needed by the body to carry out metabolic processes. When the Joper broodstock lacks energy, one of the nutrients that can be converted into energy other than fat is protein. Statistical analysis showed that there was no interaction between the energy level factor and the dose of vitamin E-selenium supplementation on protein consumption, but protein consumption was significantly ($P < 0.01$) influenced by energy level and significantly ($P < 0.05$) by supplementation dose of vitamin E-selenium. The single factor of energy level has a more dominant influence than the dose of vitamin E-selenium supplementation. The average protein intake obtained ranged from 21.90 - 22.20g/bird.

Energy level has a dominant influence on protein consumption than vitamin E-selenium supplementation dose because energy is a limiting factor in feed consumption. Protein consumption is strongly influenced by feed consumption because in this study all treatment feeds had the same protein content but different energy. This causes protein consumption to decrease when feed consumption decreases. The same opinion was reported by [21] who stated that metabolic energy in feed is positively correlated with feed intake and protein consumption. Feeds with high energy content will reduce feed consumption and protein consumption.

3.3 Energy intake

Energy level had a very significant effect ($P < 0.01$) and vitamin E-selenium level had a significant effect ($P < 0.05$) on energymetabolism intake, but the interaction between the two gave no significant effect ($P > 0.05$). High energy intake is obtained in E2 treatment. E2 is a treatment with a higher metabolic energy than E1. This shows that energy consumption is directly proportional to the amount of energy in the feed so that the energy intake also increases along with the increased energy in the feed. The energy intake obtained in this study ranged from 308-318kcal/bird and was lower than the results reported by [22] of 326-332kcal/bird in laying hens of the Lohman brown strain.

Table 2 shows that vitamin E-selenium supplementation has an impact on lower energy intake. Metabolic energy intake in broiler breeder hens is utilized for maintenance (HP), ME retained in the body and egg mass produced. Broiler breeder hens require a higher energy intake than laying hens so that surplus energy is stored as body fat mass and utilized for the synthesis of lipoproteins in egg yolks [10]. Hatching eggs require high lippoprotein to maximize embryo development during hatching.

3.4 Hen Day Production (HDP)

Energy level, vitamin E-selenium dose and the interaction between them had a very significant effect ($P < 0.01$) on HDP. The average HDP obtained in this study ranged from 76.70 - 83.00% and was lower than the [23] of 93.30% at 35 weeks of age. This is due to differences in the purpose of rearing which in the [23] is a guide for laying hens, not broiler breeders. The HDP of broiler breeders is lower than that of laying hens due to high metabolic stress. [22] reported that egg production increased with an increase in feed energy from 2,400 kcal/kg to 2800 kcal/kg. The same thing was reported by [24] that an increase in metabolic energy from 874kj to 1,034kj can increase egg production. [25] also reported that the most important factors that must be considered in hens' feed are energy and protein.

Evaluation of reproductive performance in broiler breeders is prioritized on fertility, hatchability and quality of day old chick (DOC) produced. Hatchability related with lipoprotein in egg yolk which is used as the main source of nutrition in embryonic development. High lipid requirements cause high lipid metabolism. The oxidation reaction of fatty acids will release oxygen which is called reactive oxygen species (ROS). The increased amount of Ros can cause oxidative stress, namely an imbalance between oxidants and antioxidants. Vitamins E is a natural antioxidants that play a role in the breakdown of fat chains and prevent cell damage caused by free radicals. Vitamin E-selenium supplementation can reduce oxidative stress so that egg production can increase [26].

Energy in feed is the main factor in the source of body fat mass. Body fat mass will then be used in the process of forming reproductive hormones [10]. Increased energy in broiler breeder feed will improve egg production, egg weight, egg white and yolk weight, reduce feed intake and reduce FCR [13].

The interaction between energy levels and vitamin E-selenium supplementation is very effective in controlling stressors that trigger metabolic stress in broodstock. Vitamin E is a natural antioxidant that can counteract free radicals which are stressors in Joper breeders.

3.5 Egg Weight

Statistical analysis showed that egg weight was significantly affected ($P < 0.01$) by a single factor of metabolic energy level in the feed. The average egg weight in this study ranged from 61.70 to 64.10 g/egg. This average egg weight is in accordance with the [23] which states that the standard weight of chicken eggs at 35 weeks is 62.80g.

Metabolic energy in feed plays an important role in egg formation. High energy intake will result in large egg and yolk weight. This is because 75% of the yolk component comes from feed energy. [25] and [13] reported that an increase in metabolic energy in feed can increase egg weight, egg white and yolk weight. [4] reported that the dominant factor influencing egg weight is the nutrients in the feed, especially energy and protein. The main constituent of an egg is 73.7% water; Protein 12.9%; 11.2% fat and 0.9% carbohydrates.

3.6 Feed Conversion Ratio (FCR)

The calculation of the FCR value is used to describe the effectiveness of the use of nutrients in producing eggs. FCR can be used as a parameter in measuring the efficiency of livestock business. FCR is a parameter in estimating the profit rate of livestock business. Increased energy in broiler breeder feed will improve FCR [13]. Increased energy in feed and vitamin E-selenium supplementation can reduce FCR because it can reduce feed consumption, increase egg production and increase egg weight.

Feed Conversion Ratio (FCR) was significantly ($P < 0.01$) influenced by energy level and significantly ($P < 0.05$) influenced by vitamin E-selenium dose, but there was no interaction between the two factors. The average FCR obtained in this study ranged from 2.22-2.48. The FCR obtained is higher than the [23] at the age of 35 weeks of 1.93.

Metabolic stress in broiler breeders can cause a decrease in hypothalamic function due to cell damage and result in low egg production and quality [8]. Vitamin E works synergistically with selenium to protect cells from damage that interferes with metabolism. Vitamin E-selenium supplementation will have an impact on better health and high egg production [26]. High energy feed can be used to meet the high energy requirements of broiler breeders. Energy homeostasis in the hypothalamus affects the optimal work of the hypothalamus in regulating the body's metabolism so that egg production becomes optimal. Increasing feed energy will improve feed utilization efficiency so that feed intake decreases [11, 13].

3.7 Mortality

Energy levels, vitamin E-selenium levels and the interaction between the two factors had no significant effect ($P > 0.05$) on mortality. The mortality rate in this study ranged from 0.00-0.02%. This result is lower than the standard [23] of 0.10% at the age of 35. The low mortality indicates that the maintenance system in this study is good and meets the requirements.

Necropsy on the carcass of the dead Joper broodstock found a swollen heart and ruptured blood vessels. This is probably caused by heat stress. Heat stress is the impact of oxidative stress that occurs when Joper brooders produce higher lippoproteins than laying hens in general. Oxidative stress in Joper broodstock encourages the pituitary to form the hormone cortisone. This hormone is a glucocorticoid group that plays a role in carbohydrate metabolism including the processes of glycolysis, gluconeogenesis and lipolysis. Corticosterone levels in the blood increase when oxidative stress occurs which triggers a higher rate of carbohydrate metabolism. This causes the body of the Joper broodstock to become hotter and under conditions of unbalanced energy supply and metabolic rate, it will disrupt all regulatory systems in the body.

4 Conclusion

The conclusion of this study is that the combination of energy levels and dose of vitamin E-selenium in feed can improve the performance of Joper broodstock. The best results were obtained metabolic energy at 2800 kcal/kg and a dose of vitamin E-Selenium at 100 ppm.

First of all. I thank Allah for His great love for me, giving me patience and strength to complete my doctoral studies. Second. I thank the Ministry of Education and Culture for the BPPDN scholarship for funding my doctoral studies. Third. I am very grateful to my supervisor Dr. Eko Widodo for all his knowledge, advice and evaluation on my paper.

References

1. V. Tenrisanna, S. N. Kasim, IOP Conf. Ser. Earth Environ. Sci, **492**, 1-7 (2020)
2. C. Sumantri, I. Khaerunnisa, A. Gunawan, IOP Conf. Ser. Earth Environ. Sci, **492**, 1-13 (2020)
3. C. Ferlito, H. Respatiadi, *Policy reform on poultry industry in Indonesia: center for indonesian policy studies,* (Jakarta, Indonesia, 2019)
4. Muharlien, I. W. Nursita, V. M. Pangestu, IOP Conf. Ser. Earth Environ. Sci, **478**, 1-7 (2020)
5. J. K. L. Kalangi, A. K. Rintjap, J. Lainawa, IOP Conf. Ser. Earth Environ. Sci, **478** (2019)
6. T. Immawan, A. Puruhita, W. N. Cahyo, IOP Conf. Ser.: Mater. Sci. Eng, **598**, 1-6 (2019)
7. J. Wang, C. Zhang, S. Zhao, X. Ding, S. Bai, Q. Zeng, K. Zhang, Y. Zhuo, S. Xu, X. Mao, H. Peng, Z. Shan, Poult. Sci. J., **100**, 1-7 (2021)
8. M. J. Zuidhof, M. V. Fedorak, C.A. Ouellette, I. I. Wenger, Poultry Science, **96**, 2254-2263 (2017)
9. P. F. Surai, I. I. Kochish, Poultry Science, **98**, 4231–4239 (2019)
10. S. H. Hadinia, P. R. O. Carneiro, D. R. Korver, M. J. Zuidhof, Poultry Science, **98**, 6721-6732 (2019)
11. S. H. Hadinia, P.R.O. Carneiro, C. J. Fitzsimmons, G. Y. Bedecarrats, M. J. Zuidhof, Poultry Science, **99**, 2215-2229 (2020)
12. N. Mellouk, C. Ramé, A. Barbe, J. Grandhaye, P. Froment, J. Dupont, Int. J. Endocrinol, **2018**, 1-19 (2018)
13. M. Liu, Y. Lu, P. Gao, X. Xie, D. Li, D. Yu, M. Yu, Poultry Science, **99**, 2196-2202 (2020)
14. J. Heijmans, M. Duijster, W. J. J. Gerrits, B. Kemp, R. P. K.Wakkel, H. V. D. Brandz, Poultry Science, **100,** 101131 (2021)

15. J. Ren, W. Tian, K. Jiang, Z. Wang, D. Wang, Z. Li, F. Yan, Y. Wang, Y. Tian, K. Ou, H. Wang, X. Kang, H. Li, X. Liu, BMC Genom, **22,** 1-14 (2021)
16. J. J. Lees, C. Lindholm, P. Batakis, M. Busscher, J. Altimiras, Scientific Reports **7**, 1-11 (2017)
17. Z. Bu, P. Xie, S.Y. Fu, H.B. Tong, X. Dai, J Appl Poult Res, **24,** 371–379 (2015)
18. Y. Ding, X. Bu, N. Zhang, L. Li, X. Zou, Journal of Animal Nutrition, Chinese Association of Animal Science and Veterinary Medicine, **2**, 93-98 (2016)
19. J. I. M. Fernandes, H. L. F. Bordignon, K. Prokoski, R. C. Kosmann, E. Vanroo, A. E. Murakami, Med. Vet. Zootec, **70,** 983-992 (2018)
20. L. D. Mahfudz, U. Atmomarsono, D. Sunarti, E. Suprijatna, T. A. Sarjana, Egypt. Poult. Sci. J, **31,** 491-500 (2011)
21. A. Alderey, A, H. R. Samak, A. M. E. Nasr, H. A. Abu khashaba, J.Animal and Poultry Prod, Mansoura Univ, **8**, 435 - 442 (2017)
22. F. Li, L. M. Zhang, X. H. Wu, C. Y. Li, X. J. Yang, Y. Dong, A. Lemme, J. C. Han, J. H. Yao, J Appl Poult Res, **22,** 36-46 (2013)
23. *Isa Brown Management Guide*, (Kitchener, Canada, 2014)
24. A. M. Kingori, A. M. Wachira, J. K. Tuitoek, Int. J. Poult. Sci, **13,** 151-155 (2014)
25. J. Nashr, J. Agric. Sci, **49**, 53-57 (2015)
26. Y. N. Min, Z. Y. Niu, T. T. Sun, Z. P. Wang, P. X. Jiao, B. B. Zi, P. P. Chen, D. L. Tian, F. Z. Liu, Poultry Science, **97,** 1238–1244 (2018)

Effect of forage *Legetan* (*Synedrella nodiflora*) fresh and hay on the physical quality of forage pellets

Bambang Suwignyo[1,*], *Rifqi Danang* Subagya[2], *Andriyani* Astuti[1], *Nafiatul* Umami[1] and *Ali* Agus[1]

[1] Lecturer and researcher, Faculty of Animal Science, Universitas Gadjah Mada, Jl. Fauna No.03, Karang Gayam, Caturtunggal, Kec. Depok, Kabupaten Sleman, Daerah Istimewa Yogyakarta, 55281, Indonesia

[2] Undergraduate student at Faculty of Animal Science, Universitas Gadjah Mada, Jl. Fauna No.03, Karang Gayam, Caturtunggal, Kec. Depok, Kabupaten Sleman, Daerah Istimewa Yogyakarta, 55281, Indonesia

Abstract. This study aimed to determine the effect of raw materials *Legetan* in the form of fresh and hay on the physical quality of pellets. *Legetan* (*Synedrella nodiflora*) was used as main ingredient, in the form of fresh and hay. The research method used a Randomized Completely Block Design (RCBD). Then, if there was a significant difference, it will be analyzed further using Duncan multiple range test (DMRT). The treatments consisted of P1 = 10% rice bran + 90% fresh *Legetan*; P2 = 20% rice bran + 80% fresh *Legetan*; P3 = 10% rice bran + 90% hay *Legetan*; P4 = 20% rice bran + 80% hay *Legetan*. The variables measured were *rendement*, color, texture, and price. The results showed that the utilization of *Legetan* forage as raw material, both hay and fresh with the proportions of rice bran either 10:90 or 20:80, had no significant effect on the physical quality of *Legetan* forage pellets. However, the higher the portions of *Legetan* forage in the forage pellets, the more affordable the pellet price will be.

1 Introduction

Feed plays an important role for livestock, namely for the maintenance of basic life, production, activity and health [1]. The availability of forage feed during the dry season is an obstacle at this time. In contrast, during the rainy season, forage feed is produced in abundance. Therefore, it is necessary to find alternative ways to preserve feed other than silage or fermentation [2]. One of the processing feed is in the form of pellets. Pellet is a modification of feed from the form of mash produced by pressing the pellet machine so that it becomes harder [3]. Making pellets can be done by processing of forage feed that is available and abundant around.

* Corresponding author: bsuwignyo@ugm.ac.id

Legetan (*Synedrella nodiflora*) is often found wild in open locations, such as in pasture land, fields and gardens, and is more defined as a weed or nuisance plant. This forage bush with the local name *Legetan*, has a high crude fiber of 23.27% so it is very good if given to livestock. *Legetan* in addition to having high fiber, this plant also has a high crude protein of 20.11% so it is suitable as feed for ruminants. In addition, the nutritional content in *Legetan* is DM 92.87%, OM 76.79%, EE 7.40% and NFE 36.01%. Indonesia with a tropical climate is very suitable as a place for this forage to grow, so it is easy to be found [4]. The protein content in fresh *Legetan* is at least 16.51%. Carbohydrate content is 10.01% and fat is 0.25%. The copper content is 1500 mg/kg, lead is 600 mg/kg and hydrogen cyanide is 111.38 mg/kg [5].

Forage pellet manufacture usually uses raw materials in dry form, but by utilizing abundant local resources, forage pellets possible to be made with fresh ingredients or hay. Hay is one of the forages preserved by reducing its water content [1].

The application of forage feed technology in the form of pellets by utilizing local resources such as *Legetan* and natural adhesives in the form of rice bran needs to be studied further. As well as by making pellets from *Legetan*, it is expected to get information the quality of each different proportion by observing the physical properties of the pellet.

2 Materials and methods

This research was carried out at the Forage and Pasture Laboratory, Faculty of Animal Science, Universitas Gadjah Mada, Yogyakarta. This study used a manual method in the process of cutting (harvesting forage) and mixing, but the process of making pellets was with a pelletized machine. Proximate test equipment, grinder for chopping *Legetan*, wiley mill, and digital scales with a capacity of 5 kg (accuracy of 0.05 kg) for weighing feed ingredients. The materials used in this study were forage *Legetan* weed, rice bran which functions as an additional nutrient as well as a binder and water.

Legetan forage is divided into two types of ingredients, namely dry forage/hay and fresh forage. After all the materials were available and ready, the ingredients are chopped and then mixed with other ingredients such as binders according to the treatment. The results of mixing forage and binder were put into the pelletizing machine to be compacted, after the milling process was complete, the drying process was carried out using the drying oven. Comparison between forage pellets with different portion of rice bran and *Legetan* forage (fresh and hay) in this study is presented in Table 1.

Table 1. Ration formulation of forage pellets was used in this research

Feed ingredients	**Treatments**			
	Fresh P1	**Fresh P2**	**hay P3**	**hay P4**
Rice bran	10	20	10	20
Synedrellla nodiflora	90	80	90	80
Total	100	100	100	100

The parameters observed in this study were physical (*rendement*, color, texture) and pellet price. The data obtained were analyzed by ANOVA based on Randomized Completely Block Design (RCBD). Then, if there was a significant difference, it will be analyzed further using Duncan multiple range test (DMRT) with the help of the Statistical Package for the Social Science (SPSS) series 25 computer software.

3 Results and discussion

3.1 Rendement

Data on the yield value of forage pellets *Legetan* with different compositions and raw materials can be seen in Table 2.

Table 2. *Rendement* value of *Legetan* pellets with different formulation and raw materials (%)

Block	Rice Bran		Block Average
	K1	K2	
Fresh	23,13	23,91	23,52^{a}
Hay	82,06	85,74	83,90^{b}
Average^{ns}	52,59±34,03	54,83±35,70	

ab Different superscripts in the same column showed a significant difference (P<0.05) ns non significant

K1 = pellet composition 1 (90% Legetan and 10% rice bran)

K2 = pellet composition 2 (80% Legetan and 20% rice bran)

The results of the analysis of variance showed that different compositions did not have a significant effect on the *yield* value of forage pellets. The addition of the amount of rice bran and forage did not affect the number of pellets formed after all the ingredients were mixed. In Table 2, it can be seen that there was a significant difference (P <0.05) from the ingredients of forage pellets to the pellet *yield* value. This was because the pellets used in fresh still contain a lot of water content, so hat during the pelletizing process, some of the water was wasted (as in squeezing). The water content in the raw material affects the *yield* number [6]. The greater the *yield* produced, the more efficient the treatment applied without overriding other properties [7].

3.2 Color of pellet

In this study, it was found that the color of the pellets at P1 (the amount of fresh forage was more than the rice bran) was greener (Figure 1) than in P2 (Figure 2). At P3 and P4 (fresh forage was replaced with hay) the color of the pellets was more brown than at P1 and P2, but at P3 it was more brown (Figure 3) than P4. At P4 the color of the pellet was more of a cream color (Figure 4). Retnani *et al.*[3] states that the raw materials in the mixture for pellets affects the color of the pellets. Hidayah *et al.*[8] states that a good product color is a color that is not much different from the original material, the darker or brwon-black the color, the lower the quality of the pellet.

Fig. 1. Treatment P1 (10:90) fresh

Fig. 2. Treatment P2 (20:80) fresh

Fig. 3. Treatment P3 (10:90) hay

Fig. 4. Treatment P4 (20:80) hay

3.3 Pellet Texture

The texture of the pellets produced in this study was different, in P1 it was coarser and less solid, while P2 the pellets produced were slightly coarse and dense. At P3 and P4 the pellets were smoother, denser and solid. Based on all these treatments, the most refined, dense and solid treatment was P4. Ismi *et al*. [9] stated that the texture of the pellets is influenced by the constituent materials, especially the crude fiber content and the addition of the adhesive used. Widiastuti [10] added that the quality of the texture is influenced by the water content and crude fiber in the feed, feed that contains high crude fiber will make the texture coarse.

3.4 Pellet Price

The price of pellets is one of the considerations that need to be considered. Affordable pellet prices will attract more consumers' attention so as to increase their selling power without ignoring the nutrient content. The price of rice bran feed ingredients is approximately IDR 4,000/kg while for *Legetan* IDR 2,000/kg.

Based on this price, the price of forage pellets with different compositions (not including processing costs) was IDR 1,300/kg (composition of rice bran: *Legetan* was 10:90), and Rp. 2,400/kg (composition of rice bran: *Legetan* was 20.:80). This shows that the higher the rice bran composition in the pellet composition, the higher the pellet price produced, and vice versa.

4 Conclusions

Forage pellets made from *Legetan* with the first composition of 90% forage and 10% rice bran and the second composition with a ratio of 80% forage and 20% rice bran showed no significant difference in physical quality. The price of *Legetan* forage pellets will be more affordable if the amount of forage increases.

References

1. H. R. Kartadisastra (1997)
2. I. Susilawati, J. Ilmu Ternak Univ. Padjadjaran, **12** (2012)
3. Y. Retnani, N. Hasanah, R. Rahmayeni, and L. Herawati, J. Agripet, **10**, 13–18 (2010)
4. B. Suwignyo. 2020. *Gulma Sebagai Pakan Ternak (Weed for Feed)*, Yogyakarta: K. Media (2020)
5. C. V Nnamani, United Nations University Institute for Natural Resources in Africa (2014)
6. D. A. Ertanto, S. B. Daulay, and A. P. Munir, **5** (3), 565–570, (2017)
7. W. F. Dewatisari, L. Rumiyanti, and I. Rakhmawati, J. Penelit. Pertan. Terap., **17** (2), 197–202 (2017)
8. N. Hidayah, I. P. Retno, I. M. T. Baginda, Bul. Sint.,**21** (2), 21–25 (2017)
9. R. S. Ismi, R. I. Pujaningsih, and S. Sumarsih, Fakultas Peternakan Dan Pertanian Undip (2017)
10. R. Widiastuti, Doctoral Diss. Tesis, Univ. Diponegoro, Semarang (2013)

The physical characteristics of ice cream with the addition of goat bone gelatine as a stabilizer

Ria Dewi Andriani[1,*], *Premy Puspitawati* Rahayu[1], *Mustakim*[1], *Mulia Winirsya* Apriliyani[1], *Manik Eirry* Sawitri[1], *Abdul* Manab[1] and *Sopi Sopiatun* Nida[2]

[1]Animal Product Technology Department, Faculty of Animal Science, Universitas Brawijaya, Jl. Veteran Malang, 65145, East Java, Indonesia
[2]Student of the Faculty of Animal Science, Universitas Brawijaya, Jl. Veteran Malang, 65145, East Java, Indonesia

Abstract. The purpose of this study is to determine the appropriate concentration of gelatin bone goat to the physical quality of the ice cream. The good of ice cream making can be examined from their physical properties including overrun, viscosity, melting point, and emulsion stability. This study employs a complete random design (CRD) with the treatment that is given including, P0 (controlled), P1 (the use of goat bone gelatin by 0.5%), P2 (the use of goat bone gelatin by 1%), P3 (the use of goat bone gelatin by 1.5%). The observed variable is overrun, viscosity, melting point, and emulsion stability. The results show that ice cream with the addition of a goat bone gelatin stabilizer as much as 1.5% produces ice cream with the best quality with an average value of 65.11% overrun, the viscosity to 84.57 cP, 35.31 minutes/50 g melting point and 84.03% of emulsion stability.

1 Introduction

Ice cream is one among other various kinds of dairy products that is enjoyed by all classes of society and most favoured compared to other kinds of dairy products such as yogurt, cheese and kefir. The consumption of ice cream that increases gradually from time to time is characterized by the addition of variants and amounts of ice cream in the market [1,2]. Violisa et al. (2012) reported that the consumption of ice cream in Indonesia ranges from 0.5 L/person/ year with the potential market for ice cream in Indonesia reaching 110 million liters/year. However, the number hardly amounted to 40 million litres/year [3].

Ice cream is a type of frozen shaped food made by freezing a mixture of dairy products, sugars, stabilizers, emulsifiers and other ingredients that have been pasteurized and homogenized [4,5]. The manufacturing process of ice cream requires a stabilizer to maintain the emulsion and to improve the smooth texture of the ice cream. Bahramparvar and Therani (2011) reported that stabilizer serves to store the water inside the ice cream in order to reduce the crystallization of ice [6]. El Aziz et al. (2015) adds that the purpose of using a stabilizer is to produce a soft texture, reducing the crystalline lactose and ice

* Corresponding author: riaandriani@ub.ac.id

crystals during storage, resisting the flow, forming cohesion in the product, and increasing the viscosity [7]. The stabilizer can bind water in the state where it is not frozen, therefore capable of reducing the formation of ice crystals.

Stabilizing substances that are often used are gelatin. The excess gelatin in the manufacturing of ice cream can suppress the formation of ice crystals perfectly and produce a smoother texture for the ice cream [8]. Generally, the use of gelatin in Indonesia is largely obtained by importing from other Countries such as America. The materials to manufacture gelatin are obtained from pigs and cattle, which still raises concerns among the religious communities in Indonesia. A way to solve this is to use gelatin obtained from goats. The use of gelatin as a stabilizer in the manufacturing of ice cream produces a higher average of overrun score, texture, taste, scent and an overall reception compared to the use of CMC (carboxymethyl cellulose) stabilizers [9,10,11]. Based on the description above, a study on the use of goat bone gelatin as a stabilizing material or stabilizer for good quality ice cream is carried out.

2 Materials and methods

2.1 Materials

The raw materials used in the manufacturing of this ice cream include: fresh milk as the basic material obtained from KUD Sae Pujon, sugar, skimmed milk, creamer and quick emulsifier obtained from Primarasa Store, Dinoyo, Malang. Whereas the goat bone gelatin as a stabilizer was obtained from the Laboratory of Animal Product Technology Universitas Brawijaya.

2.2 Methods

Manufacturing of the Ice Cream

The ice cream making process includes several stages, namely the mixing of the ingredients, pasteurizing, homogenizing, aging, freezing and agitation, packaging, freezing and storing [12]. The manufacturing process carried out in this study are as follows: Manufacturing of ice cream including several stages, namely, the process of mixing with a water temperature of 40-48 °C which is used to prevent clumps in the milk, as well as to dissolve the sugar and gelatin. Pasteurization to a temperature of 80 °C for 15-minute serves to kill the pathogenic microorganisms. Homogenization serves to homogenize the fat globule and to accelerate the aeration. The cooling of the ice cream dough before aging serves to prevent heat shock. Aging serves to densify the fat, increase the viscosity of the mixture from initiating the formation of gel on the gelatin, and entrap water from the gelatin. Foaming with an ice cream maker to contain the air in the gelatin gel shrinks the size of ice crystals and dismantles the globular fat. Densifying for 24 hours at 15 °C serves to maintain the robust shape of the ice cream.

Experimental Design and Data Analysis

The method applied in this study is a experimental laboratory with four treatments and four replications, including:

P0 : Manufacturing ice cream without the use of goat bone gelatin (control)

 https://doi.org/10.1051/e3sconf/202233500038

P1 : Manufacturing ice cream with the use of 0.5% w/v of goat bone gelatin
P2 : Manufacturing ice cream with the use of 1% w/v goat bone gelatin
P3 : Manufacturing ice cream with the use of 1.5% w/v goat bone gelatin
The parameters tested including overrun [13]; the viscosity used from Brook Field Viscometer DVII model following the procedure of [14]; melting point [15] and emulsion stability [14, 16]. The data obtained were tabulated and analyzed using analysis of variance (ANOVA) using the Microsoft Excel 2007 program. If the results obtained is different then it will proceed with the Distance Test of Duncan's Multiple (UJBD) to see the difference from each treatment.

3 Results and discussion

3.1 Overrun

Results from UJBD shows that the influence of using goat bone gelatin towards overrun ice cream on each of the different treatments was highly significant ($p<0.01$). The results of the overrun ice cream with goat bone gelatin stabilizer obtained an average between 65.12-87.19% (Table 1). The test results showed that the higher the concentration of the stabilizer will decrease the value of the overrun, for the reason that the higher the concentration of the gelatin used, the dough becomes more viscous and the surface tension reaches higher, therefore, the product hardly inflates. Hartatie (2011) [9]; Ntau (2021) [17] reported that a mixture of ice cream dough should not be too thick because it will perplex the inflating process, therefore minimizing the overrun. On the inflating process of the ice cream volume, the mixing process can cause an effect resulting in a spread in the fat components and forming a web around the air that binds water. The mixing process is done at low temperature (under 10 ^{0}C) to form crystallization of the fat. The process of crystallization of the fat is essential to form fat globules into three-dimensional structures that can entrap water and air, resulting in the inflating volume of the ice cream [18]. If the mixing is performed at a high temperature, then the size of the fat globule will enlarge and be unable to form three-dimensional structures [19]. In the manufacturing of ice cream, the method applied also affects overrun. The method applied in this study is the conventional method, due to the reason that conventional methods do not require stirring and cooling that are done simultaneously, instead done alternately i.e. mixing/ homogenization, cooling, stirring and then returning to cooling. This process and the changes of temperature allows the air entrapped in the dough to become contingent on the speed, stirring duration, and temperature change, in result affecting the inflate or overrun of ice cream produced.

Table 1. The average value of the overrun, viscosity, melting power and emulsion stability

Treatment	Overrun (%)	Viscosity (Cp)	Melting Point (minutes/50 g)	Emulsion Stability (%)
P0	87.19±0.10^d	74.29±0.73^a	26.90±0.64^a	76.54±0.89^a
P1	84.85±0.32^c	79.56±1.56^b	30.82±0.67^b	80.56±0.47^b
P2	72.61±0.20^b	82.52±0.98^c	32.96±0.51^c	81.18±0.19^c
P3	65.12±0.75^a	84.51±0.56^d	35.26±0.12^d	84.03±0.33^d

3.2 Viscosity

The use of goat bone gelatin provides a significant different ($p<0.01$) that influences towards the viscosity of the ice cream. The average value of viscosity of the ice cream is 74.29-84.51 cP (Table 1). The appropriate value of ice cream ranges between 50-300 cP. The largest average value of the viscosity obtained from the P3 treatment of 84.51 cP with the use of goat bone gelatin is as much as 1.5%. This is due to the increasing amount of used percentage of goat bone gelatin, which is known to bind water. Marshall and Arbuckle (2000); Syed and Shah (2016) stated that an increase in the viscosity of ice cream is due to the process of homogenization and the presence of a stabilizer added [19, 20]. Goat bone gelatin has the ability to bind water so that the initial water moving out before the molecules of the material become entrapped, therefore the space between particles narrows and eventually lead to an increase in viscosity. This aligns with Marshall and Arbuckle (2000) who states that the increase in viscosity can be caused by particles suspended in the cream dough, such as fiber, water, and protein components associated with the stabilizing materials [20].

3.3 Melting Point

The use of goat bone gelatin provides a significant difference that ($p<0.01$) towards the melting point of ice cream. The average of melting point of the ice cream using goat bone gelatin is 26.90-35.26 minutes/50 g. The ice cream produced in this study is considered well made since it has the average melting speed value above 25 minutes. This finding in line with Widiantoko and Yunianta (2014) that states ice cream with good quality has a speed of melt in between 25-30 minutes [22]. The longest average melting time resulted from the P3 treatment, with 35.26 minutes/50 g use of goat bone gelatin as much as 1.5%. This is due to the use of goat bone gelatin that affects the strength of the ice cream's body, resulting in it to be fragile and less robust. This is in accordance with the statement made by Arachchi et al. (2015) which states that the quick melting time of ice cream is shown by the body of the ice cream that is fragile and less robust. The use of stabilizer can also slow down the melting of ice cream when served [23]. The use of goat bone gelatin can bind water in the ice cream dough, therefore it is believed that the ice crystals formed do not melt quickly because goat bone gelatin increases the viscosity of the product. This aligns with Karaman et al. (2014) who states that ice cream with a high viscosity coefficient value will have a greater resistance from melting [24]. The overrun value will affect the density of the ice cream. High overrun produces a larger amount of air cavity and the ice cream that is formed is fragile and will melt quickly.

3.4 Emulsion Stability

The use of goat bone gelatin showed the significant difference ($p<0.05$) on the emulsion stability of ice cream. The average stability of the emulsion ice cream with the use of goat bone gelatin is 76.54-84.03% (Table 1). The results show that the higher concentration of the gelatin, therefore the higher the stability of the emulsion. Gelatin can increase the stability of the ice cream. The stability of the emulsion shows the resistance of the ice cream dough towards the separation of milk proteins and milk fat. One way to maintain the stability of the ice cream is to increase its viscosity through the addition of stabilizer into the ice cream mix. Gelatin is a protein type of stabilizer that is also a material for gel making, given that proteins are polymers of amino acids. Amino acids are hydrophilic and hydrophobic. In other words, amino acids can interact with water and some others can interact with oil. If protein is dissolved into the water phase, the hydrophobic part (side chain) will have an affinity to interact with the oil phase part of the interface, therefore droplets of oil can be adjacent to the water phase. This interaction will form hydrogen bonding between molecules of water with gelatin or bonds among gelatin molecules, which can then form a web and entrap the liquid to acquire an elastic and solid phase [25,26]

4 Conclusion

This finding of this study can be concluded that ice cream with the addition of a stabilizer from goat bone gelatine as much as 1.5% produces ice cream with the best physical quality, with the value of the average overrun of 65.11%, the viscosity to 84.57 cP, melting point 0f 35.31 minutes/50 g, and the emulsion stability of 84.03%, respectively.

References

1. U. Sarofa, F. Rosida Dedin, Khadik, Jurnal Rekapangan, **8**, 2-3 (2014)
2. QA, Syed, S, Anwar, R,Shukat J Nutr Health Food Eng, **8**, 422–435 (2018)
3. Violisa, A., A. Nyoto dan N. Nurjanah, Teknologi dan Kejuruan, **35**, 103-114 (2012)
4. S.J. Yeon, J.H, Kim., G.E, Hong., W. Park., S.K, Kim., H.G, Seo., and C.H. Lee, Korean Journal for Food Science of Animal Resources, **37**, 38-43 (2018)
5. G.S. Darma. D. Puspitasari, E. Noerhartati, Media Teknologi dan Manajemen Agroindustri, **1**, 46-53 (2013)
6. M. Bahramparvar, and M.M.Tehrani, Food Reviews International. **27**, 389-407 (2011)
7. M.A. El-Aziz, H.F, Haggag., M.M. Kaluobi., L.K. Hassan., M.M. El-Sayed., and A.F. Sayed, International Journal of Dairy Science. **10**, 160-172 (2015)
8. D. Hastuti, dan I. Sumpe, Jurnal Medagro, **3**, 39-48 (2007)
9. S. E, Hartatie, Jurnal Gamma, **7**, 21-22 (2011)
10. R. Juliasti, A. M. Legowo dan Y. B. Pramono, Jurnal Aplikasi Teknologi Pangan. **4**, 5-46 (2015)
11. N. Guzeler, L. Dogdu., M. Kalender., and C. Ozbek, 2nd International Energy and Engineering Conference. Gaziantep, Turkey (2017)
12. M.A. Moulina, AGRITEPA. **3**, 32-45 (2016)
13. Achmad, F., Nurwantoro dan S. Mulyani, Animal Agricultur Journal, 65-76 (2012)
14. AOAC. *Official Methods of Analysis Assosiation of Official Analytical Chemist* (Washington D.C, 2005)
15. A.M. Roland, L. G. Philips, and K. J. Boor, J. Dairy Science. **82**, 2094-2100 (1999)
16. Priastami, C.S. Karagenan Sebagai Bahan Penstabil Pada Proses Pembuatan Melorin. Skripsi, Departemen Teknologi Hasil Perairan Fakultas Perikanan dan Ilmu Kelautan Institute Pertanian Bogor. (2011)
17. Ntau, E., G.S.S. Djarkasa., dan L.E.Lalujan, Sam Ratulangi Journal of Food Research. **1**, 10 –19, (2021)
18. M.C. Milliatti, and S.C.D.S. Lannes, Food Science and Technology. **38**, 733-739 (2018)
19. H. D. Goff, dan R.W. Hartel, *Ice Cream Seventh Edition* (Springer Science plus Bussines Media, New York, 2013)
20. R, Marshall, and W. S. Arbuckle. *Ice Cream* (aspen Publishers, Maryland, 2000)
21. QA. Syed, and M. S. U. Shah, MOJ Food Processing and Technology. **3**, 246-252, (2016)

22. R. K. Widiantoko, dan Yunianta, Jurnal Pangan dan Agroindustri. **2**, 54-66, (2014)
23. Arachchi, J. Brasathe., A. Weerakoon, Proceedings of the Faculty of Agriculture Undergraduate Research Symposium, (2015)
24. S. Karaman, O.S. Toker., F. Yuksel., M.Cam., A. Kayacier., and M. Dogan, J. Dairy Sci. **97**, 97–110, (2014)
25. S.A. Salem, M. Fardous, and M.G.H, El-Rashody, Food and Nutrition Sciences, **8**, 551-564 (2017)
26. D. Fatimah, Jurnal Kimia, **1** :12-13, (2009)

A qualitative study of small-scale layer farms on a tropical island Malang Indonesia

Nanang Febrianto[1,*], *Budi* Hartono[1]

[1]Faculty of Animal Science, Universitas Brawijaya. Jl. Veteran Malang, 65145, East Java, Indonesia

Abstract. The research was conducted at Malang Regency, East Java, Indonesia, with the research purpose of understanding the amount of small-scale income and characteristics generated by layer farmers. Data collection from February to March 2020 used the survey method. The total sample was 109 respondents determined by the purposive sampling method with the consideration of laying farmers with 2.000 – 7.666 birds for at least three years. The data collection consisted of primary and secondary data. Data were analyzed using descriptive analysis. The result showed that the success rate of the livestock business is inseparable from the characteristics of the farmers. Characteristics of respondents observed in the research included age, education, occupation, breeding experience, and the number of family members. The conclusion of this research is the livestock population was 7,667-15,333 birds, the age around 40–49 years, educational level was elementary school, occupation as farmers, the farming experience was 11-15 years and family member of farmers was 5 people.

1 Introduction

Chicken egg production in Indonesia has a high potential; in 2019, the production reached 1,485,688 tons, growing an average of 8.55% during the period from 2016 to 2019. East Java is one of the biggest chicken egg-producing areas in Indonesia. Statistical data shows that the population of laying hens in East Java is around 69.94% of the total population in Indonesia. The population of laying hens for the last five years in East Java shows that from 2014 to 2019, there has been an increase. The highest population occurred in 2019 amounted to 43.221.466 birds and the lowest in 2018 was 37.035.251 birds, while the population in 2015 was 40.268.631 birds; in 2016, it was 43.066.361 birds, in 2017 it was 41.156.842 birds [1]. The increasing population from year to year indicates that Indonesia still lacks eggs for domestic needs. Therefore, efforts to increase wide-open production and domestic egg production can fill the export market.

Malang Regency has the potential to support livestock business, starting from diverse types of livestock, abundant feed, human resources, and supporting infrastructure. Following the vision of the animal husbandry and animal health service in Malang Regency, the livestock subsector is directed to realize competitive and resource-based

* Corresponding author: nanangfeb@ub.ac.id

livestock agribusiness. These conditions are characterized by the ability to meet the needs of the community, the ability to adjust production patterns to market demand, the ability to develop the region, provide employment opportunities, income, and improve living standards and play a role in economic growth. The program can provide opportunities for laying hens to increase their livestock population, which is expected to support the government's vision.

Laying hens is a leading commodity in Malang regency in terms of population size and production. Malang's laying hen's population is the second largest of the total laying hen's population in East Java Province as many as 43.221.466 birds [1]. The population of chicken eggs in the last 3 years in Malang Regency shows 16.644.990 birds (in 2017), 17.571.738 birds (in 2018) and 27.642.192 (in 2019). The increase in population is related to the amount of egg production produced. Chicken egg production in 2017 amounted to 25.080 tons, in 2018 amounted to 27.510.13 tons and in 2019 amounted to 42.198.64 tons [2]. Based on this cased, the purpose of this research was to determine the profile of small-scale laying hens in Malang Regency.

2 Materials and methods

The research was carried out in the Malang Regency. The determination of the location was carried out using the multistage sampling method. Determination of the sample in stages starting from the provinces, districts, sub-districts and villages intentionally based on the population. Malang Regency was chosen as the location by considering the second-largest laying hens centre in East Java with a population of 5,765,796 individuals [3]. The population of laying hens in the Malang Regency is shown in the following Table 1.

Table 1. The population of laying hens in Malang Regency (birds)

Sub-District	Amount (birds)
Tumpang	1.463.720
Singosari	723.973
Lawang	624.083
Turen	418.801
Poncokusumo	348.087
Wajak	279.827
Dampit	206.661
Kalipare	181.574
Jabung	168.413
Pakis	122.962

Source: BPS-Statistics of Malang Regency (2020)

The research was conducted by a survey method, where the samples were taken from one population and used a questionnaire. The number of respondents as many as 109 people were determined purposively by considering the experience of raising a minimum of 3 years and a minimum livestock population of 2000 animal, then divided into three strata based on the scale of business. The data collected is primary data and secondary data. Primary data was obtained by observation, interviews, and giving questionnaires to respondents. Secondary data was taken from the statistical data collection agency and the local Animal Husbandry Office. The data obtained were analysed using the input-output approach and qualitative descriptive analysis, namely explaining or describing primary and secondary data both qualitatively and quantitatively.

3 Results and discussion

3.1 Livestock condition in Malang Regency

Animal husbandry in Malang Regency consists of cattle, dairy, and poultry. In 2016-2019, the largest population of beef cattle was goats, beef cattle, and sheep. In 2019 the goat population was 253,209 head (up 2.87% per year), then beef cattle 83,660 head (up 3.61% per year), and sheep 33,776 head (up 2.58% per year). Dairy cattle were recorded only 234,481 dairy cows in 2019, with an average annual increase of 5.23% (Table 2). An increase in the selling price of products that impact income encourages farmers to increase their livestock population. The category of small livestock includes goats, sheep, pigs, and poultry categories namely free-range chicken, laying hens, broilers, ducks, and ducks all have increased the population. The increase in population from livestock ownership is one indicator that livestock businesses can increase farmers' incomes.

Table 2. Livestock population of Malang Regency in 2016-2019 (head)

Type of Livestock	2016	2017	2018	2019
Cattle	199.453	212.821	223.717	234.481
Dairy	75.683	78.029	81.150	83.660
Buffaloe	1.266	1.127	1.150	1.164
Horse	626	836	861	882
Goats	235.121	240.823	248.048	253.209
Sheep	31.496	33.284	33.284	33.776
Pig	12.241	12.826	13.262	13.581
Range chicken	2.201.166	2.254.982	2 318121	2.367.744
Broilers	17.571.738	27.642.192	28.335.754	28.927.203
Laying hens	3.005.562	5 597.860	5.765.796	5.912.692
Duck	400.472	468.481	481.130	492.632
Entog	400.287	420.892	432.256	442.125
Rabbit	38.505	40.667	41.590	42.606
Quail	156.288	158.055	161.690	165.412

Source: BPS-Statistics of Malang Regency (2020)

Laying hens is a superior commodity in Malang Regency based on the amount of egg production and its population. Egg production in 2019 reached 44,730 tons (19.55%) out of total production of 228,763.59 tons (Table 3), followed by meat 41,066.70 tons (17.95%) and milk 142,966.10 tons (62.50 %). The amount of broiler egg production is related to the size of the population which reached 5,912,692 heads spread in districts in Malang Regency.

Table 3. Malang Regency animal husbandry production in 2016-2019 (tons)

Commodity	2016	2017	2018	2019
Meat	22.325,74	38.885,69	39.949,12	41.066,70
Egg	27.510,13	42.198,64	43.452,20	44.730,79
Milk	117.253,67	132.052,01	137.324,46	142.966,10

Source: BPS-Statistics of Malang Regency (2020)

3.2 Characteristics of small-scale layer chicken farmers

1. Farming Scale

The magnitude of the business scale is influenced by the economic capacity, the experience of farmers, and business cooperation established by farmers. Farmers who succeed in managing their business will gradually be able to increase the scale of their business because the profits derived from livestock farming are used to increase the livestock population.

Table 4. Business scale classification of laying hens in Malang Regency

Farming Scale	Population	Percentage (%)	Respondent (people)
Small scale	2.000 – 7.666	70,64	77
Medium scale	7.667 – 15.333	18,35	20
Large scale	> 1.5334	11,01	12

Most of the respondents have a business of chicken farmers with a population scale of 2000-7666 as many as 70.64% (77 people). The majority of laying hens farmers have a business scale of 2,000-7,666 because the size of the population is related to the amount of investment and economic capability that the farmers have. The scale of livestock business is influenced by investment factors, namely: capital, cages, land, equipment. Initial investment capital will affect the amount of population to be nurtured and will have an impact on the use of production costs. [4] stated that a large number of broilers that are raised determines the amount of income and profits of broiler breeding businesses, the larger the scale of business, the higher the level of income and efficiency.

2. Age

The age of farmers is very influential on physical strength and knowledge because the livestock sector uses more human labor in its maintenance and technological development that will support the success of the chickens that it raises.

Table 5. The age classification of laying hen's farmers in Malang Regency

Age	Percentage (%)	Respondent (people)
< 29 years	11,0	12
30-39 years	22,0	24
40-49 years	36,7	40
> 50 years	30,3	33

Laying chicken farms are mostly managed by farmers aged 40-49 years as much as 36.7% (40 farmers). At the productive age, farmers tend to have a strong physique, able to work longer and energetic in managing their business. Also, productive-age farmers are braver in taking risks in trying innovations, because they have a curiosity about new knowledge and the interest to adopt technology is getting stronger. Productive farmers have excellent morale which can affect the productivity of the laying hen's business. [5] states that farmers' age and education have a positive impact on the willingness to adopt innovations from technology in livestock raising to increase profits from their livestock businesses.

3. Education

Farmer's education level influences their way of thinking and their level of acceptance of innovation and technology. The higher the level of farmer education, the better the quality of human resources in adopting innovations and applying the latest techniques in the management of laying hens.

Table 6. Classification of laying hen's farmers education in Malang Regency

Education	Percentage (%)	Respondent (people)
Elementary school	52,3	57
Junior high school	31,2	34
Senior high school	7,3	8
Bachelor	7,3	8

The education level of respondents is still relatively low because most of them are elementary school graduates as much as 53%. Respondents reasoned that the high cost of education and the location of the school from where they live is reluctant to continue to higher education. The majority of farmers are poorly educated so that they will be a barrier to accepting and adopting innovations or technologies. This situation will have implications for the development of laying hens and result in low productivity. This is reinforced by the [6] that the education of the agricultural workforce is usually characterized by low levels of education and productivity. In 2013, 39.85% of the workforce in the agricultural sector were elementary school graduates. [7] added that the higher the education level of the farmer, the input from the livestock business could increase because the farmer was more likely to be able to adopt new technology in his livestock business. The character of a farmer's high education tends to make it easy for farmers to accept the latest innovations.

4. Occupation

Occupation is an activity carried out by farmers to meet their needs and desires. The work is divided into two, namely the main work and side jobs. The main work is work whose routine is prioritized by respondents with more time.

Table 7. Occupation classification of laying hens farmers in Malang Regency

Occupation	Percentage (%)	Respondent (people)
Farmers	56,0	61
Seller	17,4	19
Farmer larbor	12,8	14
Government employees	13,8	15

The majority of respondents work as farmers as 61 respondents (56%). The topographic conditions and tropical climate of Malang Regency are very supportive of agriculture and animal husbandry. Besides, the existence of abundant natural resources caused the livelihoods of respondents to be mostly farmers and carried on for generations. The economic aspects of work can improve the economy based on financially, and socially the job can increase the needs and desires.

5. Farming Experience

The experience of raising livestock shows the length of time a farmer does his business. Experience is a farmer's knowledge gained through routine daily activities or events that have been experienced. Farmers who have long experience in managing their businesses have better knowledge, attitudes, and skills when compared to new farmers.

Table 8. Classification of experience in raising laying hens in Malang Regency

Farming experience	Percentage (%)	Respondent (people)
< 5 years	15,6	17
6-10 years	27,5	30
11-15 years	33,9	37
> 16 years	22,9	25

Farmers have experience of raising between 11-15 years of 33.9% (37 farmers). The loyalty of respondents in raising livestock is very high. Farmers who are experienced in dealing with business problems know how to overcome them, while those who lack experience will have difficulty in overcoming business problems. The experience of breeding influences the yield of production because the longer the experience of raising livestock, the farmer will get used to the problems that occur during the maintenance. The results of the research are reinforced [8] findings that the longer the experience of raising livestock, it tends to make it easier for farmers to make decisions related to the technical implementation of their livestock business. That is because the experience is used as a guideline and adaptation to a problem that is sometimes faced by farmers in the future.

6. Number of family members

The number of family members is an essential factor in the livestock business. Family members as human resource assets, especially those of productive age and help in the livestock business, on the contrary, family members can become a burden on their families' lives if they are not actively working.

Table 9. Classification of the number of members of laying hens in Malang Regency

Numbers of Family Members	Percentage (%)	Respondent (people)
2 members	7,3	8
3 members	15,6	17
4 members	16,5	18
5 members	39,4	43
> 6 members	21,1	23

The majority of research respondents had a total of 5 family members of 39.4% (43 people). A large number of family members is due to the average age of the farmers who are still in productive age, and some children from the family are still joined with their parents. A large number of family members has a positive value that is a lot of labour available to help work. On the contrary, the negative aspect of a large number of family members is the expenditure and family dependents will be higher. This is consistent with the opinion of [9] states that the number of family members is related to the labour used in the livestock business. The number of family members of farmers can affect livestock business activities because it will supply labour that helps farmer activities.

4 Conclusion

The conclusion of this research showed that the farmer have livestock population 7,667-15,333 birds, the age around 40–49 years, educational level of farmers was elementary school, occupation as farmers, the farming experience was 11-15 years and family member which the most members as farmers have were 5 people.

References

1. Dirjennakkeswan, *Statistik Peternakan dan Kesehatan Hewan* (Direktorat Jenderal Peternakan dan Kesehatan Hewan, Kementerian Pertanian, 2017)
2. BPS. *Kabupaten Malang Dalam Angka Tahun 2017* (Badan Pusat Statistik, Malang, 2020)
3. BPS. *Jawa Timur dalam Angka 2019* (Badan Pusat Statistik Provinsi Jawa Timur, Jawa Timur, 2020)
4. A. Gusasi, M.A. Saade. Jurnal Agrisistem. **2**, 1-7 (2006)
5. A. Triana, T. Salam, M. Muis. Jurnal Agrisistem. **3**, 11-15 (2007)
6. Kementan. *Rancangan Rencana Strategis Kementerian Pertanian 2010-2014* (Kementerian Pertanian, Jakarta, 2009)
7. C. B. Prayitno, K. Khotimah, Jurnal Gamma, **7**, 13-19 (2011)
8. R. Dewanti, G. Sihombing. Buletin Peternakan, **36**, 48-56 (2012)
9. P. Maspique, R. E. Sawe. Scale Livestock Regency Sidrap Journal, **10** (2011)

The different encapsulants of lileum leaves (*Clerodendrum Minahassae* L) and Avocado Seed (*Persea Americana* Mill) extraction on Physical Quality and Microscopic Structure

Mursye Nataly Regar[2], *Yuli Frita Nuningtyas*[1], *Muhammad Halim Natsir*[1*]

[1]Faculty of Animal Science, University of Brawijaya, Malang 65145, Indonesia
[2]Doctoral student, Faculty of Animal Science, University of Brawijaya, Malang 65145, Indonesia

Abstract. This research aimed to examine the differences encapsulant of leilem leaves and avocado seed (1 : 1) using a microwave oven to the microcapsule product physical quality and microscopic structure. The method was used laboratory experimental with 5 different encapsulants consist of gum arab (A1), whey (A2), chitosan (A3), maltodextrin (A4) and zeolit (A5) with 4 replications in every treatment, respectively. The Variables observed in this research consist of physical quality (dry matter, density, yield, solubility of solids) and microscopic structure. The data of physical quality was evaluate the size and structure, then the microscopic structure was analysed by analysis of variance (ANOVA) of the experiment using Completely Randomized Design (CRD). The results showed that the use of different encapsulates showed a significant ($P < 0.05$) effect on density and the solubility of solids in the product but it were not significant effect on dry matter and yield. Furthermore, different encapsulants also showed significantly different on the microscopic structure. Therefore, it can be concluded that the encapsulation process of the mixed extract of leilem leaves and avocado seeds using a microwave oven and chitosan encapsulation can produce the best physical quality and microscopic structure.

1 Introduction

The régulation of Ministry of Agriculture, number 14/2017 article 16 paragraph 1, feed and feed additives must not contain Antibiotic Growth Promoter (AGP). Therefore, the solution to overcome these problems is to use natural feed additives or Natural growth promoters (NGP). NGP can be obtained from herbal plants and an agricultural by-product as feed additives for poultry.

Leilem (Clerodendrum minhassae L.) is a widely grown plant used as food and traditional medicine in Minahasa, North Sulawesi Province. Leilem plant is part of the genus Clerodendrum L. (family: Lamiaceae) which contains bioactive compounds such as phenols, flavonoids, terpenoids, and steroids[1]. This is effective as an antimicrobial compound against microorganisms[2]. The avocado industry also produces waste in the

* Corresponding author: emhanatsir@ub.ac.id

form of avocado seeds. Avocado seeds are rich in polyphenols with potent antioxidant and antimicrobial properties [3] such as carotenoids, minerals, phenolics, vitamins, fatty acids, glycosides, phenols, alkaloids, cyanogenic, steroids, tannins and saponins [4].

This research aimed to examine the differences encapsulant of leilem leaves and avocado seed (1 : 1) using a microwave oven to the microcapsule product physical quality and microscopic structure

2 Materials and methods

This research was conducted in Animal Nutrion Laboratory, Animal Science Faculty Universitas Brawijaya, Malang, East Java started from July 15 to August 16, 2021.

2.1 Materials

The material used in this research were gum arab, whey, chitosan, maltodextrin and zeolit. The tools and materials used during the study were microwave oven extraction and microwave oven encapsulation.

2.2 Methods

The method was used laboratory experimental with 5 different encapsulants consist of gum arab (A1), whey (A2), chitosan (A3), maltodextrin (A4) and zeolit (A5) with 4 replications in every treatment, respectively. The variables observed in this research consist of physical quality (dry matter, density, yield, solubility of solids) and microscopic structure.

Encapsulation process of leilem leaves and avocado seed are at ratio 1 : 1 as many as 25% as encapsulant. At the same time, leilem leaves and avocado seed extract and encapsulant treatments are stirred into ultrasonic mixing. Furthermore, this product put into a microwave with low temperature for 4 minutes. Temperature of microwave is controlled at less than 60 °C with thermostat and blower. The protected leilem leaves and avocado seed extract is cooled in the tray. Finished protected leilem leaves and avocado seed are milled and packed[5].

The structure of the encapsulated product was evaluated by Scanning electron microscopy (SEM) type TM 3000 with 500x magnification.

2.3 Data analysis

Data were analyzed using one-way ANOVAs based on Completely Randomized Design, if there was significant effect then continued with Duncan Multiple Range.

3 Results and discussion

The results showed that the used of differences encapsulant in the encapsulation process showed highly significant (P<0.05) effect on physical quality of encapsulation product were dry matter, density, yield, solubility of solids completely can be seen in Table 1.

Table 1. Effect of differences encapsulant in the encapsulation process of leilem leaves and avocado seed extract with using a microwave oven on the product physical quality

Variables	Treatments				
	A1	A2	A3	A4	A5
Dry matter (%)	92.02±0.06	92.72±0.16	93.48±0.03	93.50±0.06	91.56±0.08
Density (g/cm^3)	0.36±0.01^a	0.39±0.01^b	0.35±0.01^a	0.43±0.01^c	0.42±0.01^{bc}
Yield (%)	46.05±1.01	45.12±0.56	45.28±0.09	42.18±1.06	43.11±0.26
Solubility of solids (%)	73.58±0.66^b	78.67±0.82^{bc}	81.64±0.56^{cd}	85.11±0.86^d	65.24±1.8^a

A1= gum arab, A2= whey, A3= chitosan, A4= maltodextrin and A5= zeolit
Different superscript in the same column indicate significant effect (P<0.05)

3.1. Effect of Treatment on Dry Matter of Microcapsule Product

The results of the analysis of variation regarding the dry matter content of various encapsulants used to coat leilem leaves and avocado seed extract were not significantly different (P > 0.05) to the dry matter of encapsulated products. Microwave temperature is the same, and the materials have the same physical characteristics. According to [6] showed that the different temperature of extraction and encapsulation process have significantly effect on the phenolic content, antioxidant activity and antimcrobial activity. In the 40 °C temperature was the optimun for obtained the TPC and chemical properties. The dry matter content of encapsulated products from various encapsulation types shows high dry matter content above 90% and water content less than 10%. High dry material indicates that the water content in the encapsulated product is difficult to spread in water because the material tends to be sticky so that no pores are formed. The study results [7] explained that the encapsulation of probiotics with dextrin and triacylglycerol ingredients resulted in water content of 7-12%. The results of this study were that chitosan and maltodextrin encapsulants showed a higher average dry matter. Therefore, microencapsulated dry matter is affected by the use of encapsulation.

3.2. Effect of Treatment on Density of Microcapsule Product

Data of analysis variance the encapsulant type to coat lileum leaves and avocado extract did not significantly different (P<0.05) on the density of the encapsulated product. The lowest value of density showed good density. The different encapsulants have other properties in coating leilem leaves and avocado seed extract. The differences in density in gum arabic and chitosan are because gum arabic has a more considerable molecular weight (±500,000) and a more complex molecular structure. Chitosan has a molecular weight of 171,790[8] compared to maltodextrin. Maltodextrin has a lower molecular weight (less than 4000) a more straightforward molecular structure. The density of the encapsulated product is lower than the material encapsulated because the encapsulated product and the materials encapsulated have bound to the encapsulated material. Therefore, it can be produce encapsulated particles with greater cohesion and adhesion forces [9].

3.3. Effect of Treatment on Yield of Microcapsule Product

Data of analysis variance the encapsulant type to coat lileum leaves and avocado extract was not significantly different ($P>0.05$) to the yield of encapsulated products. The higher yield of the product showed higher efficiency of the encapsulated product. Based on the higher number, the yield is the use of gum arabic and chitosan. This is because gum arabic, chitosan, and whey are better to be able holding the material, so the yield is increased compared to other encapsulants. Gum arabic has good emulsion-forming properties and low viscosity so that it is easy to dry and obtain high encapsulation yields. The use of gum arabic at high concentrations will form an emulsion with very high viscosity. Whey functions as an emulsifier and film-forming; whey protein has a viscosity that is not too high, making it easier for the drying process to produce dry ingredients and high microencapsulated yields. The results of this study are not in line with research [10] which stated that maltodextrin-whey (2:1) encapsulated yields more elevated than the use of arabic-whey gum (2:1) in the encapsulation process of pumpkin extract..

3.4.Effect of Treatment on Solubility of solids of Microcapsule Product

The variance analysis of the data showed that the type of encapsulation was significantly different ($P<0.05$) on the physical indicators of solid solubility. The solid solubility of the encapsulated product is the one with the highest solubility. Maltodextrin encapsulants have a high solubility level because maltodextrin has good solubility properties [11]. The good encapsulated product is yield, dry matter, high solubility, and low density [8].

3.5.Effect of Treatment on Microscopic Structure

The results of scanning the structure of the encapsulated product with black SEM can be seen in Figure 1. The results of microscopic structure observations on the leilem leaves and avocado seed extract encapsulated products showed that the treatment using arabic gum, whey, and chitosan encapsulants did not show any cavities, however, maltodextrin and zeolite show that the product has a cavity on its surface. In the SEM image, it can be seen in the encapsulation process that there are holes from the encapsulation product, which indicate structural damage to the product. In Figure 1 it can be seen that the morphological structure of the leilem leaves and avocado seed extract encapsulated products by processing using a modified microwave oven is irregular in shape. The irregular shape of the encapsulation process of a natural acidifier - phytobiotic mixture with a microwave oven is because in the encapsulation process, it does not pass through the spool as a shaper, unlike in the encapsulation process that uses a spray drier.

Figure 1. Microscopic structure of encapsulation product with Scanning electron microscopy. A1= gum arab, A2= whey, A3= chitosan, A4= maltodextrin and A5= zeolit

4 Conclusion

The conclusion of this research showed that the encapsulation process of the mixed extract of leilem leaves and avocado seeds using a microwave oven and chitosan encapsulation can produce the best physical quality and microscopic structure. This finding maybe can be use as the best product as a feed additive for poultry.

References

[1]. S. O.A Bontjura,. Waworuntu, and K.V. Siagian. Pharmacon Jurnal Ilmiah Farmasi **4** (4): 96 – 101 (2015).
[2]. Parubak and A. Sulu. Chem. Prog. **6** (1): 34 – 37 (2013)
[3]. I.Antasionasti, S. Riyanto and A. Rohman. Res. J. Med. Plants **11**(2): 55-56 (2017)
[4]. J. Tabeshpour, B.M. Razavi and H. Hosseinzadeh. Phytother. Res. **31**(6): 819-837 (2017)
[5] M.H. Natsir, O. Sjofjan, E. Widodo, I. Ardiansah and E.S. Widyastuti. Livestock Research for Rural Development **31** (1) (2019)
[6] C.,Onyebuchi, and D.Kavaz, *Scientific Reports*, *10*(1), 1–11 (2020)
[7] Seveline. Tesis. Institut Pertanian Bogor (2005)
[8] R. Yulina, W. Winiati, C. Kasipah, W. Septiani, A.S. Mulyawan, and T. Wahyudi. Arena Tekstil **29** (2) : 81-90 (2014)
[9] T.Rahayuni, Sutardi and S, Umar. J. Agrosains **15:** 391-402 (2002)
[10] M.Gardjito, A. Murdiati and N. Aini. Jurnal teknologi pertanian **2**:13-18 (2006)
[11] B.R. Bhandari, and B.R. D'Arcy. Review of Food Australia 4892:547-551 (1996)

Variation of qualitative and quantitative traits in day old chick crossbred of Bangkok and Lohman laying hens

Yuli Frita Nuningtyas[1,2] , *Muhammad Halim* Natsir [1*] , *Veronica Margaretha Ani* Nurgiartiningsih[1], *Suci Puji* Lestari[3], and *Muhamad Nadhif* Athariq[4]

[1]Lecture Faculty of Animal Science, Universitas Brawijaya, Malang 65145, Indonesia
[2]Doctoral student, Faculty of Animal Science, Universitas Brawijaya, Malang 65145, Indonesia
[3]Magister student, Faculty of Animal Science, Universitas Brawijaya, Malang 65145, Indonesia
[4]Bachelor student, Faculty of Animal Science, Universitas Brawijaya, Malang 65145, Indonesia

Abstract. The purpose of this research was to evaluate the body weight based on feather color and shank color in one-day-old (DOC) crossbred of Bangkok and Lohmann laying hens. The research material used was 373 Crossbred Day-Old Chick (DOC). The variables measured were body weight as quantitative traits, therefore feather and shank color as qualitative traits. The results showed that the feather color of Crossbred chicken varied, namely brown, grey, white, yellow, and black, with a percentage of 17.69%, 1.87%, 6.87%, 42.62%, and 30.83%, respectively. The shank colors were yellow and black, with a rate of 67.56% and 32.42%, respectively. The mean body weight for brown, grey, white, yellow, and black feather color was 40.12 ± 4.38 g; 42.43 ± 2.76 g; 40.65 ± 3.03 g; 39.82 ± 3.38 g; and 39.61 ± 3.67 g, respectively. The analysis showed that body weight was not different between feather colors. The highest percentage of color for feather and shank was yellow. The quantitative traits measured in crossbred DOC of Bangkok and Lohmann were above the DOC national standard. The results showed that the dominant qualitative trait in Crossbred chicken for feather color was yellow at 42.62%, shank color was yellow at 67.56%.

1 Introduction

Bangkok chicken is the local chicken family because it has long adapted to the environment in Indonesia and is cultivated by the Indonesian people. Bangkok chicken is usually used as a male because it has various features; a slim body shape and high fighting endurance. Laying Lohmann chickens is a potential egg-producing breed; the egg productivity is relatively high, around 60-70%. Crossed chicken is a local Indonesian chicken from a cross between a Bangkok male and a laying Lohmann hen.

Qualitative properties are traits that cannot be measured but distinguished [1]. The qualitative properties can be classified into one or more groups, and the groupings are distinctly different from each other. This difference phenotypic in the population is

* Corresponding author: emhanatsir@ub.ac.id

important to evaluate because it provides crucial information for improving and developing an effective breeding program. Qualitative characteristics will be affected in the adaptation and suitability of livestock to their habitats, especially in extreme tropical. For example, consumer preference of the color feathers laying hens is brown-black, white with black spots, and white had a percentage of 20%, 16.7%, and 46.7%, respectively, while the female chickens were 50% and 23.3% for black and grey feather. Permadi *et all.*, [2] reported that 73,3% of consumers prefer chicken with a combination color of red and black feathers. Feather color and shank color are used as trademarks that can be used to increase consumer preferences for crossbred chickens.

Quantitative properties, including body weight, chest circumference, and DOC shank length, are measures of crossbreeding chicken productivity. Body size is essential because it is a good indicator and has a reasonably close correlation value with live weight parameters [3]. Therefore, one primary research to explore genetic information is a phenotypic observation by observing feather color and measuring body weight, chest circumference, and shank length.

This study aims to determine the quantitative (bodyweight) and qualitative (feather color and shank color) characteristics of day-old crossbreed. The results of this study are expected to provide information regarding the optimal standard of crossed DOC chickens in terms of feather color, shank color, and body weight.

2 Materials and methods

This research was conducted in Berline Farm, Jl. Maguan No.11, Darungan Maguan, Maguan, Ngajum District, Malang, East Java started from March 16 to May 14, 2021.

2.1 Materials

The material used in this research was 373 crossbreed DOC chickens without sex selection (unsexed). The tools and materials used during the study were: DOC box, measuring tape, digital scales, stationery, and calliper.

2.2 Methods

The method used in this study was a field experiment and observation by measuring body weight, feather color and shank color of DOC Crossed chickens. Samples were selected by healthy DOC, consisting of standing straight, dry hair, bright eyes, and not wet nose.

1. Body weight

   Body weight were carried out on hatched DOC using a digital scale.
2. Feather color

   Qualitative data collection is carried out by observing and taking pictures to feather color
3. Shank color

   Qualitative data collection is carried out by observing and taking pictures to shank color.

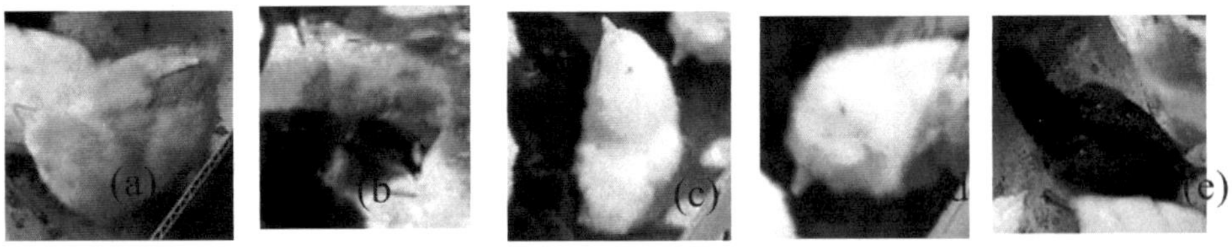

Fig. 1. Color of the feather.

(a) brown; (b) grey; (c) white; (d) yellow; (e) black

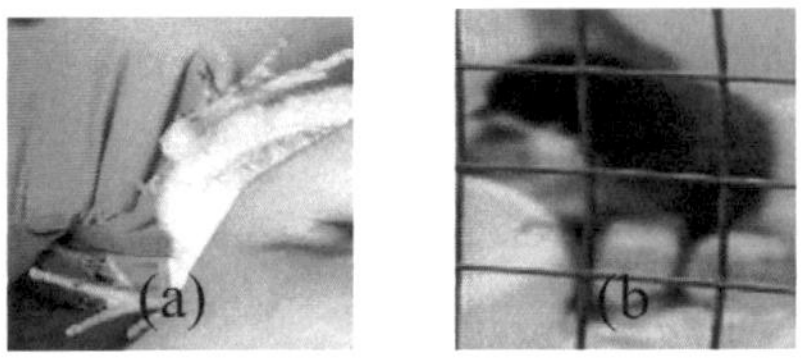

Fig. 2. Color of crossbred shank.

(a) yellow; (b) black

2.3 Data analysis

Qualitative trait data in the form of feather color and shank color were analyzed descriptively based on the frequency of the phenotype. The frequency of feather color and shank color is calculated based on the number of phenotypes that appear divided by the total number of individuals observed multiplied by 100% [4]. The formula is as follows

$$\text{Frequency Phenotype A} = \frac{\text{number of chicken trait A}}{n} \times 100\%$$

Qualitative data of DOC were analyzed based on feather color and shank color. The data were analyzed using one-way analysis of variance (One Way Anova) with the Completely Randomized Design (CRD) method with the following mathematical formula:

$$Yij = + i + ij$$

Yij = the value of observations on the color of the feathers i repetition j
μ = general mean
i = effect of i feather color
ij = test error (error) on feather color

3 Results and discussion

3.1 Color percentage of DOC crossbred

The color of feathers in poultry varies greatly; the pattern of feather color in chickens is one of the main factors determining the identification process [5]. Based on the research data, the percentage of DOC feathers of Crossed chickens was obtained as presented in Table 1.

Table 1. Color percentage of DOC crossbred

Feather color	N	Percentage (%)
brown	66	17,69
grey	7	1,87
white	26	6,97
yellow	159	42,62
black	115	30,83
total	373	100

The results showed that the feather color of the crosses varied, namely brown, gray, white, yellow, and black, with a percentage of 17.69%, 1.87%, 6.87%, 42.62%, and 30.83%. Mahmudi [6] states that crossbred chickens have non-uniform feather colors; brown, black, and white. The non-uniform coat color of the crossbred chicken is probably influenced by the feather color of the parents, namely Bangkok chickens, which have a dominant black coat color, and Lohman Brown laying hens which have brown feathers. In contrast, inherited the white color from the elder Lohman Brown. Thus, the appearance of coat color is determined genetically by either dominant or recessive genes. Edowai [1] states that the color of the feathers of male free-range chickens has various colors: plain white, colored Columbian patterns with glittering gold feathers, colored Columbian patterns with gold and silver feathers, while females have striated feather colors, Columbian type colored patterns with glittering gold feathers, and plain black.

Based on the frequency of appearance of color in crossbreed chickens, it is known that the yellow and black colors show a high percentage, namely 42.62% and 30.83%, respectively, due to the use of Bangkok males, which have colored and black feathers of 70% and 25%, respectively. The highest percentage of coat color is yellow fur (42.62%) which follows DOC Lohmann Brown's coat color, which is also yellow. It was supported research by Sitanggang [4] that the color of Bangkok chicken feathers is 99.20% colored.

The color of the feathers of this Crossed chicken is different from that of the Gaok chicken as stated by Sartika, [7] that the specific characteristic of the Gaok chicken is the color of the female feathers varies with a black and white striated neck. The color of the rooster's feathers is more uniform; it has a greenish base color with silver-yellow covering and neck feathers, black-yellow-green tail feathers (wido). It was because of the genetic difference between the Cross and Gaok chickens. The feather color of Lohman laying hens showed more dominant than bangkok.

3.2 DOC crossbred based on shank

The different color of the shank is caused by the presence of various pigments [2]. Based on the research data, the percentage of DOC shank color of Crossed chickens was obtained as presented in Table 2.

Table 2. Shank color percentage of DOC crossbred

Shank color	N	Persentase (%)
Yelow	252	67,56
Black	121	32,43
total	373	100

The results showed that the color of the crossbreed chicken shank varied, yellow and black, with a percentage of 67.56% and 32.42%, respectively. According to the opinion of Edowai [1] showed that the colors of the free-range chicken shank are white, yellow, black, and gray. Ikasari [8] showed that the color of Kedu chicken shank is white or yellow, sometimes it is black and has a fairly high egg production, and also the opinion of Sartika [7] which states that the color of the Gaok chicken shank is yellow or white.

The highest percentage of shank color is owned by yellow (67.56%) because most Bangkok chickens used as males have yellow shank color. Sitanggang [4] states that Bangkok chickens have a yellow shank as much as 69.05%, which has a yellow shank color. One of the factors that cause yellow shank in crossbreed chickens is the lack of melanin in the outer skin. This is in accordance with the opinion of Amlia [9] that the yellow shank color is caused by a lack of melanin content in the dermis (albino) layer. Gene of lohman laying hens more dominant than bangkok chicken, it can be shown in the shank color of crossbreed chicken.

3.3 Bodyweight of DOC crossbred based feather color

DOC body weight is also known as hatching weight. Lestari [10] stated that body weight is influenced by the quality and quantity of feed consumed. Therefore, the amount of feed consumption will have an effect on the resulting body weight gain. Nutrients in the feed are balanced and sufficient to promote the growth of chickens. Based on data analysis, the average and coefficient of body weight diversity of Crossed chickens were obtained as presented in Table 4.

Tabel 3. Average value ($\overline{\mathbf{X}}$), Standart deviation (SD) dan Coefficient of diversity (KK) DOC crossbred body weight

Feather color	N	$\overline{X} \pm$ SD g/bird	KK (%)
brown	66	40,12 ± 4,38	10,91
grey	7	42,43 ± 2,76	6,50
white	26	40,65 ± 3,03	7,45
yelow	159	39,82 ± 3,38	8,49
black	115	39,61 ± 3,67	9,27
total	373	40,52 ± 3,44	8,53

Based on the analysis of variance, the body weight did not differ between coat colors ($P > 0.05$). The difference in body weight between coat colors comes from elders who have uniform body weights. The results showed that the average body weight of DOC crosses was 40.52 ± 3.44 g/bird. This is referred to Trisiwi's [11] which states that Super free-range chickens have faster growth than local free-range chickens. The bodyweight of DOC Crossed chickens was 37 g/bird until harvest time (60 days) with a weight of 0.9 kg/bird. The results of the study are also higher than the provisions in the Regulation of the Minister of Agriculture Number: 49/Permentan/OT.140/10/2006, that the DOC body weight of Crossed chickens is 29-36 g/bird and minimum DOC weight in hatching is 35 g based on the SNI 7353-1:2019. The cross-program caused the results of this research applied using Bangkok chicken and selected Lohman chickens.

The bodyweight of DOC Crosses from this study was higher than the DOC of free-range chicken. Kususiyah (2011) reported that crossbreed was 25.75 g, and the Kedu chicken from the analysis of [10] that is equal to 33.02 g/bird. Due to better management and comes from superior elders.

4 Conclusion

The conclusion of this research showed that the qualitative traits of the DOC has the variety feather color with the highest percentage of feather color is yellow, shank color is black and yellow, therefore the quantitative traits of DOC crossbred has the higher body weight than the national standard.

References

1. E. Edowai, E. L. S. Tumbal, F. M. Maker. Jurnal Fapet ternak. **4**,50-58 (2020)
2. A. N. N. Permadi, E. Kurnianto, Sutiyono. Jurnal Peternakan Indonesia. **22**, 11-20 (2020)
3. A. Suparyanto, H. Martojo, P.S. Hardjosworo, et all. Jurnal Ilmu Ternak dan Veteriner. **9**: 87-97 (2004)
4. E.N. Sitanggang, Hasnudi, Hamdan. Jurnal Peternakan Integratif. **3**, 167-189 (2015)
5. A.G. Nataamijaya. Buletin Plasma Nutfah. **11**, 1-6 (2005)
6. M. Mahmudi. Doctoral dissertation, Universitas Brawijaya (2019)
7. T. Sartika, S. Sulandari, M.S.A. Zein, Pusat Penelitian dan Pengembangan Peternakan, Bogor (2006)
8. N. Ikasari, E. Kurnianto, I. Sumediana. Animal Agriculture Journal. **1**, 198-207 (2012)

9. M.A. Amlia, R. Pagala, JITRO. **3**, 31-39 (2016)
10. Lestari, R. Maskur, Jan, JITPI. **6**, 24-32 (2020)
11. H.F. Trisiwi, Jurnal Ilmiah Peternakan Terpadu. **4**, 256-262 (2016)

Relationship of motility and acrosome reaction on sexing sperm in Ongole Crossbred bull

Aulia Puspita Anugra Yekti[1*], *Rifai* Mustofa[1] and *Muhammad* Lutfi[2]

[1]Faculty of Animal Science, Universitas Brawijaya, Malang, Indonesia
[2] Beef Cattle Research Station, Grati, Indonesia

Abstract. Artificial insemination using sexing semen is expected to produce calves with the desired sex. One sexing sperm method is the percoll density gradient centrifugation method. This study aimed to determine the changes and the relationship between motility and acrosome reaction after sexing process using percoll density gradient centrifugation. The material used was semen of ±5 years old Ongole crossbred bull with a bodyweight of ±700 kg as many as three bulls with mass motility 2+ and individual motility 70%. The method used was to compare fresh semen with sexed semen after the cooling process. Parameters measured were motility characters using CASA analysis, which included motility parameters, progressive motility, capacitation, and no acrosome reaction. Statistical analysis used paired T-test to distinguish among fresh semen, after sexing and cooling process. In comparison, regression and correlation were used to analyze the relationship of capacitation and hyperactivation sperm with no acrosomal reaction with motility and progressive motility. The results showed that motility and progressive motility decreased after the sexing and cooling process. Meanwhile, the acrosomal reaction, capacitation, and hyperactivity increased.

1 Introduction

Artificial insemination using frozen semen sexing can increase the chances of producing male or female offspring [1, 2]. It was revealed by Susilawati et al. [3] that sexing spermatozoa using the Percoll density gradient centrifugation method resulted in a decrease in the quality of spermatozoa. It is because of damage to the spermatozoa membrane in the structure and function of the spermatozoa. Before the fertilization process, spermatozoa experience the biochemical and physiological structural modifications in the female reproductive organs, namely capacitation and acrosomal reactions [4].

Frozen semen has a lower fertility rate than fresh semen because the freezing process can damage the sperm plasma membrane and induce premature capacitation, increasing calcium ions [5]. The sexing process involves a centrifugation process that can reduce spermatozoa's quality to improve the capacitation of spermatozoa. Therefore, this research is conducted to see how the relationship between motility and capacitation and the acrosome reaction in semen after sexing.

* Corresponding author: auliapay@ub.ac.id

2 Materials and methods

2.1 Materials

The semen used in this study was semen from 3 Ongole Crossbred bulls aged ±5 years with a bodyweight of ±700 kg. The semen met the minimum motility requirements of mass 2+, individual motility 70%.

2.2 Methods

The research method used was experimental method. Parameters measured were motility characters using CASA on parameters of motility, progressive motility, and hyperactivity. Observations of acrosome reactions included sperm that had not undergone acrosomal reaction (intact) and spermatozoa that had capacitation [6].

2.3 Percoll density gradient centrifugation sexing

The sexing method used was percoll density gradient centrifugation using 10 gradients of percoll. The diluent used was Andromed (Minitube Ref 13503/0200) with a ratio of 1:4 with aquabidest. Gradients were arranged in every 0.5 ml from the bottom 65% to the top 20%. Then 1 ml semen was added into the tube gently. Then, it was centrifuged at 2250 rpm for 5 minutes. After centrifugation, 2 ml of the bottom and 2 ml of the top were taken and then put into different test tubes containing 3 ml of wash andromed diluent. Furthermore, wash by centrifugation at 1500 rpm for 3 minutes. Finally, remove the supernatant, and leave 1 ml of precipitate in the tube. Then diluted to the total concentration of 25 million/straw and cooled at 5°C [7, 8].

2.4 Dilution and cooling of semen

The sexed semen has then calculated the concentration using a hemocytometer and diluted to a concentration of 25 million/straw[9]. Then followed with the process of cooling to a temperature of 5°C in the refrigerator [7].

2.5 Variable of research

The variables observed in this study were:

1) Motility, progressive motility, and hyperactivation
Motility, progressive motility and hyperactivation were observed using CASA with Sperm Class Analyzer (SCA). Semen of 3-4 ul was placed on a warm object-glass with 37°C covered with a cover glass. The microscope was set at a magnification of 10 x 10 and phase contrast of pH1. The sperm was observed at 5 points of view. The results of the analysis will show the average value of the images obtained[10].

2. Capacitation Assay
The CTC stock solution containing 750 μM CTC-HCl (Sigma), 130 mM NaCl, 5 mM L-cysteine, 20 mM Tris acid (pH 7.8) was prepared daily, wrapped in foil to protect against light, and stored at 4°C until required. 10 μL of sperm were mixed with 15 μL of CTC solution on a slide at room temperature. Then, 0.3 μL of 12.5% glutaral¬dehyde in 2.5 M

Tris base was added as a fixative solution. Samples were covered with coverslips and stored in the dark room at 4°C [3]. The sperm were observed using an Olympus BX 53 fluorescent microscope. Non-capacitation spermatozoa will show fully bright fluorescence on the entire head. The capacitated spermatozoa will show fluorescence in the upper half of the head and dark in the lower half. In contrast, Spermatozoa with acrosomal reaction will show the minimal fluorescence signal with a bright signal in the equatorial region [11].

2.6 Data analysis

The statistical analysis used paired T-test to distinguish between fresh semen after sexing and cooling process. In comparison, regression and correlation were used to analyse the relationship of capacitation and hyperactivation sperm with no acrosomal reaction with motility and progressive motility.

3 Results and discussion

3.1 Motility, capacitation, no acrosomal reaction, and hyperactivity in fresh semen and after sexing

The quality of sperm in the fresh semen and after sexing, including motility, capacitation, no acrosomal reaction, and hyperactivation are presented in table 1.

Table 1. Characteristics of motility and acrosome reactions in fresh semen and after sexing and cooling

Parameter	Fresh semen	Sexing semen
Motility (%)	95,05 ± 5,16 %	71,02 ± 10,08**
Progresif Motility (%)	71,13 ± 5,82 %	47,68 ± 8,71**
Hiperactivation (%)	11,30 ± 4,83 %	12,96 ± 6,35%^{ns}
Capacitation	21,51± 11,47 %	24,90 ± 9,38^{ns}
no acrosomal reaction (%)	74,28 ± 13,15 %	69,60 ± 9,59*

Note : ** In the same row showed the very significant differences (P<0.01)
* In the same row showed the significant differences (P<0.05)

The percentage of motility of fresh semen was 95.05± 5.16%, while after sexing 71.02 ± 10.08%, the percentage of motility experienced a very significant decrease (P<0.01) after the sexing and cooling process. The percentage of progressive motility in fresh semen was 71.13 ± 5.82%, after sexing and cooling was decreased significantly (P<0.01) to 47.68 ± 8.71%. This is because the sexing process, which involves centrifugation, causes damage to the sperm membrane, resulting in decreased motility after sexing [3] [7]. The cooling process also results in changes in the protein structure of the sperm. Cold temperatures cause a restriction of the lateral movement of phospholipids from the bilayer membrane, which induces a change in shape from the liquid to the gel phase, which causes the membrane to become more rigid and brittle. This causes the separation of the lipid phase resulting in irreversible protein buildup [12]. Protein has a crucial role in the life of sperm, the fertilization process, and energy metabolism [13].

Spermatozoa capacitation in fresh condition was 21.51 ± 11.47%, while after sexing and cooling to 24.90 + 9.38, the capacitation value increased not significantly (P>0.05). The picture of capacitation using CTC staining showed an increase in calcium ions at the equator and the neck of the spermatozoa.

Hyperactivation of spermatozoa in fresh semen was 11.30 ± 4.83%, whereas after sexing and cooling, it became 12.96 ± 6.35%, the hyperactivated spermatozoa had no significant increase ($P > 0.05$). The centrifugation process in sexing and the chemicals present in the diluent induce the membrane to influx calcium ions and capacitation occurs. The cooling and freezing process causes a decrease in membrane integrity, increased capacitation and calcium ion levels [5]. Increased calcium leads to increased hyperactivation by activating the signal transduction cascade regulating hyperactivation. Calcium ions will interact with the axoneme of flagellum to trigger hyperactivation [14].

From the study results, spermatozoa that sperm with no acrosomal reaction in fresh semen were 74.28 ± 13.15% and after cooling and sexing were 69.60 ± 9.59%. This indicates that during the sexing and cooling process, there was an increase in the acrosome reaction significantly ($P<0.05$). In mammals, capacitation is followed by a direct acrosomal reaction, and then the spermatozoa are ready to fertilize the egg [15].

3.2 Relationship between capacitation and hyperactivity

In fresh condition, the capacitation value was 21.51 ± 11.47%, while hyperactivity was 11.30 ± 4.83%, the relationship between capacitation and hyperactivity was $Y = -0.0134 X + 12,179$ with $r = 0.0008$. This equation shows a positive relationship between capacitation and hyperactivity. This shows that the capacitation value is inversely proportional to the hyperactivity value. When the capacitation value increases, the hyperactivity value will increased. After being capacitated, spermatozoa can bind to the zona pellucida through the acrosome reaction (AR), followed by penetration and fertilization of the oocyte. Capacitation is characterized by actin polymerization and will develop hyperactive motility (HAM) [16]. However, the value of $r= 0.0008$ indicates that the correlation between the capacitation value and the hyperactivity value in fresh spermatozoa is weak because normally semen will undergo capacitation on the female reproductive tract before fertilization process[4].

After sexing and cooling process, the capacitation value was 25.8 ± 8.6%, while hyperactivity was 12.96 ± 6, 35%, the relationship between capacitation and hyperactivity was $y = 0.1297x+9.604$, $r=0.0311$. The same as fresh condition, the correlation equation between capacitation and hyperactivity in conditions after sexing and cooling showed a positive value. That is, an increase in in capacitation will be followed by an increase in hyperactivity. However, the value of $r = 0.0311$ still shows that the correlation between capacitation and hyperactivity value is very weak. This is not different from the correlation between capacitation value and hyperactivity value in fresh semen. The cooling process of semen causes sperm damage, including ion imbalance, protease activation, cellular acidosis, energy deficiency, membrane phase transition, and the production of free radicals or reactive oxygen species (ROS) [12]. ROS are free radicals that affect the physiology of spermatozoa during capacitation and hyperactivation [17]. Capacitation of sperm is followed by the increase of ROS and leads to a redox signaling cascade. Redox modifications of proteins accompany bull sperm capacitation [11].

3.3 The relationship between the acrosome reaction with motility and progressive motility

In the fresh semen of spermatozoa that have not reacted acrosome 74.28 ± 13.15% with motility 95.05 ± 5.16%, has a regression relationship $y = 0.0499x + 91.35$, $r = 0.0162$. The regression equation shows a correlation between acrosome reaction and motility. It is interesting that with the acrosome reaction, the motility of spermatozoa is higher. Otherwise, a decreased value of the acrosome reaction will be followed by a decrease motility. However, the correlation equation showed a very weak relationship between the acrosome reaction and the motility value of the fresh spermatozoa. This is indicated by the value of r = 0.0162.

Fresh semen also showed a similar correlation between spermatozoa with no acrosomal reaction with progressive motility of 71.13 ± 5.82%. The regression equation shows y=0,1069 + 63,186 with r= 0,05833. The positive value in the regression equation showed that an increase of acrosome reaction would follow an increase in progressive motility. However, the value of r= 0.05833 indicates a very weak relationship between the acrosome reaction and the progressive motility of fresh spermatozoa in this study.

In sexing sperm, spermatozoa that have no acrosome reaction 69.60 ± 9.59%, with motility 71.02 ± 10.08%, have a regression relationship Y= -0.7322x+121.98, r =0.48456 . This equation shows a negative value that indicates an opposite relationship between the value of spermatozoa without acrosome reaction and the motility of sexing spermatozoa after cooling. This indication shows that with the increasing value of the acrosome reaction, the value of motility will decrease, ultimately reducing the ability to fertilize [5]. The equation also shows a strong relationship between changes in the acrosome reaction value and the motility value of sexing semen after cooling. The value of r = 0.48456 indicates a fairly strong relationship between changes in the value of the acrosome reaction and the value of motility.

A negative regression relationship was also shown in the semen value after sexing and cooling between spermatozoa that had no acrosomal reaction with progressive motility 47.68 ± 8.71%, Y = -0.1295x + 56.69, r = 0.02033. Thus the value of spermatozoa with no acrosomal reaction is inversely proportional to the progressive motility in conditions after sexing and cooling. The value of progressive motility will decrease if the acrosome reaction increases, and conversely, the value of progressive motility will increase if the acrosomal reaction decreases. However, in contrast to the relationship between acrosomal reaction values and motility values, which showed a strong relationship, the regression equation for acrosomal reaction values and progressive motility did not show a strong relationship. This is indicated by the value of r = 0.02033.

4 Conclusion

Motility and progressive motility decreased after the sexing and cooling process. The process of sexing and cooling sperm causes increased capacitation, acrosomal reactions, and hyperactivation.

References

1. M.A. Prakash, Advances in Animal and Veterinary Sciences, **2**, 226-232 (2014).
2. P. Boro et al., International Journal of Applied Research, **2**, 460-462 (2016).
3. T. Susilawati et al., Advances in Health Sciences Research (AHSR), 1st International Conference in One Health (ICOH), **5**, 1-5 (2017).
4. D. Ickowicz, M. Finkelstein, and H. Breitbart, Asian J Androl, **14**, 816-821 (2012).
5. S. Collin et al., Journal of Andrology, **21**, 938-943 (2000).
6. R. Mustofa et al., Jurnal Agripet, **20**, 210-215 (2020).
7. E.D. Kusumawati et al., American Journal of Animal and Veterinary Sciences, **14**, 111-114 (2019).

8. T. Susilawati, *Sexing Spermatozoa* (2014).
9. T. Susilawati, *Pedoman Inseminasi Buatan Pada ternak*, (2013).
10. D. Ratnawati, N. Isnaini, and T. Susilawati, Jurnal Ilmu-Ilmu peternakan, **27**, 80-95 (2017).
11. A. Mostek et al., Int J Mol Sci, **22**, 1-21 (2021).
12. J.G. Baust, JD. Gao, and J.M. Baust, Organogenesis, **5**, 90-6 (2009).
13. A.A. Moura and E. Memili, Animal Reproduction, **13**, 191-199 (2016).
14. S.S. Suarez and H.C. Ho, Reprod Domest Anim, **38**, 119-124 (2003).
15. N. Hirohashi and R. Yanagimachi, Biol Reprod, **99**, 127-133 (2018).
16. S.B. Itach et al., Dev Biol, **362**, 154-161 (2012).
17. R.J. Aitken, K.T. Jones, and S.A. Robertson, J Androl, **33**, 1096-106 (2012).

Associations between Lactoferrin (LTF) gene polymorphism exon 4 and milk compositions in Senduro goat

Tri Eko Susilorini[1*], *Aswah* Ridhowi[1], *Wike Andre* Septian[1], and *Ahmad* Furqon[1]

[1]Faculty of Animal Science, Universitas Brawijaya, Malang 65145, East Java, Indonesia

Abstract. This research aimed to analyze the polymorphism of Lactoferrin (LTF) gene exon 4 associated with milk compositions in Senduro goats. A total of 42 DNA samples and milk compositions from Senduro goats were used in this study. The DNA sequence was amplified using Polymerase Chain Reaction (PCR) with a pair of primers. Genotyping was carried out using DNA sequencing and analyzed using FinchTV 1.4.0 and MEGA 6.0. In this research, the results showed that there were three genotypes (CC, CT, and TT) and two alleles (C and T). The frequencies of CC, CT, and TT genotypes were 0.381; 0.452; and 0.167, respectively. Furthermore, the frequencies of C and T alleles were 0.607 and 0.393, respectively. The genotype polymorphism did not affect on milk compositions. In conclusion, there was no association between polymorphism of LTF gene exon 4 and milk compositions in Senduro goats.

1 Introduction

Senduro goat is one of local goats in Indonesia. This breed is commonly raised for two purposes; milk and meat production [1]. As a dairy goat, the milk production of Senduro goat is 0.80-1.80 liter/head/day based on The Indonesian National Standard [2]. The variation of milk production could be affected by genetic and environmental factors. The animal selection could be conducted to reach optimal genetic quality that increase milk production. Commonly, the selection can be performed based on the phenotype and genotype.

The animal selection based on genotype can be conducted using molecular genetic approach in DNA level. Some functional genes associated with economical traits can be investigated through this approach. One of gene controlled specific protein in milk composition is Lactoferrin gene (LTF). Lactoferrin is a protein found in milk and colostrum that have biological functions in iron transport, antibacterial, antiviral, antimicrobial, and antiparasitic effects [3]. Some studies reported that polymorphism of LTF gene was analyzed and associated with milk composition [4,5]. The previous study explained that the variation of SNP in exon 4 (c.407C>T) related to milk composition on goats in Baghdad [6].

The genetic selection related to milk production and composition was not performed on Senduro goat. Regarding this issue, the genetic molecular approach can be conducted to

* Corresponding author: triekos@ub.ac.id

improve the genetic quality. Furthermore, this research aims to analyze the polymorphism of Lactoferrin (LTF) gene exon 4 associated with milk compositions in Senduro goats.

2 Materials and methods

2.1 Data Collection and Animal Management

A total of 42 Senduro goats were used in this study consisted of 9, 22, 11 ewes on 2.0; 3.0 and 4.0 years old, respectively Milk productions and compositions of Senduro goats were collected at Senduro, Lumajang Regency, East Java from the first lactation to the fourth lactation. Milk compositions were analyzed in Laboratory of Dairy Sciene, Faculty of Animal Science, Universitas Brawijaya. Genetic molecular and data analysis was conducted in Laboratory of Animal Biotechnology, Faculty of Animal Science, Universitas Brawijaya. Senduro goats were fed and maintained under similar management and feeding system. Goats had free access to consume water and were housed in a barn individually.

2.2 DNA Isolation and Amplification

DNA was isolated from blood samples collected from Vena jugularis using vacutainer needles and Ethilenediaminetetraacetic Acid (EDTA) tubes. Goat genome was isolated using Genomic DNA Mini Kit for blood and cultered cell (Geneaid Biotech Ltd., China) and according to [7] method. The DNA quality and quantity were analyzed using 1.5% agarose gel electrophoresis and NanoDrop ND-1000 Spectrophotometer (Thermo ScientificTM, Massachusetts, USA) respectively.

DNA sequence of LTF gene exon 4 was amplified using PCR method. A pair of primers were used in this study consisted of forward and reverse primer (F: 5'- GCT AAG AGC TCT CGA TTC TC-'3 and R: 5'- CAG AGA GAA ATG AAC CAG GG-'3). These primers were designed to amplify a 541 bp fragment according to the goat genomic sequence in GenBank (accession number FJ609300). A solution mixture used in this study consisted of 1 μL DNA template (50 ng/μL), 0.3 μL of each primer (10 pmol/ μL), 15 μL Go Taq Green Master Mix (Promega, USA), and 14.4 μL Nucleus Free Water (NFW). The PCR method was conducted in Bio-Rad T100™ Thermal Cycler (Bio-Rad, USA) and the machine condition followed [8]. The machine condition was set in 5 steps; a pre-denaturation temperature at 94°C for 5 min, 35 cycles of 94°C for 10 s (denaturation), 60°C for 20 s (annealing), 72°C for 30 s (extension), and a final extension at 72°C for 5 min. The PCR products were checked using electrophoresis in 1.5% agarose gel and visualized using Glite 965 GW imaging system (Pacific Image Electronics Co., Ltd.)

2.3 Genotyping

Genotyping was performed based on the result of targeted sequencing. A total of 20 μL PCR product was used for targeted DNA sequencing using forward and reverse primers. The DNA sequencing was conducted using Applied Biosystem Genetic Analyzer platform through 1st BASE DNA sequencing services (Selangor, Malaysia). The sequencing chromatogram and alignment were analyzed using FinchTV and MEGA 6.0 respectively. The double peak found in the sequencing chromatograms indicated that the sample was heterozygous.

2.4 Data Analysis

The genotype and allele frequency were analyzed following the formula of [9]:

Genotype frequency:

$$xii = \frac{nii}{N} \quad (1)$$

Allele frequency:

$$xi = \frac{2nii + \sum nij}{2N} \quad (2)$$

Where:
χii = frequency of ii genotype
χi = frequency of i allele
nii = number of individuals with ii genotype
nij = number of individuals with ij genotype
N = number of samples

The Hardy-Weinberg equilibrium was analyzed using Chi-square test [10]:

$$\chi 2 = \sum \frac{(O-E)2}{E} \quad (3)$$

Where:
$\chi 2$ = HWE test
O = the observed number of genotypes
E = the expected number of genotypes
α = 5% significance level
df = number of genotype probabilities – number of alleles

LTF gene association with milk compositions. The association study between LTF gene polymorphism and milk compositions were analyzed using General Linear Model procedure in SPSS ver. 26.0. The mathematical model used in this research was:

$$Yijk = \mu + Gi + Pj + \varepsilon ijk \quad (4)$$

Where:
Yij = observation value (milk compositions) of jth goat and i^{th} genotype
μ = overall mean
Gi = the effect of ith genotype
Pj = the effect of jth parity
εij : random error

3 Results and discussion

Lactoferrin (LTF) is an 80 kDa glycoprotein and had capability to bind and transfer irons. This protein is closely related to the serum transferrin gene family [4]. The previous studies were reported the biological functions of LTF to against bacteria [11], viruses [12], inflammation [13], and tumors [12]. LTF gene is a multifunctional gene for milk protein compositions, skeletal structure, and resistance against mastitis infection [4, 6, 14, 15].
In this research, a 541 bp of LTF gene sequence in exon 4 was successfully amplified using PCR technique at fixed annealing temperature of 60°C. The visualization of PCR product was performed in 1.5% agarose gel electrophoresis. In this research, a Single Nucleotide Polymorphism (SNP) was successfully investigated at c.407C>T (rs636609419). This SNP

was a missense mutation and changed the amino acid from Threonine to Methionine (Thr136Met).

According to patterns of PCR product cut by MspI enzyme, there were three genotypes found in current study. The frequencies of CC, CT, and TT genotypes were 0.381; 0.452; and 0.167, respectively. Furthermore, the frequencies of C and T alleles were 0.607 and 0.393, respectively. Based on these allele frequencies, the SNP c.407C>T in this research was polymorphic. The calculated chi-square was lower than the chi-square table of 3.84, descriptively. This indicated that the population of Senduro goats in Senduro village was in Hardy-Weinberg Equilibrium. According to the breeding management of Senduro goats in Lumajang regency, the goats allowed to have random mating and there was no selection program. This situation probably affected the equilibrium that there was no change of allele frequencies in a gene pool from one generation to the next.

Table 1. Genotype frequencies, allele frequencies and chi-square value

Breed	N	Genotype Frequency			Allele Frequency		χ^2
		CC	CT	TT	C	T	
Senduro	42	0.381	0.452	0.167	0.607	0.393	0.112^{ns}

Table 2. Milk production and compositions of Senduro goat

Parameter	Genotype		
	CC	CT	TT
Milk production (L)	974,85±381.50	1121.69±537.50	716±457.17
Fat (%)	26.42±1.77	25.15±3.60	26.83±2.06
Lactose (%)	4.79±0.38	4.69±0.31	4.85±0.33
Solid non-fat (%)	0.78±0.05	0.77±0.06	0.80±0.06
Solid (%)	8.27±0.55	8.07±0.70	8.45±0.61
Protein (%)	4.53±0.34	4.40±0.61	4.70±0.36

According to Table 2, the results showed that there was no association between LTF gene polymorphism and milk production and composition in this study (P>0.05). This case was probably affected by some factors such as the variation of parity, a few numbers of samples and different animal management. The goats with CT genotype had the highest milk production among genotypes, descriptively. The goats with TT genotype showed the highest milk compositions (fat, lactose, SNF, solid, and protein) although they had the lowest milk production. It indicated the negative correlation between milk production and milk composition.

4 Conclusion

In this research, the polymorphism of LTF gene exon 4 (SNP c.407C>T) was not associated with milk production and its composition. The SNP was polymorphic and needed further investigation on larger number of samples. This SNP was potential to be developed as a molecular marker to support the animal selection program that will improve the genetic quality.

The authors gratefully acknowledge the team of Dairy Science for supporting the data and the member of Genomic and Proteomic Research Group, Faculty of Animal Science, University of Brawijaya for data analysis. This research was supported by Faculty of Animal Science, Universitas Brawijaya trough Doctoral Research Grant in 2019.

References

1. T.E. Susilorini, Kuswati. *Budi daya kambing dan domba* (UB Press, Malang, Indonesia, 2019)
2. National Standardization Agency of Indonesia. *Indonesian National Standard 7352-3:2018: Bibit Kambing Senduro* (National Standardization Agency of Indonesia, Jakarta, Indonesia, 2018)
3. L. Adlerova, A. Bartoskova M. Faldyna. Vet. Med. (Praha). **53**, 457-468 (2018)
4. B. L. Guo, Y. Jiao, C. He, L. X. Wei, Z. H. Chang, X. P. Yue, X. Y. Lan, H. Chen, C. Z. Lei. Genet. Mol. Res. **9**, 2199-220 (2010)
5. I. Akisa, F.E. Gursela, C. Unb K. Oztabaka. Thai J. Agric. Sci. **45**, 83-88 (2012)
6. H.Y. Ali, W.I. Al-Samarai. Iraq. J. Agri. Sci. **49**, 200-204 (2018)
7. J. Sambrook, E. F. Fritschi, T. Maniatis. *Molecular cloning: a laboratory manual* (Cold Spring Harbor Laboratory Press, New York, USA, 1989)
8. A. Furqon, A. Gunawan, N. Ulupi, T. Suryati, C. Sumantri. Int. J. Poult. Sci. **16**, 462-466 (2017)
9. M. Nei, S. Kumar. *Molecular evolution and phylogenetics* (Oxford University Press, New York, USA, 2000)
10. D. Hartl, A. G. Clark. *Principles of Population Genetics, 3rd edn.* (Sinauer Associates: Sunderland, UK, 1997)
11. S. A. González-Chávez, S. Arévalo-Gallegos, Q. Rascón-Cruz. Int. J. Antimicrob. Agents. **33**, 301 (2009)
12. J. Małaczewska, Z. Rotkiewicz. Medycyna Weterynaryjna. **63**, 136-139 (2007)
13. H. Wakabayashi, K. Yamauchi, M. Takase. Int. Dairy. J. **16**, 1241-1251 (2006)
14. T. Kutila, S. Pyorala, H. Saloniemi, L. Kaartinen. Acta Vet. Scand. **44**, 35-42 (2003)
15. I. A. García-Montoya, T. S. Cendón, S. Arévalo-Gallegos, Q. Rascón-Cruz. Biochim Biophys Acta. **1820**, 226-236 (2012)

A review chromosome analysis comparison of swamp buffalo (*Bubalus bubalis*) with different phenotypic characters in East Java, Indonesia

Ardyah Ramadhina Irsanti Putri[1], *Gatot* Ciptadi[1,*], *Agus* Budiarto[1], *Sri* Rahayu[2], *Helly Nurul* Karima[3], *Susiati*[4], *Perdana* Finawati[4], *Mudawamah*[5], and *Yudit* Oktanella[6]

[1]Faculty of Animal Science, Brawijaya University, 65145 Malang, Indonesia
[2]Faculty of Mathematics dan Science, Biology Department, Brawijaya University, 65145 Malang Indonesia
[3]Central laboratoium (LSIH), Brawijaya University, 65145 Malang, Indonesia
[4]Biosains Institute, Brawijaya University, 65145 Malang, Indonesia
[5]Faculty of Animal Husbandry, Islamic University of Malang, 65144 Malang, Indonesia
[6]Faculty of Veterinary Medicine, Brawijaya University, 65145 Malang, Indonesia

Abstract. Chromosomal data on swamp buffalo especially in East Java province, is still lacking. For breeders, chromosomal analysis needs to be done to detect the possibility of hereditary genetic defects. In Malang Area, farmers lack of care and breeding programs for swamp buffalo also caused a decline in the genetic quality and performance. The purpose of this study was to compare the results of the chromosome analysis of swamp buffalo in Malang and another area in East Java province, which were selected based on differences in phenotypic characters. Chromosome analysis is very important for breeders because the results of this study can be used as a consideration for strategies to improve the genetic quality of swamp buffalo. The method used is standard karyotyping using whole blood with G banding staining. Chromosomal preparation using Karyo MAX medium, Colcemic Solution, Giemsa Stain, and KCl solution. Cell culture was performed according to the standard karyotyping method in mammals. The minimum number of 5 spreading chromosomes is the best was chosen, microphotographed, and then chromosome analysis is performed using Cytovision Image Analysis software. The results of this study did not find swamp buffalo with an abnormal number of chromosomes in East Java Province, Indonesia.

1 Introduction

Most of the livestock animals in Indonesia are categorized as a non-identified breed and practically have no sufficient genetic quality information, pedigree, lower productivity, and limited recording availability. There is still limited research that studies local Indonesian livestock Animals, especially buffalo, based on the information or data of cellular and

* Corresponding author: ciptadi@ub.ac.id

molecular [1]. Analysis of the history and origins of local Indonesian animals, especially local buffalo (swamp buffalo), are still compiled and described mainly based on information on phenotypic characters. This condition occurs because almost all of the buffalo are kept traditionally by small farmers in the countryside, not yet raised for commercial purposes. Buffalo farming faces serious problems because the population is decreasing by 3% per year due to the availability of specific habitats for buffalo, which is getting narrower. Genetic analysis at the molecular and cellular level needs to be done through chromosome analysis, which can reportedly be used as a device for identifying and tracking animal origins [2].

Indonesia has high animal biodiversity, one of which is ruminant livestock, namely swamp buffalo. The main type of buffalo in Indonesia is the swamp buffalo, which varies phenotypically; the hair or skin colour includes black, greyish, reddish, and white, or their body condition score (BCS) [3]. The cause of this variation is not known with certainty, whether due to hormonal problems, feed conditions, climatic influences, or other causes. It is essential to preserve genetic resources to maintain and improve the quality of phenotypic and genotyping continuously so that breeds can survive [4]. The purpose of this study was to identify the swamp buffalo in Indonesia based on the emergence of the profile of the chromosome base on the varieties of phenotypic character.

2 Materials and methods

2.1 Animals

The livestock used in this study were taken from several places the East Java region, where male buffaloes / bulls were found by showing phenotypic variations in coat and skin color, located in Malang district, Sumenep, and Kangean Island. It was taken by purposive sampling of buffalo with red, white, and blackish grey skin color, which every different color taken one whole blood sampling. This Swamp buffalo in Malang and another area in East Java province were selected based on differences and variation in phenotypic characters.

2.2 Chromosome analysis and karyotyping

The method used is standard karyotyping using whole blood with G-banding staining. Chromosomal preparation using Karyo MAX medium (GIBCO/BRL), Colcemic Solution, Giemsa Stain, and KCl (Potassium chloride) solution. Cell culture was performed according to the standard karyotyping method in mammals. The minimum number of 5 spreading Metaphase II chromosomes is the best was chosen, microphotography is performed, and then chromosome analysis is performed using Cytovision Image Analysis software.

3 Results and discussion

3.1 Animals with phenotypic traits that deviate from the standard

Swamp buffaloes selected for chromosomal analysis were characterized by phenotypic deviations of fur or skin color from the standard blackish gray colour, known from 4 regions, namely Sumenep, Kangean Island, Malang, and Banyuwangi. In general, the productivity of the swamp buffalo in East Java varies greatly. The color aberrations include red (Kangean Island), White (Kangean Island and Sumenep district), and blackish grey colour (Malang district). Black was excluded from the chromosome analysis because it was later discovered that the black colour of buffalo on Kangean Island was used as a buffalo contest and race. The breeder makes it during the Buffalo race (Fig. 1).

Fig. 1. Variations in fur/skin color are found in the fields; (A) Kangean Island, (B,C) Sumenep, and (D) Malang

Previous research results, we interviewed community leaders on Kangean Island, the incidence or frequency of the appearance of this white or fur-coloured buffalo is quite high, several or even dozens on Kangean Island, but then decreased because it is said that these buffaloes were sacrificed for ceremonies [5]. Certain traditional or religious events in Kangean Island Square. Currently, there are only three white buffalo left, while the red ones are increasing in number. Swamp buffalo with white hair color abnormality was also found in some young calves [6], for example, in Malang and Ngawi regencies, but then changed to normal as they got closer to sexual maturity; this is thought to be due to hormonal factors.

3.2 Chromosome analysis of swamp buffalo

The chromosomes are considered normal or abnormal based on standard karyotyping. The results of this study did not find swamp buffalo with an abnormal number of chromosomes, so it could be interpreted that there was no chromosomal abnormality from a phenotypic comparison of swamp buffalo in East Java Province, Indonesia. The results of the preparation of the chromosomes of the swamp buffalo using the Giemsa staining technique (G-banding standard) resulted in a fairly good spreading so that it was possible to calculate the number of chromosomes easily [7]. The results of the analysis showed that both buffaloes with normal blackish grey skin colour, or those with reddish-grey and white skin colour, showed a normal number of chromosomes, as many as 48 chromosomes. The results of the karyotyping analysis, when compared with the existing standards, both body and sex chromosomes, also showed normal results [8, 9]. Red and white buffalo on Kangean Island and Sumenep, Madura also showed a normal number of chromosomes [9]. Also in previous study have done chromosome analysis of PE and Senduro goats and found that the X-Y chromosomes of the two goat breeds did not differ based on the morphological shape of the chromosomes [11].

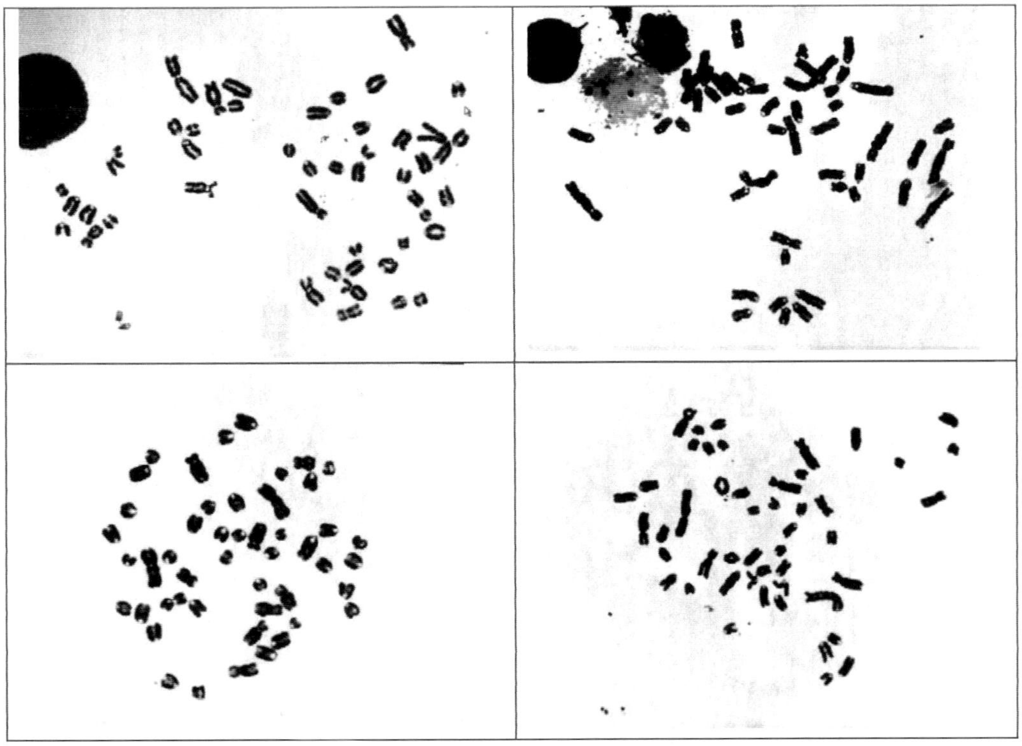

Fig. 2. The results of karyotyping of East Java swamp buffalo with various phenotypic variations that show genetically normal results

The results of the preparation of the chromosomes of the swamp buffalo using the Giemsa staining technique (G-banding standard) resulted in a fairly good spreading so that it was possible to calculate the number of chromosomes easily. The results of the analysis showed that both buffaloes with normal blackish grey skin colour, or those with reddish-grey and white skin colour, showed a normal number of chromosomes, as many as 48 chromosomes. The results of the karyotyping analysis tested in this research were normal categories. The karyotype analysis showed that the chromosomes of one cell and different individuals in each breed varied in the centromere's size, shape, and position. The standard chromosomal preparation technique for buffalo with G banding still needs to be improved to obtain better chromosome spreading results and facilitate detailed karyotyping. It seems necessary to suggest molecular studies to prove the possibility of mutations in skin colour control genes [12]. Based on genetic analysis based on the number of chromosomes from this study are that all swamp buffalo have a normal genetically. Further research needs to be done using other staining techniques such as R-banding or even FISH (Florescent In Situ Hybridization) to obtain better and specific karyotyping results.

4 Conclusion

The conclusion of this study did not find swamp buffalo with an abnormal number of chromosomes (2N=48), so it could be interpreted that there was no chromosomal abnormality in their number based from a phenotypic comparison of swamp buffalo in East Java Province, Indonesia.

This research is part of research on Analysis of Local genetic Animals funded by the grant from Brawijaya University of HPU (Hibah Penelitian Unggulan) 2021 Grant, No.r: 023.17.2.677512/2021. I wish to thank all young assistance researchers, especially Helly Nurul Karima, Dinda H.K. Wijaya, Ardyah Ramadhina I. P., who is involved and works preparing chromosome analysis in most of the work is related

References

1. G. Ciptadi, A Budiarto, M Nasich, Mudawamah, S Rahayu, Dyah Ayu O. A.P, A I Putri, H N Karima, Y Oktanella, Y Saynandya and R G Almaida, IOP Conference Series: Earth and Environmental Science, **743** (2020)
2. R. Parada, M. Kawka, M. Sacharczuk, P. Urbański, K. Jaszczak, J. Biol Sci., **7** (2017)
3. G. Ciptadi, M Nasich, A Budiarto, V.M.A. Nurgiartiningsih. Pakistan Veterinary Journal, **32** (2012)
4. A Budiarto, G Ciptadi, M Nasich, and Fatich, IOP Conf. Series: Earth and Environmental Science, **478** (2020)
5. G. Ciptadi, M. Mudawamah, V.M.A. Nurgiartiningsih, S. Wahjuningsih, Rr. F. D. Listiani, Susiati, L. Hakim, and A. Budiarto, AIP Conference Proceedings 2021, 070009 (2018)
6. A. Budiarto, G. Ciptadi, L. Hakim, Susiati, N.K. Karima, M.H. Hisam, IOP Conf. Series: Earth and Environmental Science, **247** (2019)
7. A.R.I. Putri, G. Ciptadi, A.P. Warih, IOP Conference Series: Earth and Environmental Science, **119** (2018)
8. A Budiarto, L Hakim, G. Ciptadi, M. Nasich, A.R.I. Putri, N.K. Karima, Y..D..A Prasetyo, IOP Conference Series: Earth and Environmental Science, **743** (2021)
9. G. Ciptadi, M.N. Ihsan, V.M.A. Nurgiartiningsih, I.P. Ardyah, A. Budiarto and Mudawamah, Biodiversitas Journal of Biological Diversity, **18** (2017).
10. G. Ciptadi, A. Budiarto, M. Nasich, Mudawamah, S. Rahayu, D. Ayu, A.I. Putri, H.N. Karima, Y. Oktanella, Y. Saynandya and RG Almaida, IOP Conference Series: Earth and Environmental Science, **743** (2020)
11. G. Ciptadi, M.N. Ihsan, S. Rahayu, V.M.A. Nurgiartiningsih, Mudawamah, and A.R.I. Putri, Research Journal Of Life Science, **4** (2017)
12. Mudawamah, I. Ratnaningtyas, M. Fadli, G. Ciptadi, Journal of Physics: Conference Series, **1146** (2019)

The real potential of semen production of Bali Bull: over year observation at Singosari National Artificial Insemination Center (SNAIC), Singosari-Indonesia

Chairdin Dwi Nugraha[1], *Nashi* Widodo[2], *Kuswati* Kuswati[1] and *Suyadi* Suyadi[1,*]

[1] Faculty of Animal Science, University of Brawijaya, Jl. Veteran, Malang 65145 – Indonesia
[2] Department of Biology, Faculty Mathematics and Natural Sciences, University of Brawijaya, Jl. Veteran, Malang 65145 – Indonesia

Abstract. Increased reproduction is closely related to increased bull productivity. The high productivity of these bulls will increase the amount of straw. The productivity capability of Bali bull is essential in producing straw. High productivity will result in high straw production as well. So the research aims to determine the productivity of Bali bull at SNAIC in the last year. 31 Bali bulls aged 4 – 11 years were used in this study at SNAIC. Total 1.273 ejaculations collection from these during one year. The parameters used in this study include the number of successful holding of each bull, total productivity, and estimation of straw production in one year. From 31 Bali Bulls, the ranking was grouped into two based on values of sperm productivity. The two groups were analyzed using an unpaired T-test. The results showed that the pH, volume, individual motility, concentration, and abnormality were significantly different ($P < 0.05$) between the high and low groups. The percentage of successful shelters, total TSM, and the number of straws produced in one year in the high productivity group were higher than the low productivity group. High productivity groups are the best with high productivity potential and potential as a source of semen.

1 Introduction

Bali cattle (*Bos sondaicus*) is one of the local cattle breeds with the highest population among other local cattle in Indonesia. Bali cattle are resistant to parasites and tropical environments in Indonesia and have better reproductive efficiency than other local cattle [1]. Bali cattle have been established through the Decree of the Minister of Agriculture No. 325/Kpts/OT.140/1/2010, as native Indonesian cattle and have an essential role in the national meat supply [2].

Superior bulls have been selected based on their lineage, production, and reproductive capabilities [3]. Good semen must be above the minimum standard of fresh semen to be processed into frozen semen. The minimum motility value is 70% [4], while the maximum

* Corresponding author: suyadi@ub.ac.id

abnormality is 20%, the volume ranges from 7–10 ml with a spermatozoa concentration of 1000–1500 x 10^6/ml [5, 6]. Bull fertility is influenced by age, race, body condition, feed, ejaculation ability, and environment [7].

Efforts to develop and preserve Bali bull from all aspects need to be carried out sustainably. One crucial aspect of doing is to improve the quality of reproduction. Increased reproduction is closely related to spermatozoa fertility and increased bull productivity. Reproduction is an essential step in the management system that impacts the economic aspect [8]. The high productivity of these bulls will increase the amount of frozen semen production. The government regulation of the minister of agriculture No. 10/Permentan/PK.210/3/2016 in section 4, namely the provision of frozen semen for domestically produced ruminants, must be 60% from native and/or local ruminants. This is an excellent opportunity for Bali cattle if they produce higher productivity [9].

On the other hand, the quality of Bali cattle semen varies highly. For example, the motility value of Bali cattle in several studies includes 62.07% ; 70.00%; 80.00%; 89.62% [10 - 13].

The productivity capability of Bali cattle is essential in producing frozen semen. Productivity is closely related to the value of volume, individual motility, and concentration of spermatozoa. Higher productivity will result in high straw production as well. So the research aims to determine the productivity of Bali cattle at BBIB Singosari in the last year.

2 Materials and methods

2.1 Animals

The study was carried out at Singosari National Artificial Insemination Center (SNAIC), Singosari-Indonesia (SNI.ISO 90001:2015). A total of 31 Bali bulls aged 4–11 years were used in this study. Total 1.273 ejaculations collection of these during one year. The bulls were reared under similar management based on the standard procedure in SNAIC.

2.2 Semen Collection and Evaluation

Semen collection was conducted once a week every morning. Semen was collected using an artificial vagina as a Standard Operating Procedure (SOP) of the SNAIC. The artificial vagina type is Double-Walled type. Semen quality data were determined in laboratories through direct observation after semen collection. The microscopic evaluations were individual motility, sperm concentration, and viability. The macroscopic evaluations were volume, pH, and color.

The volume of semen (ml) is determined directly after the semen is stored on the scale of the tube used to hold the semen [5]. Individual motility (%) was observed with a microscope with 400x magnification at a constant temperature using a cover glass, then determined the proportion (percentage) of progressively moving spermatozoa. The sperm concentration was observed by using a spectrophotometer [14]. Sperm viability used the eosin nigrosine staining method. The head of living spermatozoa will remain transparent and the dead spermatozoa will be red in color [15].

2.3 Data Classification

The total motile spermatozoa from each of Bali cattle for one year were calculated. The Total Spermatozoa Motil (TSM) value obtained becomes a reference for determining the rank of bull. Based on the productivity of Bali bulls in a year is obtained were divided into two groups. The top three bulls became the high group, and the lowest three bulls became the low group.

2.4 Statistical Analysis

From 31 Bali Bulls, the ranking was grouped into two based on values of sperm productivity. The two groups were analyzed using an unpaired T-test and were processed by using SPSS 24. Data were presented with the mean ± standard deviation (SD). Bull productivity was assessed from the Total Spermatozoa Motile using the following formula:

$$\textit{Total Spermatozoa Motile} = \textit{volume} \times \textit{concentration} \times \textit{individual motility} \quad (1)$$

Standard Operational Procedures (SOP) at SNAIC in a dose of 1 straw or 0.25 ml total spermatozoa concentration are $25\text{x}10^6$. So, the estimated number of frozen semen straws produced in one year is calculated by the formula:

$$\textit{Number of straws} = \textit{Total Spermatozoa Motile} \,/\, 25 \times 10^6 \quad (2)$$

3 Results and discussion

Macroscopic examination (Table 1) showed that the pH and semen volume were significantly different ($P < 0.05$) between the high and low groups. The highest pH was in the low group, while the high volume was in the high group. The color of the semen of the high group was milk-white, while the low group clear white. The range of semen volume ranged from 4.07-8.79 ml. Suyadi *et al.* [16] stated that the semen volume in Bali bulls was constant from 4.7 to 4.9 ml/ejaculate. Semen volume increases with age [14]. This is because increasing age will cause the scrotal circumference to grow so that the volume of semen increases.

Table 1. Characteristics of Bali bull semen in the two productivity groups

Parameter	Productivity Group	
	High	Low
N ejaculate	141	67
Color	Milk White	Clear White
pH	$6{,}49 \pm 0{,}40^a$	$6{,}62 \pm 0{,}22^b$
Semen Volume (mL)	$8{,}79 \pm 1{,}68^b$	$4{,}07 \pm 1{,}15^a$
Individual Motility (%)	$82{,}32 \pm 6{,}94^b$	$79{,}17 \pm 9{,}15^a$
Sperm Concentration (10^6 / mL)	$1320{,}21 \pm 292{,}95^b$	$757{,}49 \pm 369{,}54^a$
Sperm Abnormality (%)	$4{,}40 \pm 2{,}31^a$	$10{,}37 \pm 6{,}38^b$

Different superscripts within line indicate significant differences (P<0,05)

The results (Table 1) showed that microscopic examination (individual motility and concentration) of the high group was significantly ($P < 0.05$) higher than that of the low group. Meanwhile, the abnormality of the high group was lower than that of the low group. Individual motility ranged from 79.17% - 82.32%, sperm concentration per ejaculate was between 757.49 x10^6 - 1320.21 x10^6, and sperm abnormalities were between 4.40% - 10.37%.

The average individual motility in this study was higher than that of Nugraha *et al.*[14] at 62.5% and Nugraha *et al.* [17] at 65 %, while a suitable bull for breeding must have an individual motility value of at least 70% (SNI 4869-1:2017). Bali bulls in this study were included in the category of good semen quality, which was more than 60% - 75% for sperm motility [5]. The sperm concentration in this study was lower than that in PO cattle [18], which was 2250 x10^6 / mL and higher than Madura bull that is 797.526 x10^6 / mL [19]. The semen of the Bali bull in this study is included in the criteria for good quality semen because the requirements for good semen are sperm concentrations of more than 500 million/ml [5]. At the same time, the abnormality is higher than Suyadi [16], which is 2.99%. According to Ax *et al.* [5], good spermatozoa abnormality is less than 20%.

Table 2. Number of semen successes that were successfully accommodated in two different TSM groups

Productivity Group	Bull Code	Age (year)	Total Shelter (ejaculation) in 1 Year	Total Shelter (ejaculation) Succeed in 1 Year	
			(times)	(times)	(%)
High	Bull 1	10	47	47	100,00
	Bull 2	10	47	47	100,00
	Bull 3	4	47	47	100,00
Low	Bull 29	4	47	18	38,30
	Bull 30	6	47	42	89,36
	Bull 31	10	47	7	14,89

Table 2 shows that all bulls in the high group in one year were successfully accommodated 47 times out of a total of 47 shelters (ejaculation) carried out. While in the low group, no bulls were successfully accommodated 47 times in one year. Bull 29 as many as 18 times in a total of 47 shelters (ejaculation) a year. Bull 30 as many as 42 successful shelters (ejaculation) in one year. Bull 31 only seven successful shelters (ejaculation) in one year. Percentage of total shelter (ejaculation) succeed in one years in bull 29, bull 30 and bull 31 were 38,30; 89,36; 14,89, respectively.

The results of this study, bull 1, bull 2, and bull 3, which are the high productivity group, are included in the category of superior bulls because the percentage of successfully accommodated reaches 100%. This study shows that there is a wide variation in Bali bull at SNAIC-Singosari. This reflects the genetic diversity of the Bali bull, which still has the opportunity to be improved. [20] stated that the application of genetic standards needs to be considered to maximize the potential and maximum and sustainable sperm production.

Table 3. The potential productivity of Bali bulls and the estimated number of straws produced in one year

Productivity Group	Bull Code	Age (year)	N ejaculate	Total TSM in 1 Year (10^6)	∑ Straw generated in 1 Year (straw)
High	Bull 1	10	47	501.803	20.072
	Bull 2	10	47	427.069	17.083
	Bull 3	4	47	408.973	16.359
Low	Bull 29	4	18	67.688	2.708
	Bull 30	6	42	67.251	2.690
	Bull 31	10	7	37.213	1.489

The potential for frozen semen production per ejaculation has a reasonably high range between the high and low groups. The potential for frozen semen production in the high group (Bull 1, Bull 2, and Bull 3) in one year were 20,072 straws, 17,083 straws, and 16,359 straws per ejaculate, respectively (Table 3). Meanwhile, in the low group (Bull 29, Bull 30, and Bull 31), the potential for frozen semen production was 2,708 straws, 2,690 straws, and 1,489 straws.

Bali bull productivity is closely related to the amount of straw produced. Potential productivity is the ability of each bull to produce straw or frozen semen. Productivity in this study is calculated for one year. The difference in straw production between the high and low productivity groups was caused by the number of successful collections in one year. In the low group, the moderate success per individual is only 22.3 times, while in the high group, the average success per individual reaches 100% or 47 times.

The results showed that the straw produced by each bull in the high group for one year was higher than Ditjennak [21]; which is, the minimum local bull each individual produced was 7,500 straws annually. The research results are also higher than those of other Bali cattle studies Indriastuti *et al.* [13], with 11,843.17 straws per year. This shows that Bali cattle at BBIB Singosari have the potential to produce high straw. The potential for frozen semen production is influenced by volume, concentration, and the number of motile sperm [21, 23].

4 Conclusion

Based on the number of successful shelters (ejaculation), the total TSM, and the number of straws produced in one year, it can be concluded that the two groups of high and low productivity in Bali bulls were different. Bull 1, Bull 2, and Bull 3 at SNAIC-Singosari are the best with high productivity potential and potential as a source of semen.

The author thanks the Director and Team of Singosari National Artificial Insemination Center (SNAIC)-Singosari, Malang-Indonesia for providing data of reproductive and semen quality for Bali bull.

References

1. L. W. Pribadi, S. Maylinda, M. Nasich, S. Suyadi, J. Agric. Vet. Sci., **7** (2014)
2. Kementan (Kementerian Pertanian), *Decree of the Minister of Agriculture Number 325/Kpts/OT.140/1/2010 Concerning Determination of Bali Cattle Clumps,* Ministry of Agriculture (2010)
3. SNI – Standard Nasional Indonesia, *Bibit Sapi Potong – Bagian 5: Bali, Badan Standardisasi Nasional SNI 7651.4:2017* (2017)
4. SNI – Standard Nasional Indonesia, *Semen Beku – Bagian 1: Sapi, Badan Standardisasi Nasional SNI 4869.1:2017* (2017)
5. R. L. Ax, M. Dally, B. A. Didion, R. W. Lenz, C. C. Love, D. D. Varner, B. Hafez, M. E. Bellin, *Semen Evaluation In: Reproduction in Farm Animal* ed By Hafez ESE, 7th Edition (2008a)
6. R. L. Ax, M. Dally, B. A. Didion, R. W. Lenz, C. C. Love, D. D. Varner, B. Hafez, M. E. Bellin, *Artificial Insemination In: Reproduction in Farm Animal* ed By Hafez ESE, 7th Edition (2008b)
7. M. M. Rojas-Downing, A. P. Nejadhashemi, T. Harrigan, S. A. Woznicki, Clim. Risk Manag., **16** (2017)
8. S. Suyadi, L. Hakim, S. Wahjuningsih, H. Nugroho, Journal of Applied Science and Agriculture, **9** (2014)
9. Kementan (Kementerian Pertanian), *Regulation of the Minister of Agriculture of the Republic of Indonesia Number 10/Permentan/PK.210/3/2016 concerning the Supply and Distribution of Ruminant Animal Frozen Semen,* Ministry of Agriculture (2016)
10. S. Bayu, N. Isnaini, Russ. j. agric. soc.-econ. Sci., **2** (2020)
11. Yendraliza, A. E. Harap, J. Handoko, M. Rodiallah, Animal Science, **43** (2019)
12. Yendraliza, A. E. Harahap, J. Handoko, M. Rodiallah, C. Arman, Songklanakarin J. Sci. Technol., **42** (2020)
13. R. Indriastuti, M. F. Ulum, R. I. Arifiantini, B. Purwantara, Vet. World, **13** (2020)
14. C. D. Nugraha, E. Herwijanti, I. Novianti, A. Furqon, W. A. Septian, W. Busono, S. Suyadi, J. Indones. Trop. Anim. Agric., **44** (2019)
15. A. Budiyanto, M. Arif, M. P. W. Alfons, R. T. Fani, A. F. Hafid, B. Wicaksono, K. M. Insani, M. Herdinta, Acta Veterinaria Indonesiana, (2021)
16. S. Suyadi, E. Herwijanti, W. A. Septian, A. Furqon, C. D. Nugroho, R. F. Putri, *IOP Conf. Ser. Earth Environ. Sci.*, 478 (2020)
17. C. D. Nugraha, E. Herwijanti, I. Novianti, A. Furqon, W. A. Septian, W. Busono, S. Suyadi, Journal of Tropical Animal Production, **20** (2019)
18. C.D. Nugraha, N. Widodo, Kuswati and S. Suyadi, *IOP Conf. Ser. Earth Environ. Sci.,* **788** (2021)
19. I. Novianti, B. Purwantara, E. Herwijanti, C. D. Nugraha, R. F. Putri, A. Furqon, W. A. Septian, S. Rahayu, V. M. A. Nurgiatiningsih, S. Suyadi, Jurnal Ilmu-Ilmu Peternakan, **30** (2020)
20. S. Prastowo, N. Widyas, A. Ratriyanto, M. S. T. Kusuma, P. Dharmawan, I. A. Setiawan, A. Bachtiar, *IOP Conf. Ser. Earth Environ. Sci.*, **372** (2019)
21. Ditjennak (Direktorat Perbibitan dan Produksi Ternak), *Directorate of Animal Breeding and Production, Directorate General of Animal Husbandry and Health*, Ministry of Agriculture (2018)
22. S. Aisah, N. Isnaini, S. Wahyuningsih, Indones. J. Anim. Sci., **27** (2017)
23. Santoso, Herdis, R. I. Arifiantini, A. Gunawan, C. Sumantri, Trop. Anim. Sci. J., **44** (2021)

Heat shock protein (HSP) release mechanism under heat stress pressure in Goats: a review

Rafika Febriani Putri[1], *Tri Eko* Susilorini[1], *Nashi* Widodo[2] , *Kuswati* Kuswati[1] and *Suyadi* Suyadi[1,*]

[1]Faculty of Animal Science, Universitas Brawijaya, Malang, Indonesia
[2]Department of Biology, Faculty of Mathematics and Natural Sciences, Universitas Brawijaya, Malang, Indonesia

Abstract. Among the various climate variables, heat stress has been reported to be the most detrimental factor to the economy of the livestock industry. Heat stress is one of the most stressful events in the life of livestock with harmful consequences for animal health, productivity and product quality. Heat shock proteins (HSPs), also known as molecular chaperons, are prominent stress markers. Heat shock proteins consist of highly conserved protein expressed at the time of stress, and play an important role in adaptation to the environmental stress. This review discusses the scientific evidence regarding the effects of heat stress and role of HSP during heat stress on Goats.

1 Introduction

Goats have been widely used as livestock models in recent climate research. They have better thermotolerance [1], ability to survive on limited pastures, as well as their disease resistance capacity. Goats easy to mantain and can live well, even by consuming shrubs and other poor quality fiber, because goats are able to digest crude fiber [2, 3]. Rearing goats is considered more economical than large ruminants in many studies on climate change related to feed and fodder availability. Goat populations living under harsh environments have a wide diversity of traits. This proves that various environmental conditions can have different effects on goat productivity [4].

Indonesia is a tropical country because it is located on the equator. This condition causes Indonesia to experience heat throughout the year with relatively high temperatures. It can affect negatively thermoneutral zone of the animals.

Goat can easily exposed by various stress, like physical, nutritional and heat stress. Of all, heat stress is the most worrisome due to decreased in livestock performance, as well as natural immunity. Heat stress make livestock more susceptible to disease and even death. Thus, economic losses are unavoidable due to heat stress [5, 6].

When cattle are stressed, the body will maintain homeostasis through physiological mechanisms including endocrine responses and cellular heat stress responses [7]. During

* Corresponding author: suyadi@ub.ac.id

thermal stress, the body's resources including protein and energy was used to maintain homeostatis by expense the declining growth, reproduction, production, and health.

Heat shock proteins (HSPs) are responsible for maintaining the balance in organism, to acclimatize the stress and these proteins regulated Cellular tolerance to heat stress. HSPs are released intra and extra cellularly in response to various environmental stresses [8,9], beside that HSPs can be an indicator of stress in cells.

Hsp70 has a significant role in cell thermo tolerance and animal survival than other HSP family. The heat stress regulation at cellular and molecular level, include the expression pattern of Hsp70 gene was very important to understand, because it can describe the mechanism of heat stress and role of Hsp70 during heat stress adaptation in goats [10, 11].

2 Heat shock protein (HSP) and its function

Heat shock proteins (HSPs) are a specific family of proteins produced by cells, and protect the cell from damage in response to stress conditions. Subsequent studies have shown that this protein is also produced as a result of other conditions in the organism such as ultraviolet radiation [12], temperature fluctuations [13], as well as during wound healing and tissue regrowth [14].

Most of this family member act as molecular binders between the cell and some proteins in the membrane and prevent the formation of newly formed proteins to maintain their vital roles. Heat shock protein plays an important role when high temperature has an adverse effect on cell structure and cell physiology [15].

[16] noted that HSPs are various proteins that act to protect cells from different stress conditions and are classified according to their molecular weight. [17] have reported that the importance of HSPs was their direct involvement in all important protein-related processes such as protein secretion, synthesis and prevention of degradation, as well as regulation of replication processes.

3 Mechanism of HSP synthesis and release

The mechanism of action of HSPs is directly related to heat shock factor (HSF) which occurs in almost all organisms, especially in livestock. [18] demonstrated that HSF acts as a transcription factor with the gene responsible for HSP. When cells are exposed to different stress conditions, HSP genetic expression is induced.

Increased expression of these genes is considered a response to stress through various genetic and biochemical processes referred to as HSR [19]. Heat shock proteins have different molecular weights as measured by Dalton. These proteins are called HSP60, HSP70 and HSP90 [20].

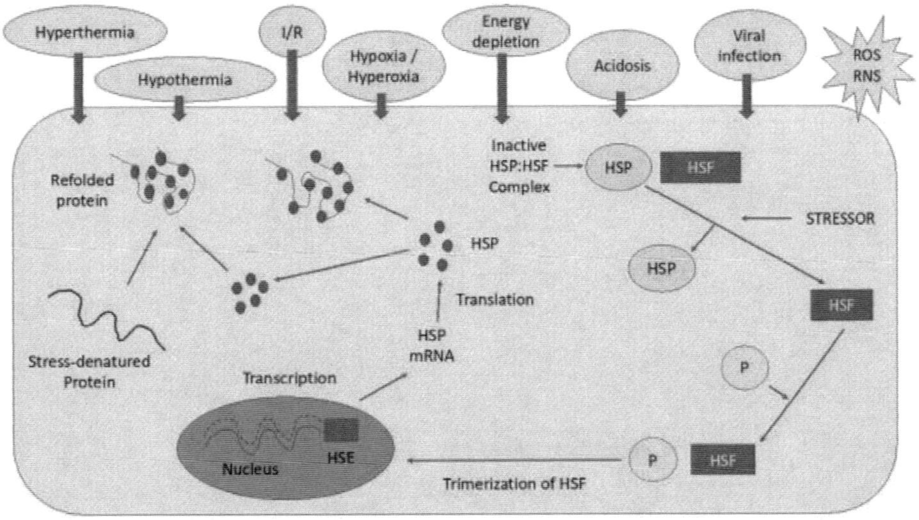

Fig. 1. Molecular Mechanism of Heat Shock Protein (HSP) Synthesis [2].

In the cytosol, the HSP connect each other with heat shock factors (HSF) to maintain an inactive state (Figure 1). Stimulation by heat stressors (HS) make HSF actived. It caused the bond between HSP and HSF is severed. Then, trimer complex (The HSF that are phosphorylated by protein kinases) will enter the nucleus and bind with heat shock elements (HSE). From the process of transcription, HSP70 mRNA was formed. The new HSP70 will be synthesised after leaves the nucleus. The newly synthesised HSP serving as a molecular chaperone. Then, facilitating the formation of new proteins within the cell. The newly synthesised HSP also repairing and refolding damaged proteins when heat stress occured [21].

4 HSPs during heat stress in goats

Stress could be defined as body's reaction to external influences that disrupt homeostasis, can affect on health and performance [22]. Small ruminant owners must be faced by climate change effects that impact to animal health and production [23, 24]. According to [24] heat stress is a problem that will continue to increase in the near future due to the development of global warming.

[25] define heat stress in which the physiological mechanisms of the animal's body are activated to maintain thermal balance, when animal was exposed by high temperatures. The effect of heat stress various on livestock according to species, growth phase, even the nutritional status. [26].

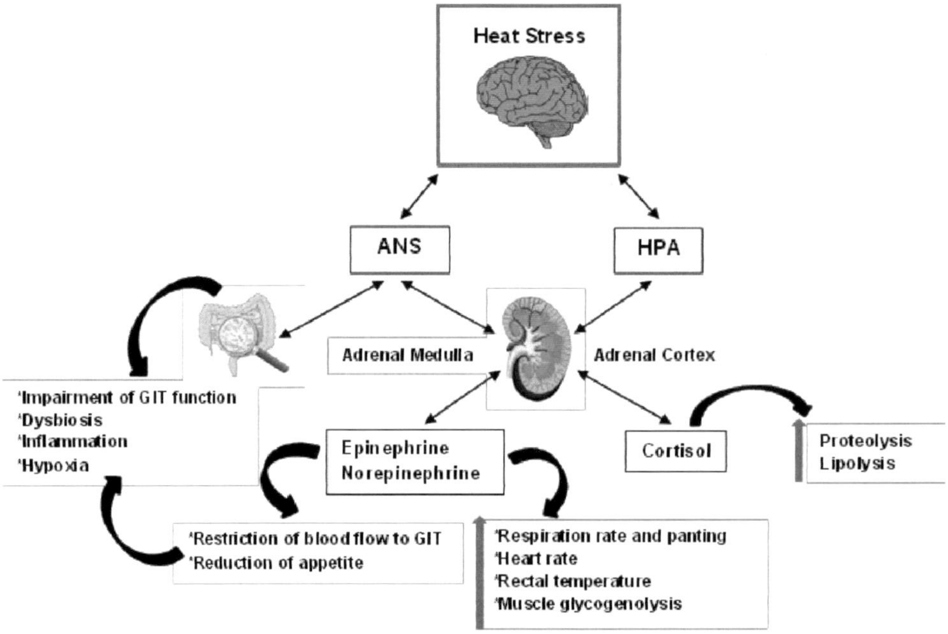

Fig. 2. Impact of Heat Stress to the Autonomic Nervous System (ANS) and the Hypothalamic-Pituitary-Adrenal (HPA) Axis [27].

Heat stress affects the all of animal's body, from their hematological until biochemical parameters [28-31]. It also negatively affects their production, growth, and natural immunity. When heat stress occured, goats more susceptible to disease and even death [31].

Heat shock protein 70 (HSP70) is used as a biomarker of cellular stress. Their expression level can show of duration of heat stress.

Heat stress denatures proteins and alters translation. Heat exerts a complex effect on autophagy [33]. Autophagy helps cell to adapt during heat stress by facilitating correct folding in response of cell adaptation [34, 35].

[36] reported that HSP70 gene expression was higher in PBMC of tropical goats during the summer season. HSP response was found in the liver and kidney of Barbari gaots which implies that higher expression of HSP70 at the tissue level protects the cells damage [36]. [38] showed that a higher expression of HSP70 also found in the kidney and heart of goats during another stress factor occured. [39] reported the lower expression of HSP genes in sheep revealed the possibility of highest adaptability. Further, [40] observed higher expression of HSP90AA1 gene during chronic heat stress while reported reduced expression of HSPA1 A and HSPA8 during summer season in Chokla, Magra, Marwari, and Madras Red sheep breeds. Seasonal variation was also established for the expression pattern of HSP70 gene family in Sirohi, Barbari, Gaddi and Chegu breeds of goats adapted to different agro climatic conditions [41]. Similarly, [42] also observed over expression of HSP90AA1 G/C-660 and A/G444 in heat stressed Manchego Spanish sheep.

5 HSPs and the comfort zone

The thermoneutral zone of ruminants is the optimal temperature range in which no additional energy above maintenance is required for thermoregulation. If environmental factors increase beyond the limits of the thermoneutral zone, the body will adapt. The process of heat loss occurs when the core body temperature increases, the ambient temperature approaches body temperature, and evaporation by sweat and panting. Under these conditions, body temperature and metabolic heat production rise, and the animal inevitably succumbs to hyperthermia and heat stress [43].

The autonomic nervous system (ANS) can be activated through a stress response mediated by catecholamines (adrenaline and noradrenaline). This results in increased respiration and heart rate, increased body temperature, redistribution of blood flow from viscera to skin for thermoregulation, and promotion of energy utilization from body reserves [27].

6 Conclusion

Stress disrupt homeostasis and affect on health and performance. Heat stress is the most worrisome due to the ever-changing climate scenario. Cellular defense to heat stress is mantained by heat shock proteins (HSPs). Among the HSPs, Hsp70 has a significant role in cell thermo-tolerance and goats survival.

References

1. Suyadi, S., W.A. Septian, A. Furqon, T.E. Susilorini, M. Nasich, IOP Conf. Series: Earth and Environmental Science, **391** (2019).
2. Shaji, S., V. Sejian, M. Bagath, G.B. Manjunathareddy, E.K. Kurien, G. Varma and R. Bhatta, Biological Rhythm Research, **48**, 65-83 (2017).
3. Nasich, M., Suyadi, G. Ciptadi, A. Budiarto, IOP Conf. Series: Earth and Environmental Science, **119** (2018).
4. Carvalho, G.M.C., S.R. Paiva, A.M. Araújo, A. Mariante, H.D. Blackburn, J. Anim. Sci., **93**, 4629–4636 (2015).
5. Sarangi S, The Pharma Innovation J, **7**, 4, 1114–1126 (2018).
6. Suyadi, S., T.E. Susilorini, W.A. Septian, A. Furqon, C.D. Nugraha, R.F. Putri, IOP Conf. Series: Earth and Environmental Science, **478** (2020).
7. Suyadi, S., S. Wahjuningsih, W.A. Septian, A.Furqon, R.F. Putri, C.D. Nugraha, 2021, IOP Conf. Series: Earth and Environmental Science, **788** (2021).
8. Chaudhary, U.B., K. Swaroop, K. Seth, P.K. Rout, G. Kumaresan, Journal of Animal Sciences, **90**, 10, 1373-1376 (2020).
9. Hecker, J.G. and M. McGarvey, Cell Stress Chaperones, **16**, 119–31 (2011).
10. Kishore, A., M. Sodhi, A. Sharma, U.K. Shandilya, A. Mohanty, P. Verma, S. Mann M. Mukesh, RRJVS, **2**, 34– 40 (2016).
11. Deb, R., B. Sajjanar, U. Singh, S. Kumar, R. Singh, G. Sengar, A. Sharma, Gene, **536**, 2,: 435–40 (2014).
12. Roh, B.H., D.H. Kim, M.K. Cho, Y.L. Park, and K.U. Whang, Ann Dermatol, **20**, 4, 184 –189 (2008).
13. Qiao, L., J.X. Wu, D.Z. Qin, X.C. Liu, Z.C. Lu, L.Z. Lv, Z.L. Pan, H. Chen, and Li, G.W, Journal of Insect Science, **15**, 1, 1-12 (2015).

14. Peterson, C.W., R.T. Carter, E. Bentley, C.J. Murphy, and H.L. Chandler, Vet Ophthalmol, **19**, 3, 262-266 (2016).
15. Tiwari, S., R. Thakur, and J. Shankar, Biotechnology Research International, (2015).
16. Richter, K., M. Haslbeck, and J. Buchner, Mol Cell, **40**, 2, 253–266 (2010).
17. Enzolifesciences, Heat Shock Proteins and The Cellular Stress Response. http://www.Enzolifesciences.com (2010).
18. Pirkkala, L., P. Nykanen, and L. Sistonen, FASEB J., **15**, 7, 1118-1131 (2001).
19. Shamovsky, I. and E. Nudler, Cell Mol. Life Sci., **65**, 6, 855-861 (2008).
20. Li, Z. and P. Srivastava, Curr Protoc Immunol (2004).
21. Rovelli, G., S. Ceccobelli, F. Perini, E. Demir, S. Mastrangelo, G. Conte, F. Abeni, D. Marletta, R. Ciampolini, M. Cassandro, U. Bernabucci, E. Lasagna, Italian Journal of Animal Science, **19**, 997-1014 (2020).
22. Gupta, M. and T. Mondal, Biological Rhythm Research, 1744-4179 (2019).
23. El-Tarabany, M.S., K.M. El-Bayoumi, Theriogenology, **83**, 3, 444–448 (2015).
24. Silanikove, N., and N. Koluman, Small Ruminant Research, 123, 1, 27–34 (2015).
25. Marai, I.F.M., A.A.M. Haeeb, Livestock Science, **127**, 2, 89–109 (2010).
26. Das, R., L. Sailo, N. Verma, P. Bharti, J. Saikia, Veterinary World, **9**, 3, 260–268 (2016).
27. Gonzalez-Rivasa, P.A., S.S. Chauhana, Minh Haa, N. Feganb, F.R. Dunsheaa, R.D. Warnera, Meat Science, 192 (2020).
28. Ocak, S., O. Güney, Anadolu Tarım Bilimleri Dergisi, **25**, 2, 113–119 (2010).
29. Phulia, S.K., R.C. Upadhyay, S.K. Jindal, R.P. Misra, Indian Journal Of Animal Sciences, **80**, 4, 340–342 (2010).
30. Alam, M.M., M.A. Hashem, M.M. Rahman, M.M. Hossain, M.R. Haque, Z. Sobhan, M.S. Islam, Program Agriculture, **22**, 1, 37–45 (2011).
31. Sharma, A.K., N. Kataria, Indian Journal Of Animal Sciences, **81**, 3, 293–295 (2011).
32. Rojas-Downing, M.M., A.P. Nejadhashemi, T. Harrigan, S.A. Woznicki, Climate Risk Management, **16**, 145–163 (2017).
33. Dokladny, K., O.B. Myers, P.L. Moseley, Autophagy, 11, 2, 200-2013 (2015).
34. Oberley, T.D., J.M. Swanlund, H.J. Zhang, K.C. Kregel, J. Histochem Cytochem, **56**, 615-27 (2008).
35. Kim, J.N., H.S. Lee, S.H. Ryu, Y.S. Kim, J.S. Moon, C.D. Kim, I.Y. Chang, S.P. Yoon, Gut and Liver, **5,** 513-20 (2011).
36. Dangi, S.S., M. Gupta, D. Maurya, V.P. Yadav, R.P. Panda, G. Singh, N.H. Mohan, S.K. Bhure, B.C. Das, S. Bag, R. Mahapatra, Trop. Anim. Health Prod., **44**, 1905-1912 (2012).
37. Rout, P.K., R. Kaushik, N. Ramachandran, Cell Stress Chaperon, **21**, 645-651 (2016).
38. Zulkifli, I., B. Norbaiyah, Y.W. Cheah, A.F. Soleimani, A.Q. Sazili, Y.M. Goh, M.A. Rajion, Animal, **4,** 973-976 (2010).
39. Singh, K.M., S. Singh, I. Ganguly, A. Ganguly, R.K. Nachiappan, A. Chopra, H.K. Narula, Small Rumin. Res., **141**, 113-117 (2016).
40. Singh, K.M., S. Singh, I. Ganguly, R.K. Nachiappan, A. Ganguly, R. Venkataramanan, A. Chopra, H.K. Narula, Cell Stress Chaperon, **22**, 675-684 (2017).
41. Banerjee, D., R.C. Upadhyay, U.B. Caudhary, R. Kumar, S.V. Singh, R.G.J. Ashutosh, S. Polley, A. Mukherjee, T.K. Das, S. De, Cell Stress Chaperon, **19**, 401-408 (2014).
42. Salces-Oritz, J., C. Gonzalez, M. Martinez, T. Mayoral, J.H. Calvo, M.M. Serano, BMC Evol. Biol., **15**, 7 (2016).
43. Renaudeau, D., Collin, A., Yahav, S., De Basilio, V., Gourdine, J. L., & Collier, R. J, Animal, **6,** 5, 707-728 (2012).

In-vitro nutrient degradability of complete feed containing myristic acid and tannins addition

Poespitasari Hazanah Ndaru, *Siti* Chuzaemi*, and *Muhimmatu* Mufidah

Faculty of Animal Science, Universitas Brawijaya, Malang 65145, Indonesia

Abstract. The purpose of this research was to determine the influence of Myristic acid and the source of condensed tannin on corn straw based-complete feed on nutrient content and in vitro digestibility. The method which was used in this experiment was randomized block design with four treatments and three replications. The treatment consisted of T0 complete feed (40% corn straw + 60% concentrate), T1 complete feed (40% corn straw + 60% concentrate + myristic acid 30 g/Kg DM), T2 complete feed (40% corn straw + 50% concentrate + calliandra leaf meal 10%/Kg DM and myristic acid 30 g/Kg DM), T3 complete feed (40% corn straw + 45% concentrate + calliandra leaf meal 15%/ Kg DM and myristic acid 30 g/Kg DM), T4 (40% corn straw + 40% concentrate + calliandra leaf meal 20%/Kg DM and myristic acid 30 g/Kg DM). Based on the result, it could be concluded that calliandra leaf meal 15%/Kg DM + Myristic acid 30 g/Kg DM on the complete feed (T3) was the best treatment based on nutrient content and increase of dry matter digestibility, but not an effect of organic matter digestibility.

1 Introduction

Corn is a type of agricultural crop whose main results are utilized by humans, while by-products include corn straw, cob, corn seeds that can be used as animal feed [1]. The productivity of corn crops every year has increased significantly. National corn production in 2014 was 19.0 million tons. The increase in corn production continued to grow in 2016, which was 23.6 tons, then in 2017, it reached 28.9 million tons, and in 2018, there was a significant increase of 30 million tons. In Indonesia, corn straw frequently becomes one of the feed compositions in complete feed. The complete feed consists of a mixture of forages and concentrates high in nutrients with a balanced ratio and arranged according to livestock needs [8].

At the time, the production of methane gas from cattle become a global issue because cattle consume a high of crude fibre. Therefore, efforts are needed to reduce methane gas by manipulating animal feed strategy and adding active compounds into the feed [2,3]. The addition of myristic acid to complete feed can affect the increase in the nutrient content of the feed and reduce CH_4 [9]. At the same time, the supplementation of concentrate with

* Corresponding author: schuzaemi@ub.ac.id

calliandra can affect the increase in the value of crude protein content and feed degradation [10].

2 Materials and methods

2.1 Materials

The material used in this study was a complete feed consisting of corn straw and concentrate composed of rice bran, copra meal, palm oil meal, soybean meal, coffee husk, cassava, urea, and molasses. Complete feed is composed of iso-protein with a CP content of 14-15%. At the same time, the additional ingredients include calliandra leaf meal, myristic acid, and cow rumen fluid. Complete feed was formulated become five treatments, and then it was analysed to determine the nutrient content of the feed and the value of feed degradation.

2.2 Method

This study used an experimental method with RBD (Randomized Block Design) with five treatments repeated three times. The complete feed formulation is as follows:

T0 : complete feed (40% corn straw + 60% concentrate)
T1 : complete feed (40% corn straw + 60% concentrate + 0% calliandra leaf meal) + myristic acid 30 g/kg DM diet.
T2 : complete feed (40% corn straw + 50% concentrate + 10% calliandra leaf meal) + myristic acid 30 g/kg DM diet.
T3 : complete feed (40% corn straw + 45% concentrate + 15% calliandra leaf meal) + myristic acid 30 g/kg DM diet.
T4 : complete feed (40% corn straw + 40% concentrate + 20% calliandra leaf meal) + myristic acid 30 g/kg DM diet.

Table 1. Nutrient content of complete feed ingredients

Feedstuffs	DM (%)	OM*	Ash*	CP*	CF*	EE*
Coffee husk	94.14	89.42	10.58	10.11	34.00	1.50
Rice bran	90.63	87.40	1.60	10.15	16.20	13.00
Cassava waste	92.59	82.87	7.13	1.76	25.39	0.44
Soybean meal	93.53	91.62	8.38	47.53	4.04	2.58
Palm kernel meal	95.39	94.97	5.03	14.24	20.91	10.01
Copra meal	95.69	92.23	7.77	22.12	21.78	2.45
Urea	99.88	99.93	0.07	244.60	0.00	0.00
Molasses	78.47	84.57	15.43	4.55	7.70	0.00
Corn straw	94.46	89.83	10.17	5.13	36.43	0.63
Calliandra leaf meal	93.67	92.37	7.63	23.16	12.08	3.90

Feed Analysis result from Laboratory of Animal Feed and Nutrient, Faculty of Animal Science, University of Brawijaya (2019).
*) Based on 100% Dry Matter

2.3 Data analysis

The research data were analysed using a Randomized Block Design (RBD) and continued with Duncan's Multiple Range Test (DMRT).

3 Results and discussion

3.1 Proximate profile

Based on the analysis of the nutrient content of the feed, the results were not too significant for the range of DM, CP, CF, EE, and OM. The different analysis results were caused by corn straw, calliandra leaf meal, and concentrate in each treatment. Corn straw has a low CP content, so it is necessary to apply technology such as preparing complete feed from feed ingredients containing high nutrients to cover the low nutrient content of corn straw. This is supported by Khoiriyah *et al.* [11], who stated that the addition of feed ingredients that contain high nutrients could increase the nutrient content of the feed. Table 2 shows the results of the complete feed nutrient content analysis for each treatment. The highest DM content was found in T1 of 93.64% compared to T0 of 92.53%, T2 of 93.11% and T3 of 92.60%, and T4 of 92.72%. Daning and Foekh [12] explain that the high and low DM content is affected by the harvesting time of the fodder plant. While the OM content showed the following results: T0 was 93.38%, T1 was 93.06%, T2 was 93.15%, and T3 was 93.33%, and T4 was 93.27%, with the highest percentage of content organic matter was in the T0 (control diet) which was 93.38%.

Table 2 shows the results of the complete feed nutrient content analysis for each treatment. The treatment feed with the highest CP content was found in treatment T3, which had the highest crude protein value of 15.95% compared to T0, T1, and T2 and T4 with 15.41%, 15.65%, and 14.35%, and 15.03%, respectively. The increase in crude protein content in complete feed can be affected by the ingredients of complete feed and calliandra leaves [13]. The crude fibre content of complete feed based on corn straw in this study showed results ranging from 17.123% to 19.98%, the high and low content of CF was affected by the content of lignin, cellulose, and hemicellulose owned by forage fodder as feed ingredients for complete feed [10].

In this study, the addition of 3% or 30 g/kg DM of myristic acid was based on the research of [3]. The higher addition of myristic acid in the diet significantly affected the decrease in the value of feed degradation and increased the crude fat content in the treated feed. Compared with the results of this study, it showed that the highest EE content in T1 was 3.72% compared to T0 (3.71%) and T2 (3.65%), while the lowest crude fat content was in treatment T4, which was 2.78%.

Table 2. Nutrient content of diet treatments

Nutrient composition (% DM)	Diet Treatments				
	T0	T1	T2	T3	T4
Dry matter (DM)	92.53	93.64	93.11	92.60	92.72
Crude Protein (CP)	15.41	15.65	14.35	15.95	15.03
Crude Fiber (CF)	17.12	19.19	19.98	19.97	19.64
Extract Ether (EE)	3.71	3.72	3.65	2.89	2.78
Organic Matter (OM)	93.38	93.06	93.15	93.33	93.27

Feed Analysis result from Laboratory of Animal Feed and Nutrient, Faculty of Animal Science, University of Brawijaya (2019)

Effect of myristic acid supplementation and calliandra leaf meal substitution (*Calliandra calothyrsus*) as a tannin condensate source in corn straw-based complete feed on dry matter degradation (DMD). Based on the results of the analysis showed that the addition of myristic acid and calliandra leaf meal (*Calliandra calothyrsus*) as a source of tannin content in corn straw-based complete feed (*Zea mays*) gave a significant difference ($p<0.05$) to the value of dry matter degradation (DMD). Further tests using Duncan's test showed that treatment T0 was not significantly different from T1 and T4, while T1 was significantly different at the 5% level ($p<0.05$) with T2 and T3. Dry matter degradation in T2 and T3 treatments gave no significant effect. The result indicates that differences in the composition of calliandra leaf meal affect the value of dry matter degradation (DMD). Table 3 explains the average dry matter degradation value (DMD) of the four treatments showing that T3 treatment has the highest average degradation value of 60.18%, when compared to the average dry matter degradation value of T0 (48.37%), T1 (49.69%), and T2 (57.15%) and T4 (50.41%). The results of dry matter degradation in each treatment gave different values caused by the high and low crude fibre and crude protein content in each treatment. According to Wijayanti [14], CF content affects the degradation value because feed containing high CF content reducing feed digestibility.

Table 3 shows the results of the analysis of the dry matter degradation value of the feed. It was known that the highest dry matter degradation value (DMD) was in the T3 treatment feed (complete feed + 15% calliandra leaf meal + 30 g/kg myristic acid). Increasing the level of giving calliandra leaf meal to the feed can increase the value of feed degradation. Lawa *et al.* [15] about the effect of adding white kabesak leaves on feed consumption and weight gain for Kacang goats, degradation, body weight gain, and feed conversion. Based on the study results, it was found that the crude fat content of each treatment did not differ much; this was presumably due to the myristic acid content added at the same level. Soy oil calcium levels in Bali cattle rations with proportions of 2.5%, 5%, and 7.5% did not significantly affect the value of dry matter and organic matter degradation [16].

3.1 Nutrient degradability profile

Based on the results of the analysis showed that the addition of myristic acid and calliandra leaf meal (*Calliandra calothyrsus*) as a source of tannin content in corn straw-based complete feed (*Zea mays*) gave no significant effect ($p>0.05$) on the value of organic matter degradation. The range of organic matter that cannot be digested properly results in the value of organic matter degradation not being significantly different ($p>0.05$).

Table 3. Average value of dry matter degradability (DMD)

Treatments	Dry Matter Degradability (%)
T0	48.37 ± 2.38^{a}
T1	49.68 ± 2.66^{a}
T2	57.14 ± 3.75^{b}
T3	60.18 ± 1.89^{b}
T4	50.41 ± 4.35^{a}

Note: Different superscripts in the same column show significant differences at the 5% level ($p<0.05$).

Table 4. The average value of organic matter degradability

Treatments	Organic matter degradability (%)
T0	57.33 ± 8.19
T1	60.58 ± 8.97
T2	63.63 ± 5.25
T3	61.38 ± 6.63
T4	59.27 ± 8.22

Note: The treatment showed that it gave no significant difference at the 5% level (p>0.05)

Based on the results of the analysis showed that the addition of myristic acid and calliandra leaf meal (*Calliandra calothyrsus*) as a source of tannin content in corn straw-based complete feed (*Zea mays*) gave no significant effect (p>0.05) on the value of organic matter degradation. The range of organic matter that cannot be digested properly results in the value of organic matter degradation not being significantly different (p>0.05).

Table 4 shows the average value of organic matter degradation (OMD) from the four treatments, namely the average value of organic matter degradation in T2 treatment, which was the highest result of 63.63%. While for treatment T0 57.33%, T1 60.58%, T3 61.38% and T4 59.27%. The average value of organic matter degradation is higher than that of dry matter degradation. Table 4 explains that the average value of organic matter degradation has increased along with calliandra leaf meal. The highest average value was found in the T2 treatment with the composition of complete feed (40% corn straw + 50% concentrate + 10% calliandra leaf meal) + myristic acid 30 g/kg DM diet). The average dry matter degradation value and the average organic matter degradation value (OMD) has a not too significant difference. The opinion of Trisnadewi *et al.* [13] states that the degradation value of dry matter and organic matter is always in harmony or not much different; this is because most dry matter is composed of organic matter.

The addition of fatty acids in complete feed can increase the value of organic matter degradation (OMD). It is known that the average results of the degradation of organic matter (OMD) are not significantly different, presumably due to the composition of the addition of the same myristic acid. This is supported by Sany *et al.* [16] stated that the effect of fish oil and L-carnitine supplementation in fermented yellow corn feed on feed degradation and quail performance showed that the addition of lemuru fish oil and L-carnitine as much as 4% could increase the value of dry matter degradation and organic matter degradation. The value of T0 (basal feed) was 59.20%, T1 (basal feed + fermented yellow corn) was 65.75%, while the dry matter degradation value of T4 (basal feed + 4% lemuru fish oil) was 72.08%.

4 Conclusion

The replacement of concentrate with calliandra leaf meal affects the percentage of feed ingredients that make up the concentrate, affecting the treatment feed's protein content. The use of calliandra leaf meal and the addition of myristic acid to complete feed based on corn straw can increase the value of dry matter degradation (DMD) and organic matter degradation (OMD). The highest dry matter degradation value was found in T3, namely the use of 15% calliandra leaf meal and myristic acid 30 g/kg DM feed with a DMD value of 60.18%. In contrast, the highest organic matter degradation value was found in T2, namely the use of 10% calliandra leaf meal and myristic acid 30 g/kg DM diet with an organic matter degradability value of 63.63%. We suggested use of complete feed according to the best treatment, namely T3 (15% calliandra leaf meal + myristic acid 30 g/kg DM diet). It is necessary to continue with research using livestock to determine livestock productivity.

References

1. W.P. Hertawan, Jurnal Peternakan Indonesia **120** (2018)
2. S. E. Hook, A.D.G Wright, B.W. Archaea **1** (2011)
3. S. Chuzaemi, Mashudi, Livest. Anim. Res. **18** (2020).
4. Yanuartono, S. Indarjulianto, H. Purnamaningsih, A. Nururrozi, S. Raharjo. Jurnal Sains Peternakan Indonesia **14** (2019)
5. E. Gustiani, K. Permadi. Jurnal Peternakan Indonesia **17** (2015)
6. O. Sjofjan, D. N. Adli, M.H. Natsir, Y. F. Nuningtyas, T. S. Wardani, I. Sholichatunnisa, S. N. Ulpah, O. Firmansyah, Jurnal Ilmu Ternak dan Veteriner **26** (2021)
7. A. Irsyamawati, S. Chuzaemi, Hartutik, Jurnal Ilmu-Ilmu Peternakan **21** (2011)
8. Yurleni, R. Priyanto, Jurnal Ilmu-Ilmu Peternakan **19** (2016)
9. M. Hambakodu, A. Kaka, Y. T. Ina, Jurnal Ilmu dan Teknologi Peternakan Tropis **7** (2020)
10. M. Khoiriyah, S. Chuzaemi, H. Sudarwati, Jurnal Ternak Tropika **17** (2016)
11. D.R.A Daning, B. Foekh, Sains Peternakan **16** (2018)
12. A. A. A. S. Trisnadewi, I. G. L. O. Cakra, I. W. Wirawan, I. M. Mudita, N. L. G. Sumardani, Pastura **3** (2014).
13. E. Wijayanti, F. Wahyono, Surono. Animal Agricultural **1** (2012)
14. E. D. W Lawa, Marjuki, Hartutik, S. Chuzaemi, J. Indones. Trop. Anim. Agric. **42** (2017)
15. A. W. Sany, S. R. Heswantari, Sudibya, S. H. Purnomo, A. Hanifa, Buletin Peternakan **39** (2015)

Effect of fermentation on mixed rumen contents and jackfruit peel using *Aspergillus oryzae* on *in vitro* gas production and digestibility

Mashudi Mashudi[1*], and *Wahyuni* Nurmawati[1]

[1]Faculty of Animal Science, Universitas Brawijaya, Jl. Veteran, Malang 65145, Indonesia

ABSTRACT. The objective of this research was to investigate the effect of *Aspergillus oryzae* on fermentation of mixture of rumen contents and jackfruit peel on *in vitro* gas production and digestibility. The method used in this study was an experiment using a randomized block design (RBD) of 4 treatments and 3 replications. The treatments including of T0 = 50% rumen contents + 50% jackfruit peel, T1 = 40 % rumen contents + 60% jackfruit peel + 0,4% *Aspergillus oryzae*, T2 = 30% rumen contents + 70% jackfruit peel + 0,4% *Aspergillus oryzae*, T3 = 20% rumen contents + 80% jackfruit peel + 0,4% *Aspergillus oryzae*. Variables observed were gas production, dry matter digestibility (DMD), and organic matter digestibility (OMD). Data were analyzed by using Analysis of Variance (ANOVA) from Randomized Block Design, if there were significant effect between the treatments then tested with least significant different (LSD). The result showed that fermentation of mixture of rumen contents and jackfruit peel have highly significant effect ($P<0.01$) on gas production, and significant effect ($P<0.05$) on DMD and OMD. Gas production, DMD and OMD of fermented mixture of rumen contents and jackfruit peel are higher than control without fermentation. It is concluded that the higher jackfruit peels the higher gas production, DMD and OMD.

1 Introduction

Animal husbandry is the activity of raising livestock for cultivation and business. Livestock is very important as a supplier of animal product such as meat and milk. There are three important factors in the livestock business, namely breeds, feed and rearing management.

* Corresponding author: mashudi@ub.ac.id

Feed is an important factor that greatly affects livestock productivity and feed costs are the largest in livestock business. High quality of feed is feed containing high nutrients needed by livestock [1]. In ruminant farm the availability of forage is absolutely necessary. Changes in land use and climate will limit the availability of forages. Climatically Indonesia has 2 seasons, namely the rainy season and the dry season, where it is very influential on the supply of forages. In the rainy season, forage can be obtained easily, but in the dry season forage becomes scarce and difficult to obtain.

Based on this, it is necessary to find alternative feeds that contain sufficient nutrients and are abundantly available as well as lower price. Alternative feed can be obtained from waste which is generally not utilized and sometime disturbs the quality of the environment.

Two of these wastes are rumen contents derived from Slaughterhouse and jackfruit peel derived from jackfruit processing industry. Rumen contents were chosen due to their relatively high protein contents and their amino acids. [2] reported that the samples of rumen contents taken from cattle had high level of crude protein (ranged between 18.52 – 19.56%). According to [3], Jackfruit peel was an alternative feed that can be used as animal feed because it contains high carbohydrate (71.53 %), but low protein. Both of these wastes can be used together as animal feed but must be processed first to increase their feeding value. One technology that can be used is fermentation technology using tape yeast or *Aspergilus oryzae*

This study was conducted to determine the effect of a mixture of jackfruit peel and rumen contents fermented using Aspergillus oryzae on *in vitro* gas production, DMD and OMD.

2 Materials and methods

The research was conducted at the Laboratory of Animal Nutrition, Faculty of Animal Science, Universitas Brawijaya. The materials used in the study were jackfruit peel, rumen contents, tape yeast (*Aspergillus oryzae*), and rumen fluid. The method used in this study was an experiment using a randomized block design (RBD) of 4 treatments and 3 replications. The treatments were T0 = 50% rumen contents + 50% jackfruit peel, T1 = 40 % rumen contents + 60% jackfruit peel + 0,4% *Aspergillus oryzae*, T2 = 30% rumen contents + 70% jackfruit peel + 0,4% *Aspergillus oryzae*, T3 = 20% rumen contents + 80% jackfruit peel + 0,4% *Aspergillus oryzae*.

Variables measured in this study were i*n vitro* gas production (on incubation 0, 2, 4, 8, 16, 24, 36, 48, 72 hours), Dry matter digestibility (DMD), and organic matter digestibility (OMD) from *in vitro* gas production residues [4] and the values of a, b and c are calculated by the formula [5]. Data was analysed by using Analysis of Variance (ANOVA) from Randomized Block Design, if there were significant effect between the treatments then tested with least significant different (LSD).

3 Result and discussion

3.1 Nutrient content of feed ingredients and feed treatment

The results of analysis of nutrient content of feed and treated feed (%DM) can be seen in Table 1.

Table 1. Nutrient content of rumen content and jackfruit peel (% DM)

No	Feed	DM	Ash	OM	CP	CF
1	Rumen content	24,31	8,42	91,58	9,56	29,65
2	Jackfruit peel	25,45	7,68	92,32	10,01	33,01

Noted: The results of the analysis from the Laboratory of Animal Nutrition, Faculty of Animal Science, Universitas Brawijaya

Nutrient content of rumen content as in Table 1 is slightly not different as reported by [6] who reported that the rumen contents of cows about 12.50% DM, 83.92% OM, 11.58% CP and 24, 01% CF. Related to the jackfruits peel, [3] reported that the nutrient content of fresh jackfruit peel consists of, 9.35% CP, 9.3% carbohydrates, 30.52% CF and 7.31% ash. The composition of the nutrient content of the rumen contents depends on feed given to the animals.

Nutrient content of each treated feed (%DM) can be seen in Table 2. It can be seen that CP and CF content of treated feed is higher than control without treatment. The higher jackfruit peels the higher CP and CF content.

Table 2. Nutrient content of each treatment feed (% DM)

No	Treatment	DM	Ash	OM	CP	CF
1	T0	50.41	16.14	83.86	9.58	23.59
2	T1	51.00	16.36	83.64	10.06	25.33
3	T2	48.42	16.78	83.22	10.40	29.12
4	T3	49.57	19.29	80.71	11.41	35.75

Noted: Results of analysis from the Laboratory of Animal Nutrition, Faculty of Animal Science, Universitas Brawijaya

3.2 *In Vitro* gas production

The process of anaerobic fermentation of organic matter by rumen microbes will produce VFA and gas. The amount of gas produced indicates the amount of organic matter that can be digested in the rumen. According to [7], *in vitro* gas production is a parameter of rumen microbial activity in degrading feed.

The amount of gas production can be used to estimate the energy value of feed in ruminants. Average *in vitro* Gas Production at different incubation periods can be seen at Table 3.

The results of statistical analysis showed that there was a significant effect ($P<0.01$) on the gas production. The higher the jackfruit peel the higher gas production. This may be because jackfruit peel contains high in protein and energy. The highest value of gas production is T3.

Table 3. Average *in vitro* gas production (ml/500 mg DM) at different incubation periods

Treat ments	Incubation Periods (hours)							
	2	**4**	**8**	**16**	**24**	**36**	**48**	**72**
T0	3.1± 1.39	5.7± 2.02^{a}	10.6± 4.65^{a}	18.4± 8.83^{a}	27.8±1 4.63^{a}	38.7±1 9.65^{a}	50.1±2 4.54^{a}	57.8±2 9.24^{a}
T1	1.9± 0.75	4.4± 2.54^{b}	9.8±5. 86^{a}	18.4± 7.28^{a}	29.8±9 .84^{a}	41.2±1 4.89^{a}	53.1±2 0.01^{a}	61.2±2 5.60^{a}
T2	2.8± 1.86	6.6± 3.40^{b}	14.5± 4.44^{b}	28.1± 2.59^{b}	39.8±1 .37^{b}	56.7±2 .31^{b}	78.6±1 6.01^{b}	86.3±1 8.01^{b}
T3	3.8± 1.76	7.3± 1.68^{b}	14.5± 1.17b	35.2± 3.83^{c}	48.2±5 .90^{b}	71.3±1 0.43^{b}	97.0±1 4.38^{b}	109.6± 23.29^{b}

Notes: Different superscripts in the same column showed very significant differences (P<0.01).

Kinetics of gas production is estimated using the exponential equation described by [5] $p = a+b\ (1-e^{-ct})$. The p value is the cumulative gas production at time t hours, while a, b and c are the exponential constants of the exponential equation.

The constant value can be interpreted as gas production from soluble fraction (a), gas production from insoluble but fermentable fraction (b) and gas formation rate (c). The a+b value can be interpreted as the maximum gas production that can be formed during the fermentation process at time t. The value of a, b and c can be seen in Table 4.

Table 4. Parameter values a, b, and c from *in vitro* gas production

Treatments	Parameters		
	a (ml/500 mg DM)	**b (ml/500 mg DM)**	**c (ml/hours)**
T0	0.6956 ± 0.80	84.3318 ± 42.26	0.01622 ± 0.0016
T1	0.0000 ± 0.00	102.8203 ± 61.68	0.01645 ± 0.0069
T2	0.1276 ± 0.22	182.9311 ± 144.86	0.01809 ± 0.0127
T3	0.3158 ± 0.55	192.3105 ± 68.19	0.01805 ± 0.0054

Noted: Parameters a, b and c show no significant effect on that value.

The results of statistical analysis showed that there was no significant effect (P>0.05) on a, b and c value. However, there is a tendency that the higher the jackfruit peels the value of a, b and c increases and the highest value is T3.

Effect of treatment, also can be seen as Graph 1, relationship between gas production of treatment feed and incubation time.

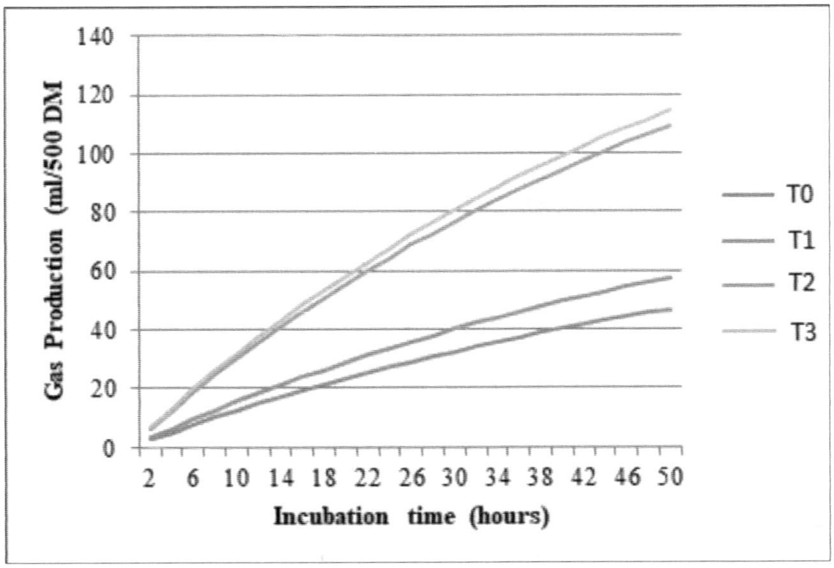

Fig. 1. Graph of the relationship between the average value of gas production from the fermented mixture of rumen contents and jackfruit skin using *Aspergillus oryzae* with incubation time using the P = a+b (1-e -ct) model.

3.3 *In Vitro* digestibility

Average value of dry matter and organic matter digestibility (DMD and OMD) based on residual *in vitro* gas production at an incubation period of 72 hours. The average DM and OM digestibility can be seen in Table 5

Table 5. Average DMD and OMD from *In Vitro* Gas production residues.

Treatments	DMD (%)	OMD (%)
T0	$39.62{\pm}2.76^a$	$37.05{\pm}2.80^a$
T1	$41.05{\pm}3.73^a$	$38.42{\pm}3.99^a$
T2	$44.45{\pm}5.00^b$	$42.36{\pm}5.06^b$
T3	$50.40{\pm}4.11^c$	$48.70{\pm}1.30^c$

Note: *Different superscripts in the same column show a significant difference in effect (P<0.05)

The results of statistical analysis showed that there was a significant effect (P<0.05) on the DMD and OMD. The higher the jackfruit peel the higher DMD and OMD. This may be because jackfruit peel contains high in protein and energy. The highest value Of DMD and OMD is T3. The digestibility value obtained based on *in vitro* gas production; therefore, the digestibility value is only the digestibility in the rumen (not including post-rumen). Low DMD and OMD values in this experiment in line with low in gas production values.

4 Conclusion

The more the proportion of jackfruit peel in the fermented (using *Aspergillus oryzae*) mixture between jackfruit peel and rumen contents, the gas production, DMD and OMD increased. Gas production, DMD and OMD were higher in the fermented mixture between jackfruit peel and rumen contents than the unfermented control. T3 treatment is the best treatment in terms of gas production value, DMD and OMD. Further research needs to be done according to the best treatment to test the response of animals to treated feed.

References

1. P. P. Ketaren, Kebutuhan Gizi Ternak Unggas di Indonesia, **20(4)**: 172-180 (2010).
2. A.E. S. Moningkey, R.A.V. Tuturoong, I.D.R. Lumenta, Pemanfaatan Isi Rumen Terfermentasi *Cellulomonas Sp* sebagai Campuran Pakan Komplit Ternak Kelinci. Zootec, **40**, 352-361 (2020).
3. S. D. Novandrini, Pengaruh Penambahan Ikan terhadap Mutu Gizi dan Penerimaan Abon Nangka, Skripsi (2003).
4. K. H. Menke, and H. Steingass, Estimation of The Energetic Feed Value Obtained from Chemical Analysis and Gas Production Using Rumen Fluid. *Anim. Res. Develop*. **28**, 7-55 (1988).
5. E. R. Ørskov, McDonald, The Estimation of Protein Degradability in the Rumen from Incubation Measurements Weighted According to Rate of Passage. *Journal of Agricultural Science*, **92,** 499-503 (1979).
6. R. Utomo, L. M. Yusiati, U. Umiyasih, Aryogi, Isnandar, Pemanfaatan isi Rumen Limbah Rumah Potong Hewan Sebagai Pakan Alternatif Pengganti Hijauan. Laporan Penelitian (2007).
7. I. Prihatini, S. Chuzaemi, O. Sofjan, Parameter Fermentasi Rumen dan Produksi Gas In Vitro Jerami Padi Hasil Fermentasi Inokulum Lignochloritik**.** Parameter Fermentasi Rumen. Jurnal Protein, **5**, 24-32 (2007).

Income analysis of integrated farming system of dairy cow breeding farm, earthworms (*Lumbricus rubellus*), and *citrus sinensis Valencia* farm. Case study in Dau District, Malang Regency, East Java, Indonesia

Puji Akhiroh[1], Hermanto[1], *Galih* Purboningrum[2], *Mentari Bertha Septina* Sase[2] and *Rizki* Prafitri[1,*]

[1] Faculty of Animal Science, University of Brawijaya, Jl. Veteran, Kota Malang, East Java, 65145, Indonesia
[2] Under Graduate, Faculty of Animal Science, University of Brawijaya, Jl. Veteran, Kota Malang, East Java, 65145, Indonesia

Abstract. The integrated farming system between dairy cow breeding, earthworms (*Lumbricus rubellus),* and citrus (*Citrus sinensis valencia*) plantations provides regular income for farmers and sustainable farming. This research aims to analyze farmers' income based on the integrated farming system that has been done on these three commodities. This study utilized quantitative and qualitative methods to analyze data. Data were collected through questionnaires to 15 farmers who are members of Gading Kulon Farmer Group in Dau District, Malang Regency, East Java, Indonesia followed by in-depth interviews with key respondents. The study showed that dairy cow breeding provided annual income which contributed up to 58% of household income. Meanwhile, earthworm farming provides weekly income and twice a year income for citrus farming. Depending on the land area, Earthworms farming contributed up to 10% of household income, and citrus plantation was 33%, depending on the land area owned by farmers. The average income of Dairy cow breeding was Rp. 163,308,000 per year; the Average Income of earthworms was Rp. 27,058,000 per year, and the average income of citrus plants was 92,480,000 per year. However, integrated farming of these commodities does not work well in the community due to farmers' lack of land and capital. Moreover, most respondents indicated that they were interested only in citrus farming because it is easier compared to dairy cow breeding and earthworm farming. Good cooperation from various parties is needed to increase the importance of integrated farming with these three commodities for sustainable farming in Indonesia.

* Corresponding author: rizkiprafitri@yahoo.com

1 Introduction

Dairy cattle are one of the leading livestock commodities in East Java Province, with a population of more than 280 thousand heads in 2019 [1]. Although the dairy cattle population in 2018 was much lower in quantity compared to 2017, in general, the dairy cattle population in East Java reached 50 percent of the total dairy cattle population in Indonesia. Likewise, the production of fresh milk in East Java is the highest production in Indonesia when compared to other regions. In 2018, fresh milk production in East Java reached more than 500,000 tons or 55.9% of Indonesia's total fresh milk production. However, fresh cow's milk production growth in Indonesia only reached 2 per cent per year.This growth is far below the growth in domestic demand for cow's milk which reaches 5% per year. Domestic raw milk needs (SSDN), which reached 3.8 million tons per year, can only be met by domestic production of around 20 percent, while the rest still has to be imported from various countries such as Australia, New Zealand, and the United States [2].

Integration of dairy cattle breeding businesses owned by farmers with other businesses, namely earthworms (Lumbricus Rubellus) and Valencian sweet orange plantations (Citrus sinensis Valencia) provide a solution for farmers toget regular income during the maintenance of dairy calves. Farming business integration will be carried out by utilizing cow dung from dairy cattle breeding as earthworms feed (Lumbricus Rubellus) kept around the cowshed. Farmers harvested Earthworms (Lumbricus Rubellus) every day and provided regular farmers' income. In addition, the use of organic fertilizers on citrus plants owned will also provide additional income for farmers every time they harvest. Dairy cattle breeding business integrated with the maintenance of earthworms (Lumbricus Rubellus) and Valencia sweet oranges (Citrus sinensis Valencia) will provide regular income in daily, weekly, monthly and annual income for farmers.Integrated dairy cattle breeding business with the maintenance of profitable earthworms (Lumbricus Rubellus) and Valencian oranges (Citrus sinensis Valencia) will increase the interest of dairy farmers to conduct dairy cattle breeding businesses and at the end affecting the number of dairy cows in Indonesia.

Integrated dairy cattle breeding business has been started by the Gading Mandiri farmer group, in Dau District, Malang Regency since 2019. Gading Mandiri livestock group started a side business of cultivation earthworms by considering economic and environmental aspects. One of the positive impacts of dairy cattle breeding is reducing environmental pollution, especially local drainage channels by cow dung waste. The most obvious impact is on the quality of cleanliness and environmental aspects. Earthworm cultivation is one of the efforts to overcome cow dung waste and increase the economic income of farmers. However, farmers have not been able to calculate the business analysis they have. Therefore, this study analyzes the economic benefit of the integrated business that farmers have carried out.

2 Materials and methods

This research was conducted on the Gading Mandiri Cattle group, Dau District. This research was conducted using survey research methods and quantitative approaches. Data Collection Methods are:

1. Primary Data
   a. Observation
   The observation method is used as a way to collect the necessary direct data, facts and information in the field. Data collection was also carried out by observing and observingdirectly observing the research object, conditions and matters relating to field observation research

b. Interview
Interview is a method of collecting data by conducting questions and answers (direct interviews) to the Gading Mandiri livestock group, one of the research subjects. Interviews must be done systematically and based on research objectives. Interviews conducted with group members aimed to obtain more in-depth information about the dairy cattle breeding business integrated with earthworms (Lumbricus Rubellus) and how this business has an economic and social impact on farmer groups and the Gading Kulon community in general.

2. Secondary Data
The method of collecting secondary data obtained from the Gading Mandiri Cattle groupand conducting this literature study includes literature studies, statistical data and other information by comparing books, journals, articles that are useful for obtaining information related to theories and concepts related to feasibility research. integration of dairy cattle breeding.

Data collected
The stages of data analysis that will be carried out include:

1. Capital Analysis
Capital or capital is raw materials or funds combined with production factors in land, tools, and labor to produce a new item.
2. Total Cost

$$\mathbf{TC = TFC + TVC}$$

- TC = Total Cost
- TFC = Total Fixed Cost
- TVC = Total Variabel Cost

3. Revenue
Revenue in this study will be calculated based on the total production generated bythe sale of dairy cattle and earthworms multiplied by the selling price:

$$\mathbf{R = P \times Q}$$

Description:
- R = Revenue
- P = Price
- Q = Quantity
If : TR > TR : then profit will be obtained
TR = TC : break even point or Break Even Point
TR<TC : will result in loss

4. Break Even Point (BEP)
Break event point or BEP is the point where production costs equal revenue. BEP will be calculated using the formula:

$$\boldsymbol{BEP\ (price) = \frac{Total\ Cost\ (Rp)}{Quantity\ Production\ (Rp)}}$$

5. Profit
   Calculation of profit or profitability value can be calculated by using a formula.
   $$\pi = TR - TC$$
   Description:
   - π = Profit on milk storage unit (Rp/year)
   - TR = Total Revenue or total revenue at the milk storage unit (Rp/year)
   - TC = Total Cost or total cost of the milk storage unit (Rp/year)

6. Profitability Analysis
   Profitability is a reflection of the company's capital ability to earn profits or profits. Profitability is a reflection of the efficiency of a company in using its working capital. Profitability can be calculated by the formula:
   $$\mathbf{R} = \frac{L}{M}$$
   Description:
   - R = Profitability
   - L = Profit
   - M = Foreign capital + own capital

3 Results and discussion

Livestock Population

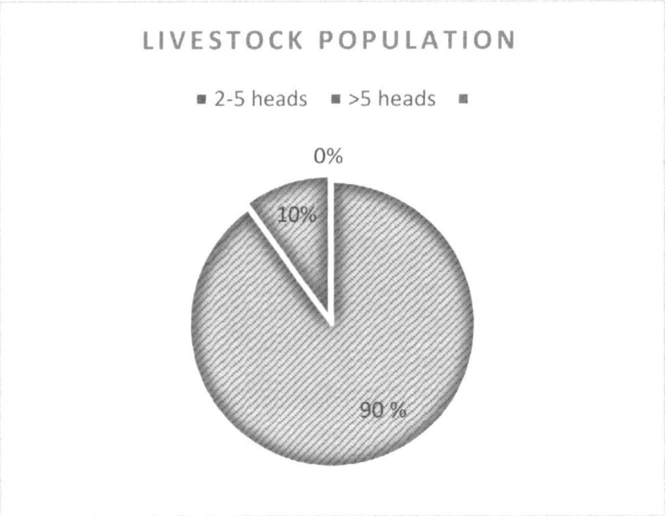

Fig. 1. Persentation livestock population

In the picture shows the experience of raising the 15 respondents who were interviewed divided into 2 groups. Respondents with a population of 2-5 cattle as much as 90% 5 and for a livestock population> 5 heads as much as 10%. In the picture shows the experience of raising the 15 respondents who were interviewed divided into 2 groups. Respondents with a population of 2-5 cattle as much as 90% 5 and for a livestock population> 5 heads as much as 10%. This is in accordance with the opinion of Widayati *et al.* [3] which states that the number of cows has a positive influence on experience.

Type and number of livestock ownership

The composition of the types of livestock kept by farmers in the Gading Mandiri livestock group, Dau District is very diverse, consisting of female dairy cows, lactating female cattle and some breeders who raise beef cattle. The number of livestock also greatly affects the income of livestock business.

Table 1. Number of Livestock Ownership in the Gading Mandiri Cattle Group

No.	Type of Livestock	Amount Cattle (Heads)	Amount Respondent (Person)
1.	Integrated Dairy Cattle Breeding (Large Scale)	18-20	1
2.	Lactation Dairy Business (Small Scale)	3-4	4
3.	Dairy Cattle Business (Large Scale)	18-20	2
4.	Beef Cattle Breeding (Small Scale	3-4	3

Source: Harmoko *et al*. [4]

Based on the data in table 1 shows that the cattle breeding business owned by respondents from the Gading Mandiri livestock group is very diverse, ranging from small to large scale. In table 1 there is 1 farmer respondent who has an integrated large-scale dairy cattle breeding business. In addition, there are 4 farmer respondents have a small-scale lactation dairy cattle business of only 3-4 cows. For breeders who have a large-scale lactation dairy cattle business of up to 18 cows, only 2 people. This is according to Novarista *et al*. [5] that the difference in the number of cows kept is divided into three levels, namely 3 which illustrates that raising livestock is still a small business and is a side business, while if the number of cows is 4-10 it illustrates that the farmer's business is still not on an economic scale even though it may have become the main business, > 10 headsillustrates that the farmer's business is already on an economic scale.

Business Capital

Business capital is operational cost capital used to facilitate running a business that runs out in one production process. The capital used in livestock groups comes from self-financing, cooperative loans or bank loans originating from the KUR (People's Business Credit) program. Asperinche *et al*. [6] stated that initial investment capital is the capital issued for a dairy cattle business, such as costs in purchasing livestock, cage costs, and equipment costs.. The production costs incurred by the dairy cattle breeding business owner in the Gading Mandiri Cooperative, Gading Kulon Village, Dau sub-district consist of 2 types, namely fixed costs and variable costs.

Table 2. Fixed costs in Dairy Cattle Breeding Business and Lactation Cattle Business

No	Fixed Cost	Quantity	Unit price (IDR)	Total (IDR)
1.	Shrinkage of Cages and Warehouses	1	250.000	250.000
2.	property tax	1	214.500	214.500
3.	Equipment Depreciation	1	41.500	41.500
Total Fixed Cost				506,000

Source: Harmoko *et al*. [4]

In table 2, the largest expenditure is on the cost of depreciation of the cage and warehouse of Rp. 250,000. So that the total fixed costs in the dairy cattle breeding business and lactating cattle business are Rp. 506,000. Fixed costs between dairy and lactating cattle breeding businesses are considered the same, the difference is the variable cost. Expenditures on land and building taxes are taken on average from 10 farmer respondents. Variable costs in dairy farming and integrated with the cultivation of earthworms (Lumbricus rubellus) can be seen on table 3.

Table 3. Variable costs of production per period Data A

No	Variable Cost	Quantity	Unit price (IDR)	Total (IDR)
1.	2 Months Dairy Cow Calf Breeding	9	3.333.333,33	30.000.000
2.	Forage/period (kg)	27.000	1.200	32.400.000
3.	Concentrate (kg)	9.000	2.400	21.600.000
4.	Wormectin	9	10.000	90.000
5.	Vitamin B-Plex	9	20.000	180.000
6.	IB	9	25.000	225.000
7.	Sales Vehicle Rental	2	200.000	400.000
8.	Household and Enclosure Electric	16	35.000	560.000
9.	Household and Cage Water	1	25.000	25.000
10.	Earthworm Seeds (Lumbricus rubellus)	5	50.000	250.000
	Total Variable Cost			85.730.000

Source: Harmoko *et al*. [4]

The data in table 3 shows the total variable costs of farmers. Variable costs are costs that will change every period. Variable costs include the cost of dairy cattle in the form of a 2-month calf, forage feed and concentrates, medicines, vitamins, vaccines, AI, electricity, water, rental of sales vehicles and worm seeds. Based on table 3, the largest expense of variable costs is the cost of feed in the form of forage with the type of dry tumpi obtained from the cooperative at a price of Rp. 1,200,-/kg and concentrate costs range from Rp. 2,400, - up to Rp. 2,600,- .kg and the cost of dairy cattle for 2 months is Rp. 3,333,333 per head or ordinary breeders make a purchase of IDR 10,000,000 for 3 calves aged 2 months. Data A is a type of calf that is 2 months old with a maintenance period of 1 period for 16 months. Data B is a type of calf with a maintenance period of 10 months in 1 period of 6 months. The earthworm cultivation business in table 3 does not use feed costs, because the earthworm feed comes from the use of dairy cow dung. Variable costs or variable costs are influenced by the livestock population, the more population you have, the greater the variable costs.

Table 4. Variable costs of production per period Data B

No	*Variable Cost*	Quantity	Unit price (IDR)	Total (IDR)
1.	2 Months Dairy Cow Calf Breeding	9	9.000.000	81.000.000
2.	Forage/period (kg)	13.500	1.200	16.200.000
3.	Concentrate (kg)	4.500	2.400	10.800.000
4.	Wormectin	9	10.000	90.000
5.	Vitamin B-Plex	9	20.000	180.000
6.	IB	9	25.000	225.000
7.	Sales Vehicle Rental	2	200.000	400.000
8.	Household and Enclosure Electric	1	35.000	560.000
9.	Household and Cage Water	1	25.000	25.000
10.	Earthworm Seeds (Lumbricus rubellus)	5	50.000	250.000
	Total Variable Cost			109,380,000

Source: Harmoko *et al.* [4]

Table 4 is a variable cost in Data B. The cost of seeds in data A and data B is very different due to differences in the age of dairy cows. Data A uses dairy cows aged 2 months at a price of Rp. 3,333,333 for Data B uses dairy cows aged 10 months at a price of Rp. 9,000,000,-. The pattern of rearing for breeders who do dairy cattle breeding on a large scale, namely 18 cows, breeders have two different types of dairy cows, this also affects the cost of animal feed. The longer the maintenance of livestock, the greater the cost of feed and production costs. High feed costs are caused by the purchase of forage and concentrates. Farmers are still buying forage, because there is not enough forage in the fields. The type of forage given is corn tumpi. Breeders prefer corn tumpi because of the higher dry matter content than fresh forage. For the most part concentrate costs are subsidized by local cooperatives.

Data A shows that the total variable cost reaches Rp.85,730,000,- this is lower than Data B, which has a total fixed cost of Rp. 109.380.000,-. This is because the cost of dairy calves is very high which reaches Rp. 9,000,000 per head. The cost of feed in Data A is higher, because the maintenance on Data A takes longer, so it requires a fairly high cost. This is done by farmers so that dairy cows that are kept at the age of 10 months can be harvested first at the age of 18 months. The results from the harvest can be used to finance the first loan of a cooperative or bank, and can be used for daily living.

The lactation cattle business in table 5. also shows the amount of costs required to purchase pregnant heifers. One pregnant heifer or lactating cow that will produce milk has a price of Rp. 17,000,000, - This is very different from a dairy cattle breeding business which buys calves aged 2 months at a price of Rp. 3,333,333, - and female dairy cows aged 10 month at a price of Rp. 9,000,000, - per head. The need for water in the maintenance of lactating cows is more, this is because the management process requires a lot of water to maintain the sanitation of the cage and dairy products. The following table 5 is a variable cost in large-scale lactation cattle business production.

Table 5. Variable Costs of Large-Scale Lactation Cattle Production

No	*Variable Cost*	Quantity	Unit price (IDR)	Total (IDR)
1.	Pregnant heifer	18	17.000.000	306.000.000
2.	Forage/period (kg)	77.760	1.200	93.312.000
3.	Concentrate/ period (kg)	19.440	2.400	46.656.000
4.	Wormectin	18	10.000	180.000
5.	Vitamin B-Plex	18	20.000	360.000
6.	IB	18	25.000	450.000
7.	Milk Sales Transportation	18	80.000	1.440.000
8.	Household Electricity and Water with Cage	18	75.000	1.350.000
9.	Labor Wage (Month)	18	600.000	10.800.000
	Total Variable Cost			472.548.000

Source: Harmoko *et al.* [4]

Table 5 represents the variable costs of large-scale lactation cattle production with the maintenance of 18 cows. The biggest variable production costs in this business are pregnant heifers at a price of Rp. 17,000,000 per head. In addition, the other biggest cost is forage feed at a cost of IDR 93,312,000 per period or for 18 months. The cost of concentrate is also quite calculated because it takes a large amount of Rp. 46,656,000, -.

Table 6. Variable Costs of Small-Scale Lactation Cattle Production

No	*Variable Cost*	Quantity	Unit price (IDR)	Total (IDR)
1.	Pregnant heifer	4	17.000.000	68.000.000
2.	Forage/period (kg)	17.280	1.200	20.736.000
3.	Concentrate/ period (kg)	4.320	2.400	10.368.000
4.	Wormectin	4	10.000	40.000
5.	Vitamin B-Plex	4	20.000	80.000
6.	IB	4	25.000	100.000
7.	Milk Sales Transportation	4	80.000	320.000
8.	Household Electricity and Water with Cage	1	75.000	1.350.000
Total Variable Cost			100.994.000	

Source: Harmoko *et al.* [4]

Based on table 6. Regarding the variable costs of small-scale lactation cattle production, which only raises 3-4 cows. Although the scale of this business is relatively small, the cost of forage feed in one period or 18 months is Rp. 20,736,000, - for concentrate feed costs, it costs Rp. 10,368,000, -. The difference in the variable costs of small and large scale lactation cattle production lies in the burden of labor costs. Lactation cattle business with a small scale, or having 3-4 livestock does not require hired labor for a lactation cattle business. Breeders in this case devote some of their time to developing cattle. The remaining labor time is allocated to other people's sideline farming businesses, so that farmers still get additional income.

Table 7. Initial Capital for Orange Business

No	Land Area (m^2)	Initial Capital (IDR)
1.	7.000	550.000
2.	7.500	500.000
3.	8.000	600.000
4.	10.000	600.000
5.	7.000	500.000
6.	10.000	600.000
7.	8.000	500.000
8.	7.000	500.000
9.	8.000	600.000
10.	7.000	550.000
11.	20.000	500.000
12.	8.000	600.000
13.	5.000	10.000.000
14.	5.000	10.000.000
15.	10.000	10.000.000

Based on table 7 regarding the business capital of oranges, it shows that the smallest capital spent to manage the citrus business by the respondents is Rp. 500,000 and the largest capital issued by the respondents is Rp. 10,000,000. In addition, respondents also use the smallest land 5,000 m2 and for the largest land 20,000 m2. There is limited land owned by respondents to be able to produce oranges on a large scale and included with the capital owned. However, a managed citrus business can provide more benefits even though there are limited land areas.

3. 1 Production Cost (Total Cost)

The calculation of production costs is the sum of the addition of fixed costs and variable costs. The determination of the cost of production is carried out by reducing production costs as low as possible and while maintaining the quality of the goods or products produced, this is in accordance with economic law to set production costs to a minimum to get maximum results, this causes the cost of goods produced by livestock to be higher. Lower than the previous one. The total cost of production (total cost) in the integrated dairy cattle breeding business in the Gading MandiriCattle Group, Gading Kulon Village, Dau District can be seen in table 7.

Table 8. Production Costs (Total Cost) of Integrated Earthworm Dairy Cattle Breeding Business

No.	Cost Tyoe	Data A (IDR)	Data B (IDR)
1	Total Fixed Cost	506,000	506,000
2	Total Variable Cost	85.730.000	109,380,000
Quantity		86,236,000	109,886,000
Total Cost			196,122,000

Source: Harmoko *et al.* [4]

The data in Table 8 shows the total cost of production (total cost) in the breeder. The data shows that the production cost in Data A is Rp. 86,236,000,- for the production costs in Data B of Rp.196.122,000. The total production cost for an integrated dairy cattle breeding business with earthworms (Lumbricus rubellus) is Rp. 196,122,000 for a period of 18 months. One period will produce 1 harvest for dairy cattle breeding and 18 harvests for earthworms. The earthworm harvest period is done once a month, this is due to the fast breeding cycle. Several farmers who are members of the Gading Mandiri Livestock Group have activities to raise both dairy and beef cattle, so that many farmers have difficulty processing and utilizing cow dung waste properly. So that a dairy cattle breeding business is created that is integrated with the cultivation of earthworms (Lumbricus rubbelus). This is done so that farmers can utilize organic waste by cultivating earthworms (Lumbricis rubellus) so that they can create new business units based on science and technology so that they can develop entrepreneurial motivation among farmers.

Table 9. Business Production Costs (Total Cost) of Lactation Dairy Cattle Breeding

Average/Period	Breeder	
	18 Lactation Dairy Cattle (IDR)	3-4 Lactation Dairy Cows (IDR)
Fixed Cost (Rp)	506,000	506,000
Variable Costs (Rp)	472.548.000	100.994.000
Total Cost (Rp)	473.054.000	101.500.000

Source: Processed Primary Data (2020)

Based on the data in table 9, it can be seen that the total cost or the total production cost of large-scale lactating dairy cattle with 18 cows reached Rp.473,054,000. For production costs that must be incurred by small-scale farmers or with the maintenance of 3-4 heads reach Rp. 101,500,000. The price of beef cattle greatly determines the quality of dairy cows. Buying cattle seeds from the animal market is very risky, because breeders have the opportunity to get low quality or sick cows. Breeders only sell cattle seeds to the animal market if the quality of their cattle is not good, unless there is an urgent need to sell good quality cattle.

Table 10. Business Production Cost (Total Cost) of Orange Business

No	Total Tree Ownership			Total Cost (IDR)
	Siam Madu	Keprok	Baby	
Quantity	1740	1960	1400	236.000.000
Average	116	130,6667	93,33333	15.733.333,33

Source: Harmoko *et al.* [4]

Based on table 10, it can be seen that the total number of ownership of Siam Honey trees is 1740, tangerines are 1960 and baby oranges are 1400 so that the total cost to produce oranges is Rp. 236,000,000. Based on the opinion of Djumadil *et al.* [11] explained that the production received is not always the same because there are parts that have large revenues influenced by sufficient production costs and good planting and maintenance processes.

3. 2 Revenue (Total Revenue)

Receipts can be divided into cash receipts and calculated receipts. Cash receipts are in the form of sales of livestock products or side businesses, while the income that is taken into account is the value of livestock products. Cash receipts for farmers who carry out integrated businesses come from the sale of by-products, namely earthworms, vermicompost and manure. For receipts that are taken into account in the form of sales of dairy cows. The average income received by farmers who have dairy cattle integrated with earthworms can be seen in Table 11.

Table 11. Acceptance of Earthworm Integrated Dairy Cattle Breeding

No.	Revenue	Total (IDR)	Contribution Percentage (%)
1.	Dairy Cattle Seed Sales	360.000.000	90.85
2.	Sales of Earthworms and their Components	36.240.000	9.14
	Total Sales	396.240.000	100

Source: Harmoko *et al.* [4]

The data in table 11 shows the total income of farmers. Farmers' income is obtained from selling dairy cattle seeds as the main product and earthworms as a by-product. On average, breeders sell live dairy cattle at a price of around Rp. 20,000,000 with a live weight of 250 kg-300 kg. The price of dairy cows depends on the quality of the dairy cows. In addition to selling dairy cattle as the main product, farmers also sell earthworms to cooperatives, catfish or fish owners with a fairly high selling price of Rp. 30.000,-/kg. Earthworms have derivative products in the form of worm droppings or what is known as vermicompost. Kascing has many benefits, namely it is used as a natural solid organic fertilizer which is fermented directly by earthworms.

The data listed in table 9 shows that the sales of dairy cattle breeds contribute 90.85% and earthworms contribute 9.14% of the revenue. The percentage contribution of earthworms and their components in table 10 is relatively small compared to the sale of dairy cattle breeds. Although the percentage of the contribution is quite small, it is very helpful in providing side income for farmers. If farmers only rely on income from dairy cattle breeding businesses, farmers only get income in the 6th and 18th months, as shown in table 12.

Table 12. Revenue Receipts From Dairy Cattle Breeding Business

No	Month Income n-th	Quantity (Ekor)	Total Income (IDR)
1.	6th Month Income	9	180.000.000
2.	18th Month Income	9	180.000.000
Total Revenue			360.000.000

Source: Harmoko *et al.* [4]

In Table 12 there are two periods of revenue receipts from dairy cattle breeding, this is because farmers keep dairy cows with different seed ages, namely dairy cows with 2 months and 10 months of seed purchase. Dairy cows are sold at the age of 18 months or at the age of 1.5 years at a price of Rp. 20,000,000 per head. Breeders prefer to buy dairy cattle from cooperatives, KUD or well-known breeders. The quality of dairy cows that come from cooperatives or better known breeders will be guaranteed. The price you get will be less expensive, than livestock in the animal market. Revenue from earthworm cultivation can be seen in table 13 below.

Table 13. Revenue from Earthworm Cultivation

No	Month Income n-th	Quantity (Kg)	Total Income (IDR)
1.	1stMonth Income	29	870.000
2.	2ndMonth Income	18	540.000
3.	3rdMonth Income	123	3.690.000
4.	4thMonth Income	67	2.010.000
5.	5th Month Income	112	3.360.000
6.	6th Month Income	179,5	5.385.000
7.	7th Month Income	152,5	4.590.000
8.	8th Month Income	39	1.170.000
9.	9th Month Income	89	2.670.000
10.	10thMonth Income	75	2.250.000

11.	11th Month Income	61,5	1.845.000
12.	12th Month Income	88	2.640.000
13.	13th Month Income	76,5	2.295.000
14.	14th Month Income	124	3.720.000
15.	15th Month Income	84,5	2.535.000
16.	16th Month Income	48	1.440.000
17.	17th Month Income	168,5	5.055.000
18.	18th Month Income	152,5	4.575.000
	Total Revenue	1639,5	36.240.000

Source: Harmoko *et al*. [4]

The data in table 13 shows the total income of farmers who sell large and small scale lactation cows. Farmers' income is obtained from daily milk sales. The daily milk production obtained by farmers from two milkings in the morning and evening is 13 liters/day. Farmers get a price of Rp. 6,300, - for one liter of milk. The finished milk will be sold to PT. Nestle Indonesia. The amount of daily cow's milk production depends on several factors such as genetics, feed and rearing management. Sales receipts of large and small scale lactation cows to farmers are shown in table 14 below.

Table 14. Sales Revenue of Large and Small Scale Lactation Cattle

No.	Revenue	Total (IDR)
1.	Large-Scale Sales of Lactation Cattle (18 heads)	796.068.000
2.	Small-Scale Lactation Cattle Sales (3-4 heads)	176.904.000

Source: Harmoko *et al*. [4]

Table 15. Orange Business Revenue

No	Orange Revenue			Revenue per year (IDR)
	Siam Madu (10.000/kg)	Keprok Batu 55 (8.000/kg)	Baby (8.000/kg)	
Quantity	278.400.000	548.800.000	560.000.000	1.387.200.000
Average	18.560.000	36.586.666,67	37.333.333,3	92.480.000

Source: Harmoko *et al*. [4]

Based on table 15, it can be seen that the average orange revenue is Rp. 92,480,000 per year. In carrying out citrus farming, it is necessary to maximize income. This is in accordance with the opinion of Pramadinata, *et al* [10] which states that the acquisition of revenue value must be reduced by farming costs in order to obtain maximum income.

Income

Revenue is the result obtained from the difference between receipts and costs or expenses. Income can be influenced by many factors, one of which is the scale of business, capital, labor efficiency and the level of production of farmers. This income is calculated by the farmer per period or for 18 months.

Table 16. Income from the Business of Integrated Earthworm Dairy Breeding

No.	Income	Total (IDR)
1.	Total Revenue	396.240.000
2.	Total Production Cost (Total Cost)	196,122,000
	Total income	200.118.000

Source: Harmoko *et al*. [4]

The data in table 16 shows that the net income of farmers for one period (18 months) is Rp. 200,118,000. Revenue is the result of revenue from the sale of dairy cattle which are integrated with earthworms. Revenues in table 10 have been deducted by production costs and land and building taxes. The amount of income of farmers depends on efficient production costs. This is in accordance with Ervina *et al*. [9] which states that the high and low income of dairy cattle is influenced by factors such as the number of ownership of lactating dairy cows, the selling price of livestock, the amount of milk production, labor wages, milk prices, and additional feed prices. The income of farmers who run dairy cattle breeding businesses and are integrated with earthworms is different from that of farmers who run lactating dairy cows. In table 17 below is the income of farmers who do lactation dairy cattle with two types, namely large and small scale.

Table 17. Business Income of Large and Small Scale Lactation Dairy Cattle

Income	Total (IDR)	
	Large Scale Lactation Dairy Cows (18 Heads)	Small Scale Lactation Dairy Cows (4-5 Heads)
Total Revenue	796.068.000	176.904.000
Total Production Cost (Total Cost)	472.548.000	100.994.000
Total income	323.520.000	75.910.000

Source: Harmoko *et al*. [4]

Based on table 17, the following is the income of farmers who come from lactating dairy cows. Large-scale lactating dairy farmers have a net income of IDR 323,520,000 per period, or 17,973,333 per month with a total production cost of IDR 796,068,000. The net income earned for small-scale lactating dairy farmers is Rp. 75,910,000 per period or for 18 months. If calculated, the farmer will get Rp. 4,217,222 per month. Expenditure for production costs for small-scale farmers is Rp. 176,904,000.

Table 18. Orange Business Income

Income	Total (IDR)
Total Revenue	1.387.200.000
Total Production Cost	236.000.000
Total Income	1.364.200.000

Based on table 18, the total revenue is Rp. 1,364,200,000 of the total revenue which is deducted by the total production cost of all types of oranges that are managed.

Business Contribution

Calculating the business contribution means knowing how much influence the contribution of a dairy cattle business that is integrated with the cultivation of earthworms (Lumbricus rubellus) has on the income of the farmer's family. The following table 21 shows the contribution of dairy cattle breeding business to livestock income per period.

Table 19. Contribution of Earthworm Integrated Dairy Cattle Breeding Business

Type of Business		Contribution (%)
Dairy Cattle Breeding Business Income	163.308.000	88,29%
Earthworm Cultivation Business Income	27,058,000	12,71%
Total Farmer's Household Income	184.966.150	100 %

Source: Harmoko *et al.* [4]

The contribution of dairy farming is 88.29% of the total household income of farmers per period. The contribution of earthworm business is 12.71% of household income. Farmer household income is generated by the main business of dairy cattle breeding with earthworm cultivation as another side income.

4 Conclusion

The research, which was conducted at the Gading Mandiri Cattle group, Dau District, became the integration of dairy cattle breeding businesses owned by farmers with other businesses, namely earthworms (Lumbricus Rubellus) and Valencian sweet orange plantations (Citrus sinensis Valencia) which is one way out so that farmers can earn income regular in the dairy calf rearing period. However,integrated farming of these commodities does not work well in the community due to farmers' lack of land and capital. Moreover, most respondents indicated that they were interested only in citrus farming because it is easier compare to dairy cow breeding and earthworm farming. Good cooperation from various parties is needed to increase the importance of integrated farming with these three commodities for sustainable farming in Indonesia.

References

1. BPS-Statistics Indonesia, *Populasi sapi perah menurut Provinsi 2009-2019*, (2020).
2. Kementrian Pertanian, *Buku outlook komoditas peternakan susu sapi* (2019).
3. T.W. Widayati, I. Sumpe, Prosiding Seminar Teknologi Agribisnis Peternakan (STAP) Fakultas Peternakan Universitas Jenderal Soedirman, **7** (2020)
4. H. Harmoko, I. Ibrahim, N. Kusrianty, M. Marhayani, Cendekia Eksakta **5** (2021)
5. N. Novarista, M. Maiyontoni, R. A. Putra, H. D. Triani, Agrifo: Jurnal Agribisnis Universitas Malikussaleh, **5** (2020)
6. S. Asperinche, A. Widyasworo, AVES: Jurnal Ilmu Peternakan, **14** (2020)
7. G. A. R. Maria, S. T. Raharjo, Jurnal Kolaborasi Resolusi Konflik **2** (2020)
8. R. Fahlewi, R. C. Amri and A. M. Sari, Journal of Economic, Business and Accounting, **4** (2020)
9. D. Ervina, A. Setiadi, T. Ekowati, SOCA, **13** 2 (2019)
10. B. Y. A. Pramadinata, N. Khoiriyah, L. R. Maula, Jurnal Sosial Ekonomi Pertanian dan Agribisnis, **9** (4) (2021)
11. N. Djumadil, Y. Syafie, Prosiding Seminar Nasional Agribisnis, **1** (1) (2020)

Evaluation of qualitative characteristics of Bali cattle at Bali breeding center

Rosidi Azis[1], *Veronica Margareta Ani* Nurgiartiningsih[2*], *Sri* Wahjuningsih[2], *Herni* Sudarwati[2], and *Ahmad* Furqon[2]

[1]Animal Science Doctoral Program, Faculty of Animal Science, University Brawijaya, Malang 65145, East Java, Indonesia
[2]Faculty of Animal Science, University Brawijaya, Malang 65145, East Java, Indonesia

Abstract. The aim of this study was to identify the qualitative characteristics of Bali cattle at Bali Breeding Center. The samples used were 136 heads. Qualitative data was analysed based on the percentage for each trait and analyzed descriptively. The results showed that the characteristic with the highest percentage were brick red for females of Bali cattle (84,81%), black for males Bali cattle (86.66%), dominant thick back line (79.74%) for females, while on males was not clear (86,67%), strict line of body color border with leg (100% for males and 91,14% for females), strict line of body color border with buttocks color (93,33% for males and 88,61% for females), white of edge lips color (95% for males and 93,67% for females), black eyelid color (100%), white ear hair color (males: 100% & females: 96,20%), dark circles on the ears (98.73%), dominant medium wattle for males (93,33%) & small wattle for females (53,16%), upwards horn (males: 100% & females: 94.93%), black tail feathers (males: 100% & females: 98.04%), and no hump (males: 96.66% & females: 100%). The qualitative characteristics of Bali cattle at Bali breeding center were in line with Indonesian National Standard. However, there were deviations and abnormality with a small percentage.

1 Introduction

One of the diversities of Indonesian livestock genetic resources is Bali cattle as beef-producer. Bali cattle are native Indonesian cattle that contribute to the fulfillment of meat in Indonesia. Bali cattle are spread in several areas in Indonesia such as West Nusa Tenggara, East Nusa Tenggara and Sulawesi [1, 2]. The distribution of Bali cattle has been widespread in tropical and subtropical environments outside of Indonesia before its distribution was limited. Throughout the wide distribution of Bali cattle, there were no significant problems in terms of reproductive function so that Bali cattle grow and develop as in their native area [3]. The high adaptability of production and reproduction despite being in a marginal environment has made Bali cattle much in demand today [3-5]. However, the genetic potential of Bali cattle developed outside Bali is not the same as that in Bali.

* Corresponding author: vm_ani@ub.ac.id

The qualitative characteristics of female Bali cattle are brick red, but the bull is black, the back of the thigh is white, the edges of the upper lip and the thigh of the lower leg start from the tarsus and carpus to the upper edge of the hoof, fur on the tip the tail is black, the fur on the ears is white and on the top of the back there is a black line [6, 7]. The most easily identifiable characteristic of Bali cattle is brick red, but at the age of 12-18 months, bulls change color from brick red to black [8].

Qualitative properties are less valuable than quantitative traits [6], but in certain circumstances have high economic value. Qualitative nature becomes important at the time of traditional and religious events. Complete information on the qualitative characteristics of Bali cattle is still limited. This study aims to identify qualitative characteristics such as coat color, horns and other specific qualitative characteristics of Bali cattle. This is done as an effort to evaluate the qualitative characteristics of Bali cattle at the Bali Breeding Center.

2 Materials and methods

The total sample was 139 Bali cattle consisting of 79 females and 60 males with the criteria of 2-5 years of age at the Bali breeding center. Qualitative characteristics include the presence of dominant skin color (dominant color in female cattle: brick red, light brown, yellowish brown, spotted; while bulls: dark brown, black, reddish black), body color border with the buttocks, body color border with the back. legs, tail feathers, black eye circles, black lines on the back, the direction of the horns, the presence or absence of wattles and the presence or absence of a hump.

Qualitative trait data were analyzed descriptively with frequency calculated and displayed in percentage form for each trait using the following formula:

$$P\ (x) = \frac{\sum Xi}{n}\ x\ 100\%$$

P : Percentage of colors that appear
xi : the property being measured
n : total sample

3 Results and discussion

The results of observations of the qualitative characteristics of bull and cow Bali cattle at the Bali Breeding Center are as shown in Table 1 and 2. Based on the two Table, Bali cattle at Bali Breeding Center have dominant black color (86,67%) of males and brick red (84,81%) of females. The thick back lines on females was clear (79,75%), while on males was not clear (86,67%). The body color border with legs have dominan strict lines (males = 100% and females = 91,14%), as well as the body color border with the buttocks have dominan strict lines (males = 93,33% and females = 88,61%). The white color on the edges of the lips was both dominant (males = 95,00% and females = 93,67%). The eyelid color was black dominant (100%), ear hair color dominant white on males and females, black ring on dominant ear black, dominant moderate and small wattle, dominant horn direction upwards, dominant black tail feathers and no dominant hump. Characteristics of qualitative properties obtained in previous studies [9-11]. These characteristics have the appearance of similar character to Banteng, but the body of Bali cattle was smaller than Banteng [12]. The phenotypic similarity between Bali cattle and Banteng was evidence that the relationship between Bali cattle and Banteng have similar qualitative characteristics [13, 14]

Table 1. Qualitative characteristics male Bali cattle at Bali breeding center

Qualitative Characteristic	n	Character	Percentage (%)
dominant skin color	7	Dark brown	11,67
	52	Black	86,67
	1	Reddish black	1,66
Back line	0	Thick	0
	1	Medium	1,67
	7	Thin	11,67
	52	Not clear	86,67
Body color border with legs	60	Strict lines	100
		Gradation (no clear)	0
		No border	0
Body color border with buttocks	56	Strict lines	93,33
	4	Gradation (no clear)	6,67
		No border	0
Lip edge color	57	White on the edge	95,00
	3	White circle	5,00
Eyelids color	60	Black	100
		Kedness	0
Ear hair color	60	White	96,67
		Brown	3,33
		Black	0
Ear black circle		Black	0
	4	No black circle	1,67
Wattle	1	Long	1,67
	56	Medium	93,33
	3	Small	5,00
Horn		Forward	0
	60	Upward	100
		backwards	0
Tail hair	60	Black	100
		White/panjut	0
Hump	2	Yes	3,33
	58	No	96,67

Table 2. Qualitative characteristics female Bali cattle at Bali breeding center

Qualitative Characteristic	**n**	**Character**	**Percentage (%)**
Dominant skin color	67	Brick red	84,81
	7	Light brown	8,86
	3	Yellowish-brown	3,79
	2	Spotted	2,53
Back line	63	Thick	79,75
	9	Medium	11,39
	7	Thin	8,86
	2	Not clear	2,53
Body color border with legs	72	Strict lines	91,14
	7	Gradation (no clear)	8,86
		No border	0
Body color border with buttocks	70	Strict lines	88,61
	9	Gradation (no clear)	10,13
		No border	1,26
Lip edge color	74	White on the edge	93,67
	5	White circle	6,33
Eyelids color	79	Black	100
		Kedness	0
Ear hair color	76	White	96,20
	3	Brown	3,79
		Black	0
Ear black circle	78	Black	98,73
	1	No black circle	1,27
Wattle		Long	1,49
	37	Medium	46,84
	42	Small	53,16
Horn	1	Forward	1,26
	75	Upward	94,94
	3	backwards	3,79
Tail hair	77	Black	97,47
	2	White/panjut	2,53
Hump	0	Yes	0
	79	No	100

Based on the results in Table 1 and 2, there were several deviant characters and were considered defective. For example a. There were spots on the body (2.53%), b. the white tail feathers (panjut) (2.53%), c. no back line (2.53%), d. the body color border with the buttocks color was not clear (1.26%), e. circular white color on the lips (males = 5.00% and females = 6,33%), f. the color of the fur was brown on the ears (3.79%), g. Long wattle (1.67%) was common in bulls, but generally wattle in medium-sized on male of Bali cattle, h. There was a small hump (males = 3,33%). The occurrence of characteristic deviations in cattle such as the appearance of spotted in Bali cattle could be caused by gene mutations [11]. The spotted character in Bali cattle appears at an adult age, while at a young age there was no spotted character. In contrast to the character of the white tail feathers that could be detected from a newborn age. The spotted and panjut characters were immediately removed (selection) and they were not given the opportunity to produce.

Some other deviant characters such as no back line, body color boundaries on the legs were not clear, body color limits on the buttocks color and others were caused by the genes controlling these traits experiencing mutations or due to the presence of genes from other cattle breeds from the ancestors [11]. The existence of a crossbreeding program for Bali cattle with other breeds of cattle could be one of the reasons for the occurrence of undesirable traits in Bali cattle [15]. This possibility could occur uncontrolled spread of Bali cattle breeds, resulting in crosses with other breeds. However, for some of these traits, they were still given the opportunity to produce and were not considered defective as the spotted character and white tail feathers.

Generally, the results of this study showed that the qualitative characteristics of Bali cattle have met the standard of SNI 7651-4:2017. It means that the Bali Breeding Center applied a good breeding program to maintain the specific qualitative character of Bali cattle.

4 Conclusion

Qualitative traits in Bali cattle have a good dominant percentage on the typical character of Bali cattle. The percentage of deviant and defective characters in Bali cattle is very small. The qualitative trait selection program was carried out with the age standards prevailing in the community, such as spotted and white tail traits being removed, while other deviant traits were ignored.

References

1. T. Saili, IOP Conf. Ser.: Earth Environ. Sci. **465** 012004 (2020)
2. S. Baco, R. Malaka, M. Hatta. IOP Conf. Ser.: Earth Environ. Sci. **492** 012101 (2020)
3. S. Rahayu, Berkala Penelitian Hayati, **20**, 28-35 (2014)
4. M. Purwanti, H. Harry, Jurnal Penyuluhan Pertanian, **1** (2006)
5. K. Amiano, W. Busono, S. Maylinda, TERNAK TROPIKA (Journal of Tropical Animal Production), **21**, 39-43 (2020)
6. M. Bahary, Perbedaan Sifat Kualitatif dan Kuantitatif Sapi Bali Tidak Bertanduk dengan Sapi Bali Bertanduk, Undergraduate Thesis, Universitas Hasanuddin Makassar, Indonesia (2017)
7. H. Hikmawaty, A. Gunawan, R. Noor, Jakaria, Jurnal Ilmu Produksi dan Teknologi Hasil Peternakan, **2**, 231-237 (2014)
8. H. Martojo, *Penyebaran Sapi Bali pada Masa Lampau. Analisis Manfaat dan Risiko Hasil Rekayasa Genetik dalam aspek: Produktivitas, perlindungan dan keanekaan*

hewan. Pusat Antar Universitas Bioteknologi (Institut Pertanian Bogor, Bogor, Indonesia, 1995)
9. A. Supriyantono, O. Yoku, T. Widayati, I. Sumpe, in Prosiding Seminar Nasional Teknologi Peternakan dan Veteriner, pp. 42-46 (2014)
10. W. Hardjosubroto, *Aplikasi pemuliabiakan ternak di lapangan* (PT Grasindo Jakarta,Jakarta, Indonesia, 1994)
11. R. Setiaji, S. Prastowo, D. Prasetiyo, N. Widyas, IOP Conf. Ser.: Earth Environ. Sci. **372** 012032 (2019)
12. Z. Zulkharnaim, S. Baco, L. Rahim, M. Yusuf, Hasanuddin, J Anim Sci, **2**,70-75 (2020)
13. K. Mohamad, M. Olsson, G. Andersson, B. Purwantara, H. Van Tol, H. Rodriguez-Martinez, B. Colenbrander, J. Lenstra, Reproduction in domestic animals, **47**, 18-20 (2012)
14. B. Purwantara, R. Noor, G. Andersson, H. Rodriguez-Martinez, Reproduction in domestic animals, **47**, 2-6 (2012)
15. R.J.W. Setiyabudi : Evaluasi Tren Fenotipik Dan Genetik Sapi Bali Di Balai Pembibitan Ternak Unggul Dan Hijauan Pakan Ternak Denpasar, Master Thesis, Institut Pertanian Bogor, Bogor, Indonesia (2016)

The impact of education and experience on profit of smallholder dairy farming at Batu City of Malang Raya, Indonesia

Hari Dwi Utami[1,*], *MB* Hariyono[1], *Umi* Wisaptiningsih[1], *Hary* Nugroho[2], and *Nur* Cholis[2]

[1]Department of Livestock Socio-Economic, Faculty of Animal Science, Universitas Brawijaya, Malang 65145, Indonesia
[2]Department of Animal Production, Faculty of Animal Science, Universitas Brawijaya, Malang 65145, Indonesia

Abstract. The research was conducted at Batu, City, Malang Raya of Indonesia. Study addressed to examine the farmer characteristics, dairy farming income, and the factors influencing on profit. The case study applied multistage sampling method to select 34 representative farmers which divided into three strata namely, stratum-1 (rearing <4 Animal Units), stratum-2 (owning 4-8 AU), and stratum-3 (controlling >8 AU). Primary data collection used survey method with structured questionnaire, whereas secondary data were available in related institutions and sources. Data analysis implemented descriptive and multiple regression technique. Results confirmed that farmers has experienced about 6-10 years in raising dairy farming and they has secondary school education. The profitable dairy farming was smallholder dairy farming that rearing more than 8 AU with daily income per Animal Unit of IDR 64,554 and structured with IDR 11,131 of revenue and IDR 47,577 of production cost. Farmer's experience was positively explaining the smallholder dairy profit, and the high school education attainment was more likely to increase venture's income. The farm return has positive and strong relationships with the more number of dairy cattle owned.

1 Introduction

Dairy farming was popular in rural areas as the favorable activity that can generate family income. As the growth of dairy farming, more family members has more participated in this farming and therefore it can contribute on family prosperity. In regard to dairy farming livelihood, previous studies discovered that this farming become imperative for villager livelihood since it can provide daily earning from selling fresh milk. This venture represents as prospective farming and contributes the good income for farmers' household [1]. Particularly, in the area where land holding is inadequate and unbalanced and therefore, smallholder dairy farming might become an alternative income enhancement [2]. Farmers who kept 3-5 AUs can contribute about 39% to household Income [3]. Profit of dairy cattle

* Corresponding author: hrdwutami4@ub.ac.id

per Animal Unit in monthly basis was about IDR 767,271 R. D. Haloho, *et al.*. [4], IDR 166,667/AU/month in West Java [5]. Study in in Budur Province of Turkey found net profit USD 188.28/AU/year [6].

Some previous studies ascertain that farmer's education and experience in rearing dairy has impacted on increasing dairy farming practices [7]. The diminishing technical inefficiency for milk production will related significant with the years of education [8]. Education of household head negatively affects technical inefficiency of milk production but, positively influence it as found in [9, 10, 11].

The presence of faster technology development can offer potential tools that related to increase milk productivity, and therefore lead to dairy income. The proper application of technology might promise the higher milk production. IOT is one of the implementation in industry 4.0. Farmer's experience or knowledge of farming were more likely to accept the modern dairy farming in which implementing modern ways of treating farm animals [12]. Also, the adoption of advance technology have a positive relationship with the profit. The farmer that have good education only can capable using the advance technology. The research will try to insight into the influence of these variables in pursuing technology, either the have similar or opposite direction in regard to dairy farming income. Hence, the study addressed to three questions: (i) what is farmer's characteristic, (ii) how much profit will be obtained from dairy farming and (iii) what is factors explaining the venture income? Two hypothesis were formulated to answer the research objectives. Hypothesis-1: It was predicted that farmer's experiences in rearing dairy cattle has positively influenced on farming profit. Hypothesis-2: It was predicted that farmer's education was positively effecting on farming income.

2 Materials and methods

The research located in Malang Raya which selected by purposive sampling. The research was conducted at Batu, City, Malang Raya of Indonesia. Study addressed to examine the farmer characteristics, dairy farming income, and the factors influencing on profit. The case study applied multistage sampling method to select 34 representative farmers which divided into three strata namely, stratum-1 (rearing <4 Animal Units), stratum-2 (owning 4-8 AU), and stratum-3 (controlling > 8 AU). Primary data collection used survey method with structured questionnaire, whereas secondary data were available in related institutions and sources. Three analysis was done to analyse the data. First, descriptive analysis was important to perform the farmer's characteristics. Second, descriptive analysis using economic formulation is employed to analyse the farming return. Finally, the multiple regression technique might appropriate to explain factors influence on dairy farming income.

3 Results and discussion

In general, farmers has experienced about 6-10 years in controlling dairy farming which majority (60%) was represented by farmers in stratum-2 (Figure-1). It implied that farmers has many solutions in dealing with various problems in operating dairy farming. Therefore, this farm can be maintain and sustain this farm in order to provide daily income for their family for long time.

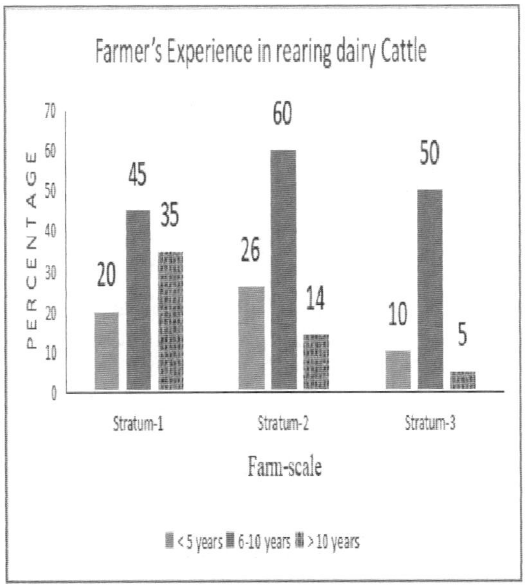

Figure 1. Farmer's experience in rearing dairy cattle

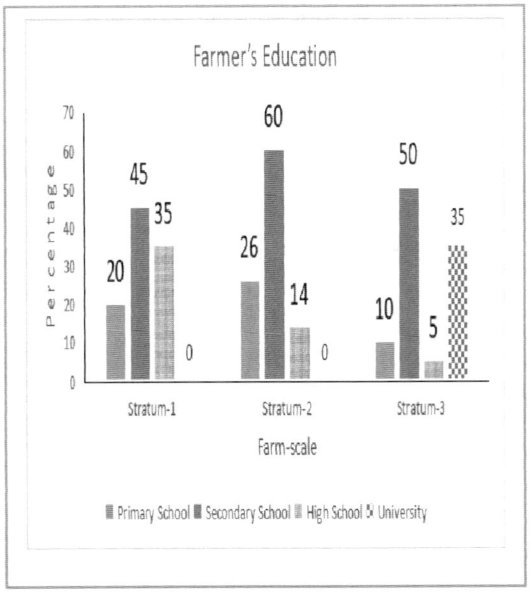

Figure 2. Farmer's Education

Figure 2 depicted that secondary school has dominated in all strata of dairy farming. Particularly, farmers in stratum-2 showed high percentage (70%) in the secondary school attainment than those stratum-3 (50%) and stratum-1 (45%). The education will be related to the capability in access new technology and even easy to adopt them in running dairy farming. Hence, farmers can improve their skill in dairy farming practice to increase farm's productivity that resulting in high income achievement.

3.1 Variable and Fixed Costs of Smallholder Dairy Farming

Daily production cost per Animal Unit ranged IDR 43,617 -IDR47,577 with the efficient cost was coming from stratum-1. The expenses structured with 11.94%-14.79% for fixed cost (Figure-3) and 85.26%-85.89% variable cost (Figure 4). The efficient of fixed cost can reach in stratum-2 about 11.94%, while stratum-3 was thriftily on the variable expenses up to 85.26%.

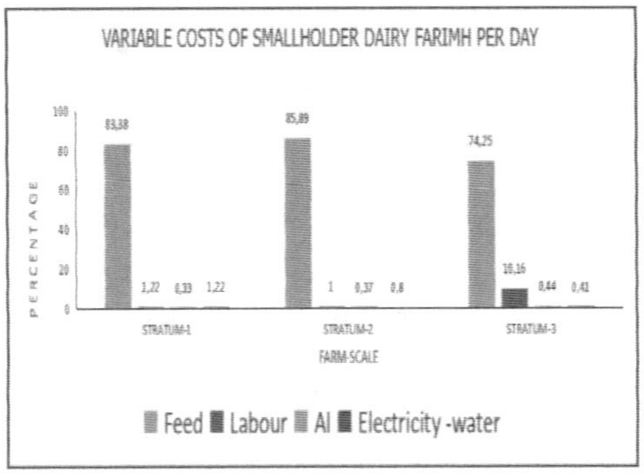

Figure 3.Daily variabel cost per Animal Unit of dairy farming

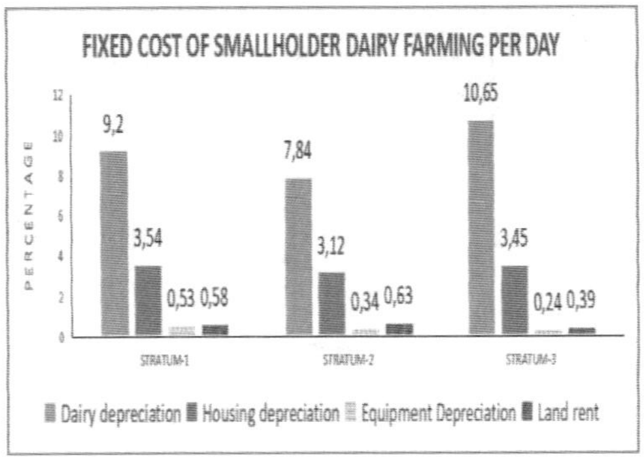

Figure 4. Daily fixed cost per Animal Unit of dairy farming

Concerning to fixed cost, dairy depreciation in stratum -2 used the little depreciation costs (7.84%) than those of stratum-1 (9.20%) and stratum-3 (10.65%). Likewise, the less fixed expenses happened in stratum-2 was about 3.12% in the comparison with stratum-3 (3.35%) and stratum-1 (3.54 %). Variable cost has counted lowest in feed concentrate expenses for stratum-3 (74.25% in comparison with stratum-1 (83.38%) and followed by stratum-2 (85.89%).

3.2 Revenue and Profit of smallholder dairy farming

The revenue of smallholder dairy farming increase as the adding number of dairy cattle (Figure 5). The daily revenue per Animal Unit started from IDR 9,931 for stratum-1 to IDR 99,978 for stratum-2 and reached the highest one about IDR 112,131 in stratum-3. The revenue of stratum-3 composed with 64.11% from selling fresh milk structure for stratum-2, whereas and 35.89% from calf yielding. Selling fresh milk dominated (69.12%) revenue structure for stratum-2 compared those from stratum-1 (65.12%). It was small different the calf revenue between stratum-2 (30.88%) and stratum-1 (34.88%).

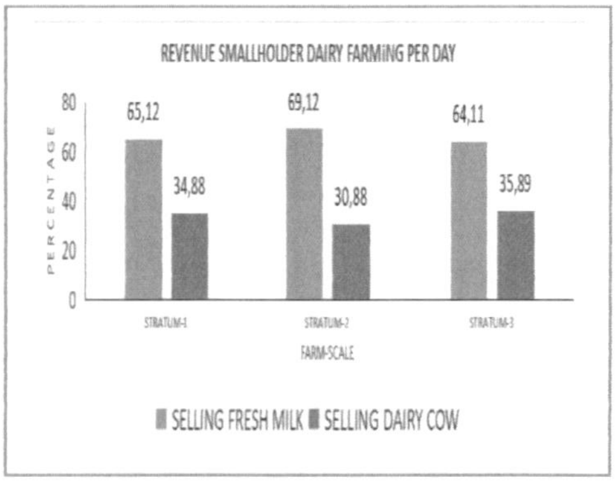

Figure 5.Daily revenue per Animal Unit of dairy farming

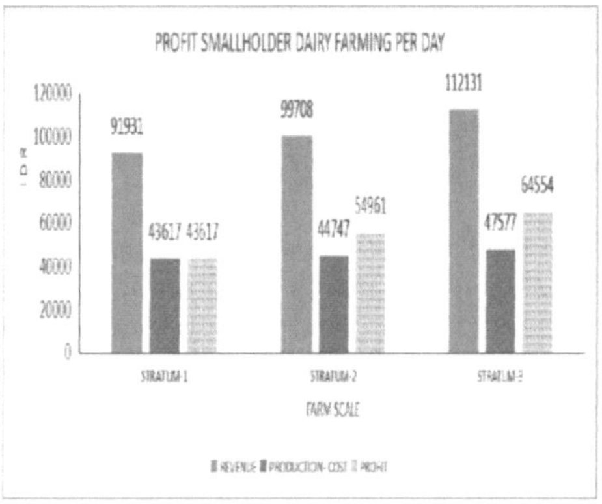

Figure 6. Daily profit per Animal Unit of dairy farming

Figure 6. exhibits the incline of daily Profit per Animal Unit aligning with the increase of the number of dairy cattle owned. The higher profit occurred on stratum-3 (IDR 64,554) in comparison with those for stratum-2 (IDR 54,961) and stratum-1 (IDR 43,617). The finding was highest than IDR 25,576 daily income per Animal Unit [4]. Smallholder dairy farming represented the prospective smallholder dairy farming in contributing the good income for farmers' household [3] and it might become an alternative in generating income in rural areas [2].

3.3 Factors explaining smallholder dairy farming income

Three of the six predicted variables can explain the profit of smallholder dairy farming (Table 1.). Coefficient determination (R2) was counted for 81%. It can be interpreted that six predictor variables explain about 81% of smallholder dairy farming income, whereas the rest of 19% was clarified by other variables.

F-test addressed to examine the influence of all variables dependent toward dependent variable. F-calculation was 32.41 and very significant (P< 0.000). Results informed that six variables namely farmer's age, education, experience in raising dairy, number of household members, land occupation and number of dairy cattle together had simultaneously affect on farming income.

Table 1. Factor influencing in Smallholder Dairy Farming

Explanation	Regression Coefficient
Constant	33.14
Age	0.469
Farmer's Education	-0.238*
Family number	-0.329
Dairy number	1.126***
Land occupation	0.256
Farmer's experience in raising dairy	0.337*
R Adjusted = 74%	
R Square (R2) = 81%	
n = 34	
Fcalculation = 32.406	
*P<0.01; ***P<0.000	

The result of t-test showed that three of six predicted variables were significantly having effect on farmers income. There were dairy number and the experience have positive influenced, whereas it was negatively affecting on farm income. The t-test shows that only three variables, experience and the number of dairy cattle were found significant positively influencing in dairy farming income, while it was the negative impact on farmer's education. The detail explanation will find in the following paragraph.

Dairy cattle number explains positively and very significant (P<0.000) in regard to farming return. Regression coefficient was 1.126. It can be interpreted that the increase 1 unit of dairy number will raise profit 1.126 unit of profit. Likewise, study of H. D. Utami, *et al.* [1] reported that the more dairy owned, the many profit will be obtained. Smallholder dairy farming, therefore, become an alternative in generating income in rural areas [2].

Farmer experience in operating dairy farming has positive and significant P < 0.01) in relation with dairy profit. The results has accepted Hypothesis-1: It was predicted that farmer's experiences in rearing dairy cattle has positively influenced on farming profit. Regression coefficient was 0.337. It can be interpreted that the incline 1 unit of dairy number will increase profit about 0.337 unit. Farmers have experienced long time about 6-10 years in operating their farm, therefore they have ability to deal with any problem in raising dairy cattle. This finding was aligning with study of Fita, *et al.* [7] that farmer's experience in rearing dairy has impacted on increasing dairy farming practices.

However, the farmer's education has negative and significant (P<0.01) impact on dairy return. The secondary school education of farmers will reduce the farm earning. In other words, the farm income will increase as the farmers has the high school education attainment. This discovering was agreeing with hypothesis-2: It was predicted that farmer's education was positively effecting on farming income. Regression coefficient was 0.238. It can be interpreted that the raising 1 unit of farmer's education will improve farming income about 0.238 unit. This finding confirmed with the study of H.Girma, Heliyon [8] in that the diminishing technical inefficiency for milk production will related significant with the years of education. Therefore, education of household head positively affects technical efficiency of milk production [11].

4 Conclusion

The research has confirmed that farmers has experienced about 6-10 years in raising dairy farming and they has secondary school education. The profitable dairy farming was smallholder dairy farming that rearing more than 8 AU with daily income per Animal Unit of IDR 64,554 and structured with IDR 11,131 of revenue and IDR 47,577 of production cost. Farmer's experience was positively explaining the smallholder dairy profit, and the high school education achievement was more likely to increase venture's income. The farm return has positive and strong relationships with the more number of dairy cattle owned

It was suggested that the utilisation of qualified forage need to be improved to increase milk productivity. Farmers need to improve the clean and healthy environment to obtain the good quality, safety, and healthy milk products.

References

1. H. D. Utami, U. Wisaptiningsih, H. Nugroho, Laporan Penelitian (2018)
2. O. M. Kocturk, J Anim Vet Adv **8**, 1685-1688 (2009)
3. H. D. Utami, A. P. Seruni, The 2nd Animal Production International Seminar (APIS) 2013 : Sustainable Livestock Production Based on Local Resources in the Global Climate Changes Era Prospect and Challenges (Universitas Brawijaya, Malang, Indonesia, 2013)
4. R. D. Haloho, S. I. Santoso, S. Marzuki, Jurnal Pengembangan Humaniora **13**, 65-72 (2013)
5. K. S. Mudikdjo, R. Pambudy, Media Peternakan **24**, 57-66 (2000)
6. V. Demircan, T. Binici, H. Koknaroglu, A. Aktas, Czech J. Anim. Sci. **51**, 8–17 (2006)
7. Fita, M. M. Trivedi, B. Tassew, J. Agric Extension Rural Dev. **4**, 69-78 (2012)
8. H. Girma, Heliyon, **5**, 1-24 (2019)
9. Bardhan, Sharma, Springer Plus **2**, 1-7 (2013)
10. Z. Adane, K. Shiferaw, B. Gebremedhin, International Livestock Research Institute (ILRI) **11**, 1777-1786 (2015)
11. A. Al-Sharafat, J. Agric. Sci. **5**, 45-53 (2013)
12. B. K. Boogaard, B. B. Bock, S. J. Oosting, J. S. C. Wiskerke, A. J. van der Zijpp, J Agric Environ Ethics **24**, 259-282 (2011)

Using quail waste as an adsorbent on biogas quality

Aurelia Aprilianty[1,*], *Arina* Manasikana[2], *Eggi Pur* Pinandita[3], *Shafa* Fa'izah[3], *Mochammad* Junus[4], and *Hary* Nugroho[4]

[1]Undergraduate program of Faculty of Animal Science, Universitas Brawijaya, Malang 65145, East Java, Indonesia
[2]Doctoral program of Faculty of Animal Science, Universitas Brawijaya, Malang 65145, East Java, Indonesia
[3]Magister program of Faculty of Animal Science, Universitas Brawijaya, Malang 65145, East Java, Indonesia
[4]Faculty of Animal Science, Universitas Brawijaya, Malang 65145, East Java, Indonesia

Abstract. The aim of this research is to determine the effect of coconut and mineral waste of quail as adsorbents on the biogas purity. This research was designed to use the Completely Randomized Design (CRD) method with 6 treatments and 4 replications and the significant influence will be tested using Duncan Test. The result form this research showed that the use of coconut waste and mineral waste of quail as adsorbents gives highly significant influence ($P<0,01$) to increase the CH4 gas concentration, highly significant influence ($P<0,01$) to decrease CO_2 gas concentration, highly significant influence ($P<0,01$) to decrease gas pressure and highly significant influence ($P<0,01$) to increase gas flow rate on biogas purity. The conclusion of this research is the use of 100% of activated coal from coconut waste, the use of 50% of activated coal from coconut waste and 50% of mineral waste from quail waste as an adsorbent are able to increase the quality of biogas. However, it is lack of effectiveness due to inappropriate application of biogas purification. It is suggested to do physical activation for both adsorbents in order to avoid saturation of the adsorbent so those absorbents be able to adsorb optimally the impurities gases on biogas.

1 Introduction

Indonesia is one of the countries facing problems related to the availability of fuel as an energy source. This is due to the large dependence of the Indonesian state on fossil fuels, if this problem is not addressed immediately it will lead to a scarcity of energy sources in Indonesia in the future. In addition, the use of fossil fuels as an energy source can lead to an increase in the concentration of greenhouse gases in the atmosphere caused by carbon emissions. Therefore, it is necessary to use alternative energy by utilizing renewable energy or renewable energy sources whose use is intended to reduce the use of fossil fuels. Indonesia is an agricultural country where the majority of the population works in the agricultural and livestock sectors so that the quantity of waste produced from this sector is

* Corresponding author: Aureliaaprlnty@student.ub.ac.id

abundant. In otherwise, the provision of feed for animals is a major contributor to land and water use greenhouse gas emission [1]. In line, overfishing and reduction in available land and water resources in huge archipelago island like Indonesia on which crops can be grown has led to an increase in animal industry. [2]

Moreover, it is unfortunate that this waste from the agricultural and livestock sectors is often not utilized and thrown away. According to [3] the presence of animal waste that is wasted on land without the involvement of water will produce types of greenhouse gases such as CO_2, N_2O, NO_2, NO_x and CH_4. Then animal waste that is dumped into the waters will produce greenhouse gases such as CO_2 and CH_4. Therefore, further handling and processing is needed so that waste from this sector does not become a source of environmental pollution. Biogas is a renewable energy, which in the energy production process utilizes non-fossil raw materials such as waste from the agricultural sector, especially livestock. Biogas is the result of anaerobic degradation of organic compounds with the help of methane bacteria which can be a substitute for natural gas and fossil fuels. Biogas contains components such as methane gas (CH_4), carbon dioxide gas (CO_2), oxygen gas (O_2), hydrogen sulfide gas (H_2S), nitrogen gas (N2), hydrogen gas (H_2), ammonia (NH_3), and gas. carbon monoxide (CO). According to [4] the quality of biogas is determined from the percentage of methane gas (CH_4) and carbon dioxide gas (CO_2). Biogas which has a high content of methane (CH_4) will have a high calorific value, and vice versa if the level of carbon dioxide gas (CO_2) in biogas is high, it will result in a low calorific value of biogas.

This is confirmed by [5] based on his research, the largest composition of biogas produced from anaerobic fermentation is methane gas (CH_4) around 54-70%, and carbon dioxide gas (CO_2) around 27-45%. Therefore, it is necessary to make efforts to reduce the percentage of carbon dioxide (CO_2) gas content in biogas to improve the combustion quality and biogas calories. Purification of biogas is one way to reduce the levels of impurity gases in biogas such as carbon dioxide gas (CO_2) and hydrogen sulfide gas (H_2S) which are corrosive, with purification causing the levels of methane gas (CH_4) in biogas to increase. Based on research [6] purification or purification of biogas aims to increase heat efficiency, increase the amount of methane gas composition between 4-20% from before and can reduce the content of CO_2, H_2O and H_2S. Purification of biogas can be carried out by chemical absorption methods and adsorption on solid surfaces. The adsorption method on solid surfaces is one of the effective methods for adsorbing impurities in biogas [7]. [4] added that one of the adsorbents that can be used in the purification of biogas is charcoal/activated carbon. Besides being easy to obtain, the use of charcoal/activated carbon is widely used because the price is relatively cheap compared to zeolite and chemical adsorbents. Coconut shell waste and quail livestock waste minerals can be used as adsorbents for gas impurities in biogas, so that the resulting biogas contains methane gas (CH_4) and a higher calorific value. Furthermore, biogas as renewable energy is able to overcome the negative impacts of livestock and fishery waste, so that a clean environment can be realized and can maintain the sustainability of domestic energy sources. According to thus condition the objective of this research to determine the effect of activated carbon and mineral quail waste as adsorbents on the biogas quality.

2 Materials and methods

Field research was conducted to determine the manufacture of biogas digester tanks (jurigen), biogas purification equipment, biogas manufacture and the application of various biogas purification treatments. The materials in this study are divided into two, namely the main material is the material observed in the biogas purification process and the supporting material is the material used in the purification process.

The research equipment included: 1). plastic balloons, 2). buckets, 3). digital scale, 4). Biogas installation from Jurgen with 35 l capacity 5). biogas purifier and 6). Stopwatch. While the materials were 1). dairy cattle manure, 2). Octopus waste 3). quail manure. 4), well water, 5) activated carbon and 6). Quail minerals. The method used were conducted using completely randomized design. The experiment consisted six treatments and four replicate. The parameters observed were biogas composition before and after purifications; flow rate; and pressure of biogas.

The analysis of variance and Duncan multiple range test were used to analyze the observation results.

3 Results and discussion

3.1 Biogas Content Before Purification

The results of measuring the content of biogas were using a GC (Gas Chromatography) tool show that before the purification process was carried out, the gas content was detected, namely, methane gas (CH_4), carbon dioxide gas (CO_2) and hydrogen gas (H_2). The measurement results show that the concentration of gas in the biogas before purification has a relatively low CH_4 gas level and a high CO_2 gas level. The following in Table 1 is the average concentration of gas content in biogas before purification is carried out.

Table 1. Biogas composition before purification

Gases	Average
Methane (CH_4)	33.99±4.41
Carbon dioxide (CO_2)	60.52±5.45
Hydrogen (H_2)	5.49±2.49

Table 1 shows the average level of methane gas (CH_4) produced in the biogas produced in the jurigen anaerobic digester, which is 33.99%, which is still quite low. While the content of impurity gases such as carbon dioxide gas (CO_2) has a fairly high content of 60.52%. The results of this biogas measurement show the low quality of the biogas produced in the jurigen anaerobic digester. The factor that causes the low concentration of CH_4 gas in the biogas can be caused by the absence of the activity of the bacteria forming the biogas. Methane gas in biogas is a by-product of the anaerobic biological process of methanogenic bacteria. Methanogenic bacteria can carry out the process of producing methane gas if it is supported by optimal environmental conditions for their production activities. The type of organic material used as raw material for the manufacture of biogas sludge and environmental temperature can affect the activity of microorganisms in forming methane gas. Methanogenic bacteria can perform methanogenesis/methane gas formation activity at the highest temperature of ±35-550C [8]. The average temperature for 1-2 months during the production of biogas in the biodigester from jerry cans is 27.50C, so the activity of methanogenic bacteria in producing methane gas is not optimal. To increase the temperature of the biodigester, it can be put in a thick cardboard or covered with a tarp and an extra heater using a light bulb is used.

The organic material of biogas sludge used in this study was a mixture of quail livestock waste and squid waste with the addition of cow feces as a starter which was fermented using a digester made of plastic jurigen following the digester design of [9] Organic matter

Quail waste and squid waste are placed in closed jurygen tanks and processed anaerobically for 1-2 months depending on the amount of material added. The advantage of making biogas using this type of digester is that it is easy to apply and the construction cost is cheaper than using other types of digesters. The weakness of this type of digester is that the temperature of the digester is influenced by the ambient temperature so that to achieve the optimal temperature in the formation of methane gas, an extra heater is needed such as a light bulb. Then, this type of digester has the potential for leakage so that the fermentation process does not take place anaerobically which causes the methanogenesis process to be hampered and biogas production decreases.

3.2 Biogas Content After Purification (CO_2, CH_4, H_2)

The results of the measurement of the biogas content which consists of the calculation results of the percentage concentrations of CO_2, CH_4 and H_2 gases. The results of the analysis of variance showed that the treatment using activated carbon adsorbents from coconut shells and minerals from quail livestock waste gave a very significant difference ($P < 0.01$) to the content of biogas, namely CO_2, CH_4 and H_2 gas concentrations. The results of the research showed that the highest to lowest concentrations of CH_4 gas were treatments T1, T3, T2, T4, T0 and T5, respectively. The average value of CH_4 gas concentration in T5 treatment was 5.13 ± 0.85a with 100% mineral adsorbent from quail livestock waste, the average value of T5 resulted in the lowest average value of CH_4 gas concentration compared to treatments T0, T1, T3, T2 and T4. Meanwhile, the concentration of CH_4 gas in T1 treatment with 100% activated carbon as adsorbent from coconut shell waste resulted in the highest average value of 63.89 ± 3.80d. It is suspected that biogas purified with 100% mineral adsorbent from quail livestock waste (T5) has a mixture of other gases such as CO_2 and H2 gases with higher concentrations than other adsorbent treatments. Figure 1 is a bar chart showing the effect of adsorbent concentration on the biogas content, the diagram shows the CO_2 and H_2 gas concentrations in the T5 treatment respectively 78.57 ± 1.141d and 16.30 ± 0.62c.

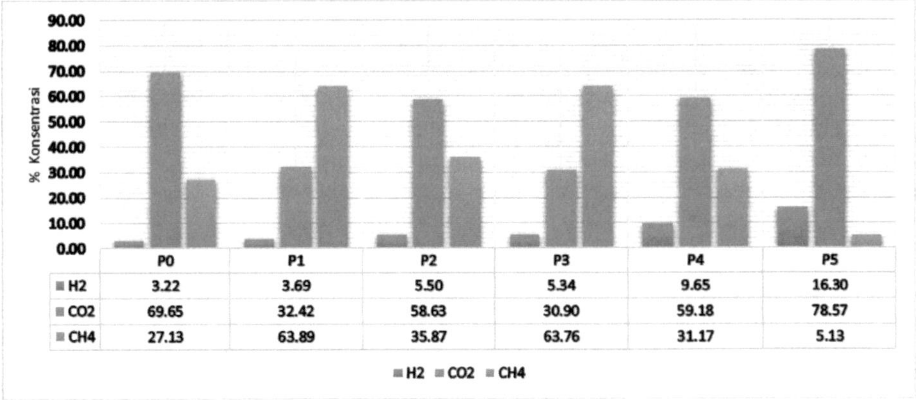

Fig. 1. Effect of adsorbent concentration on biogas content

This is also supported by research conducted by [10] that the production of CH_4 gas tends to decrease along with the increase in the amount of Fe (mineral) in the manufacture of bio gas, so the decrease in the concentration of CH_4 gas can be associated with an increase in the concentration of CO_2 gas produced during the fermentation process. The lowest to highest CO_2 gas concentrations were treatments T3, T1, T2, T4 and T5, respectively. The average value of CO_2 gas concentration in treatment T5 was 78.57 ± 1.141d with 100% mineral adsorbent from quail livestock waste producing the highest average value compared to treatments T0, T1, T2, T3 and T4. The concentration of CO_2 gas in treatment T1 (32.42 ± 5.08a) with 100% activated carbon as adsorbent from coconut shell waste resulted in the lowest average value.

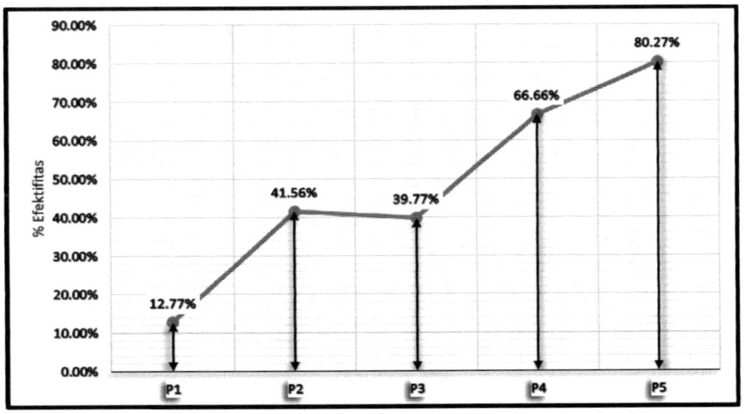

Figure 2. H_2 gas concentration improvement chart (%)

Meanwhile, the concentration of H_2 gas in biogas from the lowest to the highest were treatments T0, T1, T2, P3, P4 and T5, respectively. The average value of H_2 gas concentration in treatment T5 was 16.30 ± 0.62c with 100% mineral adsorbent from quail livestock waste producing the highest average value compared to treatments T1, T2, T3, T2 and T4. The concentration of H_2 gas in the control treatment T0 was 3.22 ± 0.139a which was the control treatment or without the adsorbent treatment resulting in the lowest average value. The data from this study indicate that the increase in the concentration of mineral adsorbents from quail livestock waste is directly proportional to the increase in the concentration of H_2 gas in the bio gas content. Therefore, the use of 100% mineral adsorbent from quail livestock waste was effective in purifying hydrogen gas (H_2) in bio gas with an increase in H_2 gas value of 80.27% in T5 treatment (Figure 2). This is supported by research conducted by [11] that the use of Fe_2O_3 and Al_2O_3 as mineral adsorbents has resulted in 99.8% purity of hydrogen gas in synthetic biogas. The purity of hydrogen gas occurs due to the decomposition of CH_4 gas as described in the thermodynamic equilibrium of the Boundouard reaction. So, it can be assumed that the decrease in the concentration of CH_4 gas in the P5 treatment biogas is caused by the decomposition of CH_4 gas into H_2 gas. Or caused by other factors such as saturation of the adsorbent used in the purification of bio gas treatment T5 so that the adsorbent is not able to capture CO_2 gas as an impurity gas in biogas. This saturation is due to the fullness of the adsorbent cavities by CO_2 gas which causes an increase in the CO_2 gas concentration in the T5 treatment.

3.3 Treatment of flow rate and pressure of purification of Biogas

The test will be carried out of treatment with different amount of carbon and mineral quail waste. The biogas water flow rate and gas pressure the test results after passing through the adsorption process is shown in Table 2.

Table 2. Water Flow rate and gas pressure after passing through the adsorption process

Treatment	Flow Rate (mL/s)	Pressure (Pa)
T0	2.363^a	99.319^e
T1	2.468^a	93.801^d
T2	11.868^b	29.019^c
T3	15.128^c	21.253^d
T4	19.943^e	5.926^a
T5	19.678^d	6.540^a

$^{a-}$values with different superscripts within a column are significantly differences (P<0.05)

The water flow rate and gas pressure are shown in Table 2. The results showed significant differences in water flow rate and gas pressure ($P<0.05$). As a result, increase in the water flow rate will decrease gas pressure. The treatments of P2, P3, P4, and P5 showed higher water flow rate than the control treatment (P0). However, P1 treatment had lower water flow rate than the control treatment even it had no significant difference. T5 and T4 had completely high-water flow rate (19.678-19.943 mL/s) than the other treatments and the control. The results were similar by [12] the effectiveness of CO_2 absorption of biogas increase along with higher biogas pressure and water flow rate. This is because the adsorbent takes more time to absorb the impurity gases in the bio gas, also due to the remodeling process the microbes in the digester are not good. However, this bio gas purification flow rate cannot be an absolute benchmark in determining the quality of bio gas. The length of time required for gas to pass through the adsorbent can be caused by various factors.

On the other hand, there were significant differences among the gas pressure of treatments. However, the gas pressure was 6.540, 5.926, 21.253, 29.019, 93.801, 99.319 for T5, T4, T3, T2, T1, and T0 respectively. The treatments of T1, T2, T3, T4, and T5 showed significant differences with lower gas pressure than the control of treatment ($P<0.05$). This is because the minerals in quail waste contain calcium (Ca) compounds which have the ability to bind CO_2 gas, resulting in an increase in the concentration of CH_4 gas which causes an increase in the calorific value of the bio gas. However, in T5 treatment (100% minerals from quail excreta) the percentage of pressure drop was lower than in T4 treatment, this is presumably due to saturation of the adsorbent used. Saturation is caused by the cavities in the adsorbent that have been filled with impurities so that they are no longer able to absorb. The decrease in the value of gas pressure in the bio gas that has gone through the purification process occurs due to the absorption of CO_2 gas in the adsorbent. The lower the resulting pressure value, the more CO_2 gas is absorbed. Gas pressure has a function to flow gas into the adsorbent tube and into the bio gas sample holder. [12] during the time of gas pressure test in columns adsorbent should be conservation as to not become over-pressured nor pull out (under-pressured). This is caused by anaerobic fermentation process is very sensitive to changes in temperature and biogas production will decrease rapidly due to changes in temperature reactor inside reactor.

4 Conclusion

Results from this study proved that various treatments with different amount of carbon and mineral quail waste had high of water flow rate and less gas pressure than the control treatment. Moreover, the high amount of carbon and minerals quail waste had the best result. These findings indicated that P4 had greatest effectiveness in water flow rate and gas pressure. 1. Efforts to purify bio gas with the best quality can be carried out using an adsorbent of 100% activated carbon from coconut shell waste (P1) and using an adsorbent of 50% activated carbon from coconut shell waste and 50% minerals from quail livestock waste (P3). The use of 100% Ca^{2+} mineral adsorbent in quail excreta to purify bio gas is considered ineffective because it is not good at applying adsorbents during bio gas purification, such as not having physical activation such as aerating and drying after using the adsorbent so that this adsorbent treatment does not can show an increasing trend in the concentration of CH_4 gas and a decreasing trend in the concentration of CO_2 gas in the purified biogas.

It to be recommend the use of both types of adsorbent needs to be considered again, especially regarding the homogeneity of the adsorbent content used so that it has the same adsorption ability. It is necessary to carry out a physical activation process first after using the two adsorbents so that the effectiveness of CO_2 gas absorption on the two adsorbents can take place optimally.

References

1. D. N. Adli, LRRD, **33**, 94 (2021)
2. O. Sjofjan, D.N. Adli, R. P. Harahap, A. Jayanegara, D. T. Utama, A. P. Seruni, F1000 Research, **10,** 183 (2021)
3. M. Junus, SAADC, **1** (2015)
4. N. Hamidi, Wardana, D. Widhiyanuriyawan, Jurnal Rekayasa Mesin, **2,** 227-231 (2011)
5. A. Yamliha, B. D. Agro, W. A. Nugroho, Jurnal Bioproses Komoditas Tropis, **1,** 67-72 (2013)
6. S. Wahyudi, M. Arif, N. A. Hidayati, Rekayasa Mesin, **11**(2020)
7. D. Widhiyanuriyawan, N. Hamidi, C. Trimandoko, Jurnal Rekayasa Mesin, **5**(2014)
8. M. Junus, F. Jaya, Ustadi, M. P. Kurniawan, Y. A. Romadhon, International Journal of Research-Granthaalayah, **9**(2021),
9. R. Jammet, *Methane Production guide- how to make biogas*, (2012)
10. I. Suraya, Tiarasti, H. Irvan, B. Trisakti. Jurnal Teknik Kimia USU, **1**(2012)
11. S. Bock, R. Zacharias, V. Hacker. Royal Society of Chemistry, **9**(2019)
12. T. M. Gantina, P. Iriani P, Maridjo, C. K. Wachjoe. J. Phys. Conf. Ser., 012011 (2020)

The potential of frozen rumen fluid for ruminant feed evaluation using *in vitro* gas production technique

Melyndra Sonya Pramita and *Hendrawan* Soetanto*

Faculty of Animal Science, Universitas Brawijaya, Malang 65145, Indonesia

Abstract. The aim of this study was to determine the potential of frozen rumen fluid (RF) for ruminant feed evaluation using *in vitro* gas production (IVGP) technique. This study used a factorial randomized block design with a 3x3x3 pattern, consisting three feeding regimes varying in forage to concentrate proportion, namely 20%:80% (feed A), 40%:60% (feed B), and 60%:40% (feed C), respectively; three storage time (2, 11, and 19 days); and RF from 3 breeds of indigenous cattle, namely Bali, crossbred Ongole and Madura cattle as groups. The results showed that the effects of feeding regimes, inoculum source, or storage time on total gas production, potential of gas production, rate of gas production, dry matter and organic matter degradability. Based on the results, it can be concluded that frozen RF may be used as inoculum source for feed evaluation using the IVGP technique and hence it supports the previous findings reported in the literature. Considering that the effects of different feeds, inoculum sources, and storage times gave the specific values, it is advisable that to use RF from slaughtered animals should be taken from the same breed to allow accurate comparison for feed evaluation using IVGP technique.

1 Introduction

The *in vitro* gas production (IVGP) technique has been widely applied as a routine feed evaluation for ruminant animals in many countries. By recording the gas at several time intervals, the kinetics of rumen fermentation can be measured [1] and the values of feed digestibility calculated correlate very strongly with the values obtained by in vivo digestibility measurements. [2]. It therefore, is understandable if the IVGP technique becomes more preferred than in vivo digestibility measurement that is more tedious, capital, and time-consuming.

The application of the IVGP technique, unfortunately, requires the use of rumen fluid as microbial inoculum from rumen cannulated animals that pose questions on the compliance of animal welfare [3, 4] This recent pressure on the ban of cannulated animals for such researches has encouraged researchers to seek other alternatives. A number of reports are available in the literature on the use of ruminant's faeces such as cattle and sheep as a source of inoculum [5, 6, 7], but the digestibility values obtained are generally lower which is

* Corresponding author: hendrawan07@ub.ac.id

attributable to the lower population of rumen microbes in the faeces [8] RF can also be collected by inserting a stomach tube via the mouth of animals, however RF may be contaminated with saliva and fluid exchange with the omasum and abomasum [9]. The alternative source of RF as microbial inoculum that is abundantly available is from slaughtered animals at the abattoir or in some instances it is available in a particular religious festival time such as Eid al–Adha in the Muslim-populated countries like Indonesia where the practice of sacrifice takes place during three days of festival by slaughtering ruminant animals.

The problem that often arises in the use of RF from slaughtered cattle is ignorance of the prior feed consumed by livestock before being slaughtered which may affect the extent of gas production during the course of studies. Another mitigating question is the feasibility usage of frozen RF for IVGP studies. Several studies had reported the agreement in gas production using fresh RF from slaughtered cattle as compared with frozen RF even though some precautions require judicious attention. Recently, [11] reported their study on the use of frozen RF for IVGP and concluded that stored RF up to 72 hours had no significant decrease in gas production, but it declined after 96 hours of storage. This notion opens for further studies on the possibility of using cryoprotectants such as dimethyl sulfoxide (DMSO) or glycerol [12] to improve the longevity of rumen microbes during the frozen phase. The existence of the majority of Indonesian people who are Muslim strongly supports the implementation of this alternative because there is an Eid al-Adha, where many ruminants are slaughtered simultaneously. It can be used to compare the yield of feed fermentation carried out by rumen microbes of different livestock breeds.

Quality of feed greatly affects the yield of rumen fermentation and it is the reason why the low quality of tropical forages seldom meets the nutrient requirement of ruminants to support their genetic potential. Therefore, it is important to add concentrates to correct the nutrient imbalance for this particular objective. The proportion of forage to concentrate in the diet mostly used by breeders in Indonesia is 60%: 40%, but it may change up to 20%:80% for fattening cattle. Involving the proportion of concentrate in feed up to 100% can increase the degradability of dry matter from 75.03g / 100g DM to 88.85g / 100g DM [13], but this practice may impose some disadvantages, namely economically inefficient and lowering the rumen pH that leads to acidosis, a type of digestive disturbance. This study aimed at elucidating the feasibility of using frozen RF taken from slaughtered cattle to evaluate three cattle feeds varying in the proportion of forage to concentrate by the IVGP technique.

2 Materials and Methods

2.1 Experimental Location and execution

The research took place from August to October 2020 at the Laboratory of Animal Nutrition and Feed, Faculty of Animal Science, Universitas Brawijaya.

2.2 Materials

A complete feed consisting of grass hay and commercial concentrate was used in this study. The hay was a mixture of signal grass (*Brachiaria decumbens*) and star grass (*Cynodon plectostachyus*) obtained from Artificial Insemination Centre (Singosasi, Malang) while concentrate was obtained from a local feed market.

2.3 Methods

This study used a factorial randomized block design with a 3x3x3 pattern. Three feeding regimes varying in the forage : concentrate ratio, namely 20% : 80% (feed A), 40% : 60% (feed B), and 60% : 40% (feed C), respectively as the first factor. The RF was stored in the deep freezer at -25^{0}C for 2, 11, and 19 days as the second factor. RF extracts from three different breed were used as a group in the design.

2.3.1 Sample preparation

Each feed sample was taken as much as 1 kg and oven-dried at 60°C to obtain air-dry matter content. The dried sample was then mashed with a grinder through a 1 mm sieve for further use and stored in plastic clips. Feed samples were analysed for proximate nutritive values [14] of dry matter (DM), organic matter (OM), crude protein (CP), ether extract (EE), and crude fibre (CF). Table 1. below describes the chemical composition of three feeding regimes

Table 1. Chemical composition of feed treatments

Feeding Regime	Chemical Composition (%)				
	DM	OM	CP	CF	EE
A	82.94	91.78	14.34	15.68	5.75
B	82.08	92.04	11.67	22.08	3.71
C	81.18	91.95	8.79	28.25	3.48

Based on %DM

2.3.2 Collection of rumen digesta

Rumen digesta was taken from 3 breeds of cattle that were slaughtered for the Islamic festival of Eid al-Adha 1441 H (31 July 2020 and 1 August 2020) within the regency of Malang. The traveling time from the onset of rumen digesta collection to reach the nutrition laboratory varied between 18 - 50 minutes. The detailed information of the sampling location is presented in Table 2.

The rumen digesta of the respective cattle breed was immediately taken after the rumen was opened, checked for the temperature, and squeezed gently while transferred into the pre-warmed flasks until full and closed with the lid. They were then transported to the Nutrition and Feed Laboratory for further process.

2.3.2.1 Preparation of inoculum

As soon as arrived in the laboratory, all rumen digesta was emptied from the flasks and transferred into plastic bottles, and stored in the freezer at -28^{0}C. The contents of the rumen of each cattle breed were transferred into 5 plastic bottles with a capacity of 330 ml. The rumen digesta in the flask was poured into a plastic bottle and mixed between respective cattle breed under continuous flushing with CO_2 prior to be sealed with a metal cap and stored until required.

Table 2. Information on the distribution site for RF intake

Cattle Breed	Pick Up Date	Location	Distance for Location to Laboratory (km)*	Duration for transport (minute)	No. of Cattle (head)
PO	31 July 2020	Pujon Sub-district	29	51	1
	1 August 2020	Moslem Boarding School Sabilurrosyad Gasek village, Malang Regency	3.4	10	2
Madura	1 August 2020	Tlogomas Village, Lowokwaru sub-district Malang City	5.8	13	3
Bali	1 August 2020	Wiagan Sekarpuro village , Pakis sub-district Malang City	8.8	18	3

*) Information was obtained from *Google Map*

2.3.2.2 Thawing procedure

The bottle containing the contents of the rumen was thawed first using pre-warmed water 39°C for ± 30 minutes. After the thawing process was done, the contents of the rumen in the bottle were filtered using a three-layered filter cloth and measured as needed. The source of inoculum from one breed of cattle was taken from three heads, so that the RF requirement per cattle was 1/3 part of the total RF requirement in one solution formula.

2.3.3 In vitro gas production analysis

The technique of IVGP has been described by Soetanto *et al* [15] and the procedure is briefly as follows. A 500 mg of feed representing respective feeding regimes was inserted into a 100 ml glass syringe according to [16]. The syringe was incubated overnight before being used for *in vitro* analysis. In the *in vitro* test, the solution is added to the syringe, with a solution formula consisting of 1095 ml distilled water, 730 ml buffer solution, 365 ml macro-mineral solution, 0.23 ml micro-mineral solution, 1 ml resazurin, 60 ml freshly prepared reductive solution, and 660 ml of RF. A syringe without a sample was included in the incubation process as a blank, to correct the production of the gas produced. The syringes were incubated in a water bath at 39°C for 48 hours. Gas production at 2, 4, 6, 8, 10, 12,18,24,36, and 48 hours incubation was recorded to allow the kinetic of respective gas production was calculated. Meanwhile, the potential and rate of gas production were calculated by the equation according to [16] as follows, $Y = b(1-e^{-ct})$, where Y: gas produced at the time 't' (ml/500mg DM), b: the potential of gas production (ml/500mg DM), c: rate of gas production (ml/hour), t: incubation time (hour), and e: exponential.

2.3.4 Rumen degradability

Rumen degradability was calculated gravimetrically according to the tare DM or OM weight of the original sample minus residue. The remaining residue from the gas production was transferred to the tube and was centrifuged at 1409 *g* for 5 minutes until the supernatant was clear and then discarded. The sediment was transferred to the crucible and oven-dried at 105 ^{0}C and weighted to obtain feed degradation in the rumen. The dried sediment was then burned at 550°C to measure the OM content of the sediment for further calculation of OM degradation. This procedure refers to [16] which was modified according to the laboratory conditions in this study.

2.3.5 Statistical analysis

This study used a factorial randomized block design with a 3x3x3 pattern. If the results of the design show no interaction between factors, each factor was tested using a randomized block design to determine its effect on individual factors [17]. The data obtained were tabulated using Microsoft Excel 2016. While the potential and rate of gas production were obtained by the NEWAY method using Microsoft Excel 2016. If the data in ANOVA showed differences due to treatment, it was continued by a Duncan's Multiple Range Test.

3 Results and Discussion

Table 3 summarizes the statistical analysis of all parameters measured during the IVGP study and clearly demonstrated that there was no significant decrease ($P>0.05$) as the storage time increased from 2 to 19 days. Furthermore, the interaction among feeding regimes, source of inoculum, and storage time of RF also revealed no significant difference ($P> 0.05$). In accordance with [18], as there is no interaction between feed, source of inoculum, and storage time, further statistical analysis was done to the main effects, namely feed, source of inoculum, and storage time only.

As shown in Table 4, There is variation in results among the main effects of feed, inoculum sources, and storage time in the respective cattle breed. The effect of feed on storage time is presented in Table 4. Storage time of RF for 2 days showed a highly significant difference ($P<0.01$) on total gas production, gas production rate, and OMD. Storage time of RF for 11 days showed a highly significant difference ($P<0.01$) in the total gas production and the rate of gas production. Meanwhile, the storage time for 19 days showed a significant different ($P<0.05$) on gas production and gas production potential, as well as a very significant different ($P<0.01$) on the rate of gas production.

The effect of feeding regime on the inoculum source is presented in Table 5. The inoculum source of PO and Madura cattle showed a highly significant different ($P<0.01$) on both the total and rate of gas production. Meanwhile, the inoculum source from Bali cattle showed a very significant different ($P<0.01$) on the rate of gas production and DMD, and had a significant different on total gas production ($P <0.05$).

Table 6 describes the effect of storage time on the inoculum source. PO cattle RF showed a very significant different ($P<0.01$) on total gas production and gas production rate, the use of Madura cattle RF also showed a significant different ($P <0.05$) on total gas production. Meanwhile, Bali cattle RF showed a very significant different ($P<0.01$) on DMD and OMD, as well as a significant different ($P<0.05$) on the rate of gas production.

Sources of inoculums obtained from different breeds of cattle indicate that there are differences in the diversity of life in the rumen which can be influenced by the type or species of the host itself, physiological status, and age of the cattle [19, 20, 21]. Information on these factors is not recorded, because the origin of livestock raising prior to slaughter is not known certainty. In addition, the feed consumed can affect the abundance of rumen microbes. The feed will be hydrolysed by rumen microbes through the enzyme activity they produce by attaching the rumen microbes to the feed particles. This is the most efficient way for microbes to prolong their residence in the rumen and carry their enzymes to adhere to other substrates [22].

Gas production describes the products produced from rumen microbial fermentation. Gas production from three feed treatments given showed the highest value in feed A. The high gas production during the 48-hour incubation was in line with the increasing proportion of concentrate in the feed. The same result was also shown by [23] who used rice straw as forage and concentrate with the same treatment ratio. This was due to the increased CP content with the increasing proportion of treated feed concentrate. In accordance with the results of previous studies [24, 25], protein will be easily fermented in the first 4-6 hours of incubation.

Based on the gas production data presented in Figure 1 the highest increase in gas production s shown at the 4th hour for the PO and Madura cattle RF groups, and at the 6th hour for the Bali cattle group with an average value range of 10-14 ml/500 mg DM. This is because the main part of the protein will be fermented at the beginning of incubation along with other soluble components [26].

The potential of gas production shows a significant difference ($P<0.05$) only occurs in the feed treatment with RF storage time for 19 days as shown in Table 4. The highest potential of gas production value is shown in feed C, while the highest gas production value is shown on feed A. The three feed treatments used contain high crude fibre but have the potential to be fermented by rumen microbes to produce gas [27]. Fibrous feed which has low digestibility will be broken down slowly because the digestion process is slow for the first time, so the enzyme action is delayed [28]. While the rate of gas production is directly proportional to total gas production, this is due to the fermentation of the protein contained in the feed that is generally easier to be fermented in the rumen and hence enhances the gas production during IVGP measurements.

The preservation of RF was carried out to determine the storage capacity of the RF so that it could be reused and facilitate the use of the inoculum from the same source in a series of studies. In Table 4. the two-day shelf life of the feed treatment gave the highest DMD value ($P<0.01$) compared to the storage time of 11 and 19 days. This decline in DMD may be attributable to the dehydrating effect of freezing and increment in storage time beginning at 11 and 19 days. This notion is supported by evidence reported by [12] where the adverse effect of freezing had been minimized by adding cryoprotectants such as glycerol and DMSO to the microbial medium (in this case: RF). It has been known that cryoprotectants act at the cell membrane level by limiting the dehydrating effect of cells during the freezing process, lowering the freezing point of extracellular and intracellular biological fluids, and encouraging the vitrification process rather than the formation of intracellular ice crystals [29, 30]. Previous research [30] has proven that addition of 1 % and 10 % DMSO in the rumen ciliate cultures after being equilibrated at 25°C for 5 minutes has a low sensitivity to potential toxicity of DMSO. Another study [31] has used glycerol to compare lyophilized and frozen rumen microbes, showing that there is no effect of proteolytic activity on microbial preservation. Furthermore, [12] compared 5% DMSO with 5% glycerol as a cryoprotectant in RF, showing that the value of DMD was higher with DMSO-preserved RF than glycerol. However, recent research suggests that the addition of cryoprotectants to organic chemicals can cause toxicity effects such as DMSO and may interfere with subsequent *in vitro* testing because they are fermentable substrates (e.g. glycerol) [32]. Based on the negative effect of the addition of cryoprotectants, this study did not use cryoprotectants and refers to [11, 33, 34].

Table 3. The effect of treatment on observed variables

Item	PO Cattle			Madura Cattle			Bali Cattle			SEM	p-value (PxLxS)	Sig.
	2 days	11 days	19 days	2 days	11 days	19 days	2 days	11 days	19 days			
Total of gas production (ml/500mg DM)												
A	126.00	130.17	113.97	122.43	131.51	130.19	104.88	77.61	95.60	3.689	0.998	ns
B	113.30	120.17	102.91	114.18	124.30	126.04	101.65	65.71	90.21			
C	110.87	110.68	92.38	102.87	105.93	116.30	88.72	59.96	87.03			
Potential of gas production (ml/500mg DM)												
A	139.39	132.03	126.69	145.21	147.91	144.01	122.28	105.17	112.63	16.059	0.798	ns
B	135.94	137.32	136.53	144.43	162.50	151.41	135.64	93.35	111.10			
C	161.83	126.94	170.06	162.15	197.99	158.88	202.33	556.96	167.87			
Rate of gas production value (ml/hour)												
A	0.050	0.064	0.045	0.044	0.047	0.049	0.041	0.029	0.037	0.003	0.999	ns
B	0.039	0.052	0.028	0.035	0.030	0.038	0.030	0.027	0.033			
C	0.025	0.041	0.016	0.022	0.016	0.027	0.012	0.002	0.015			
Rumen Dry matter degradability (%)												
A	37.58	32.80	37.03	46.47	42.40	37.88	48.38	41.86	47.35	0.984	0.993	ns
B	40.70	33.79	31.91	45.63	44.52	43.43	42.95	38.25	42.37			
C	35.51	34.87	29.03	46.47	42.33	45.17	42.11	37.68	40.38			
Rumen Organic matter degradability (%)												
A	50.67	46.20	44.87	50.93	52.39	48.18	52.66	52.75	70.00	1.535	0.945	ns
B	49.08	43.93	43.62	50.34	61.23	49.28	48.41	40.98	64.79			
C	42.24	42.05	48.16	43.45	49.95	47.49	45.34	42.18	70.53			

A, B, C= feed treatments; SEM= Standard Error of the Mean; PxLxS= interaction of feed, storage time of rumen fluid, and inoculum source; Sig.= Significance
ns) there was no significant different ($P>0.05$) in interaction of feed, storage time of rumen fluid, and inoculum source

Table 4. Main effect of feeding regime on IVGP parameters due to different storage time

Variable	2 Days Storage			SEM	p-value	Sig.	11 Days Storage			SEM	p-value	Sig.	19 Days Storage			SEM	p-value	Sig.
	A	B	C				A	B	C				A	B	C			
IVGP parameters																		
Y	117.77^{b}	109.71^{ab}	100.82^{a}	3.794	0.007	**	128.04^{b}	121.51^{b}	108.37^{a}	3.399	0.004	**	113.25^{b}	106.38^{ab}	98.57^{a}	3.361	0.017	*
b	135.62	138.67	175.44	7.786	0.097.	ns	145.71	152.78	173.01	5.676	0.070	ns	127.78^{a}	133.01^{a}	165.60^{b}	7.375	0.049	*
c	0.045^{c}	0.035^{b}	0.020^{a}	0.004	0.0002	**	0.047^{c}	0.034^{b}	0.022^{a}	0.004	0.001	**	0.044^{c}	0.033^{b}	0.019^{a}	0.004	0.003	**
Rumen degradability																		
DMD	44.14	43.09	41.36	1.449	0.467	ns	42.25	44.53	44.65	0.907	0.439	ns	40.75	39.24	38.19	2.021	0.785	ns
OMD	51.42^{b}	49.28^{b}	43.68^{a}	1.208	0.003	**	50.50	53.62	46.96	1.601	0.142	ns	54.35	52.56	55.40	3.669	0.390	ns

Y= total of gas production (ml/500mg DM); b= potential of gas production (ml/500mg DM); c= rate of gas production (ml/hour); DMD= Dry Matter Degradability (%); OMD= Organic Matter Degradability (%); A,B,C= feeding regimes; SEM= Standard Error of the Mean; Sig.= Significance
Different superscript (a-c) in the same column showed significant different among feed treatments
*) there was significant different at $P<0.05$; **) there was significant different at $P<0.01$; ns) there was no significant different ($P>0.05$) in the effect of feed composition variation between different storage time

Table 5. Main effect of feeding regime on IVGP parameters attributed to different inoculum source

Variable	PO Cattle			SEM	p-value	Sig.	Madura Cattle			SEM	p-value	Sig.	Bali Cattle			SEM	p-value	Sig.
	A	B	C				A	B	C				A	B	C			
IVGP parameters																		
Y	123.38^{b}	112.13^{a}	104.64^{a}	3.822	0.002	**	128.04^{b}	121.51^{b}	108.37^{a}	3.399	0.004	**	92.69^{b}	85.85^{ab}	78.57^{a}	5.109	0.018	*
b	132.70	136.60	152.94	5.031	0.284	ns	145.71	152.78	173.01	5.676	0.070	ns	113.36	113.36	309.05	48.652	0.224	ns
c	0.053^{c}	0.040^{b}	0.027^{a}	0.005	0.0002	**	0.047^{c}	0.034^{b}	0.022^{a}	0.004	0.001	**	0.036^{b}	0.030^{b}	0.010^{a}	0.004	0.0009	**
Rumen degradability																		
DMD	35.80	35.47	33.14	1.147	0.548	ns	42.25	44.53	44.65	0.907	0.439	ns	45.86^{b}	41.19^{a}	40.06^{a}	1.205	0.002	**
OMD	47.25	45.54	44.15	1.027	0.575	ns	50.50	53.62	46.96	1.601	0.142	ns	58.47	51.39	52.68	3.845	0.122	ns

Y= total of gas production (ml/500mg DM); b= potential of gas production (ml/500mg DM); c= rate of gas production (ml/hour); DMD= Dry Matter Degradability (%); OMD= Organic Matter Degradability (%); A,B,C= feeding regimes; SEM= Standard Error of the Mean; Sig.= Significance
Note: see Table 4 for notation explanation

Table 6. Main effect of storage time on IVGP parameter among different inoculum source

Variable	PO Cattle			SEM	p-value	Sig.	Madura Cattle			SEM	p-value	Sig.	Bali Cattle			SEM	p-value	Sig.
	2	11	19				2	11	19				2	11	19			
IVGP parameters																		
Y	116.72^{h}	120.34^{h}	103.09^{a}	3.822	0.002	**	113.16^{a}	120.58^{h}	124.18^{h}	3.399	0.030	*	98.41	67.76	90.94	5.109	0.0009	ns
b	145.72	132.10	144.43	5.031	0.490	ns	150.60	169.47	151.43	5.676	0.151	ns	153.41	251.82	130.53	48.652	0.539	ns
c	0.038^{b}	0.052^{c}	0.030^{a}	0.005	0.0003	**	0.034	0.031	0.038	0.003	0.086	ns	0.028^{b}	0.019^{a}	0.028^{b}	0.004	0.034	*
Rumen degradability																		
DMD	37.93	33.82	32.66	1.147	0.192	ns	46.19	43.08	42.16	0.907	0.199	ns	44.48^{h}	39.26^{h}	43.36^{a}	1.205	0.005	**
OMD	50.50	53.62	46.96	1.601	0.116	ns	48.24	54.52	48.32	1.601	0.116	ns	48.80^{a}	45.30^{a}	68.44^{h}	3.845	0.002	**

Y= total of gas production (ml/500mg DM); b= potential of gas production (ml/500mg DM); c= rate of gas production (ml/hour); DMD= Dry Matter Degradability; OMD= Organic Matter Degradability; SEM= Standard Error of the Mean; Sig.= Significance
Note: see Table 4 for notation explanation

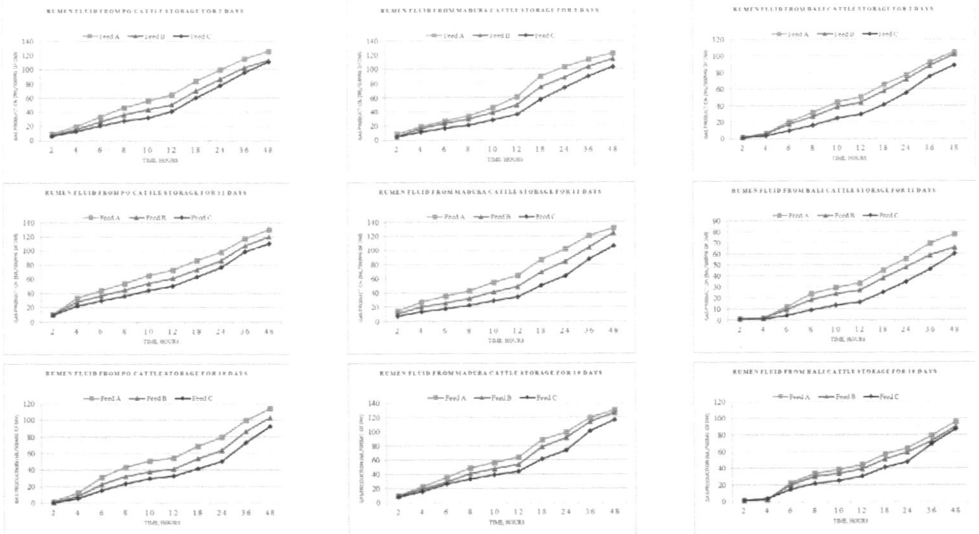

Fig. 1. Gas production patterns produced from different feeding regimes, microbial inoculum source and storage time of rumen fluid in the respective indigenous breed of cattle

4 Conclusion

Based on the data obtained, this study supports the previous findings reported in the literature on the feasibility to use the frozen RF for IVGP measurement with the special attention should be considered, that is only using the RF from the same breed of cattle. Further studies may be warranted to evaluate the benefit of adding cryoprotectants during freezing process that render the detrimental effect of freezing and hence to improve longevity of microbes in the rumen to digest feed during IVGP measurements.

The authors wish to express our sincere gratitude for tangible help given by Rizka M.A., Aprilia D.K., Roi K.J., Muhammad R.A., Lisa M.M., Ainy, Hafidz, and Arya D.S. Thanks also due to the person in charge at each RF collection site, who made the collection and transportation of RF arrived promptly at our nutrition laboratory. We are indebted to the financial support provided by Brawijaya University through a special grant for professorship to HS that enabled the completion of this study.

Reference

1. M. Blümmel, E.R. Ørskov, Anim. Feed Sci. Technol. **40**, 109-119 (1993)
2. M.S. Dhanoa, S. Lopez, J. Dijkstra, D.R. Davies, R. Sanderson, B.A. Williams, Z. Sileshi, J. France, B. J. N. **82**, 131-142 (2000)
3. A.S. Chaudhry, R.A.I. Mohamed, Czech J. Anim. Sci. **57**, 265-273 (2012)
4. M. Mutimura, C.B. Myambi, P. Gahunga, D.M. Mgheni, G.H. Laswai, L.A. Mtenga, D. Gahakwa, A.E. Kimambo, C. Ebong, Agric J. **8**, 173-180 (2013)
5. S.E. Cavalcante, V.R. Vasconcelos, A.L. Abdalla, A.A. Alves, A.S. Natel, D. Shanasekaran, L.R.F. da Silva, J.C. Siqueira, *Faecal inoculum as alternative microbial source for in vitro rumen fermentation techniques*, in Proceeding of Conference on International Research on Food Security, Natural Resource Management and Rural Development, 16-18 September 2015, Berlin, Germany (2015)
6. J.W. Cone, A.H. van Gelder, H. Bachmann, Anim. Feed Sci. Technol. **99**, 221-231 (2002)
7. C.S. Pandian, T.J. Reddy, K. Sivalah, M. Blümmel, Y.R. Reddy, Anim. Feed Sci. Technol. **16**, 271-281 (2016)
8. G.T. Bergmann, B. M. C. Vet. Res. **13**, 253 (2017)
9. T. Geishauser, A. Gitzel, S. R. R. **21**, 63-69 (1996)
10. G.A. Beyihayo, R. Omaria, C. Namazzi, A. Atuhaire, U. G. A. S. **16**, 93-98 (2015)
11. A.S. Chaudhry, R.A.I. Mohamed. Czech J. Anim. Sci., 57, 2012 (6): 265–273
12. N. Denek, A. Can, M. Avci. S. A. J. A. S. 40, 251-256 (2010)
13. S.A. Nagadi, Met. Env. & Arid Land Agric. Sci. **28**, 27-36 (2019)
14. AOAC Official Method of Analysis. 18th Edition, Association of Officiating Analytical Chemists, Washington DC, Method 935.14 and 992.24. (2005)
15. H. Soetanto, R.M.Aprilia, M.S.Pramita and I.Banna. Comparison of Three Different Rumen Fluid As A Source of Inoculum To Evaluate In Vitro Gas Production And Digestibility of Elephant Grass-Concentrate Mixture ,In Proceedings of 2nd International Conference On Animal Production For Food Sustainability, Organized by the University of Andalas and University of Udayana, held on June 16th, 2021,(2021).
16. H.P.S. Makkar, M. Blümmel, K. Becker, B. J. N. **73**, 897-913 (1995)

17. M. Blümmel, H. Steingaß, K. Becker, B. J. N. **77**, 911-921 (1997)
18. A. Sastrosupadi, Rancangan Percobaan Praktis Untuk Bidang Pertanian, Kanisius, Yogyakarta (1995)
19. E. Jami, A. Israel, A. Kotser, I. Mizrahi, The ISME J. **7**, 1069-1079 (2013)
20. S. Wang, J. Pisarčíková, M. Kreuzer, A. Schwarm, Can. J. Anim. Sci. 1-30 (2018)
21. F. Li, C. Li, Y. Chen, J. Liu, C. Zhang, B. Irving, C. Fitzsimmons, G. Platstow, L.L. Guan, Microbiome, **7**, 1-17 (2019)
22. T.A. McAllister, H.D. Bae, G.A. Jones, K.J. Cheng, J. Anim. Sci. **7**, 303-316 (1994)
23. D.V. Dung, W. Shang, W. Yao, Asian Australas. J. Anim. **27**, 797-805 (2014)
24. J.W. Cone, A.H. van Gelder, Anim. Feed Sci. Technol. **76**, 251-264 (1999)
25. A. Karabulut, O. Canbolat, H. Kalkan, F. Gurbuzol, E. Sucu, I. Filya, Asian-Aust. J. Anim. Sci. **4**, 517-522 (2007)
26. J.W. Cone, A.H. van Gelder, F. Driehuis, Anim. Feed Sci. Technol. **66**, 31-45 (1997)
27. E.R. Ørskov, I. McDonald, J. Agric. Sci. **92**, 499-503 (1979)
28. P. McDonald, R.A. Edwards, J.F.D. Greenhalgh, C.A. Morgan, Animal Nutrition, 7th Edn, John Willey and Sons, New York, USA (2010)
29. P. Mazur, A.J.P. Cell **247**, 125-142 (1984)
30. I. Tanasawa, Ann. N. Y. Acad. Sci. **858**, 227-234 (1998)
31. E. Nsabimana, S. Kisidayova, D. Macheboeuf, C.J. Newbold, J.P. Jouany, Appl. Environ. Microbiol. **69**, 3826-3832 (2003)
32. N.D. Luchini, G.A. Broderick, D.K. Combs, J. Anim. Sci. **74**, 1134-1143 (1996)
33. A. Prates, J.A. de Oliveira, L. Abecia, M. Fondevila, Anim. Feed Sci. Technol. **155**, 186-193 (2010)
34. M. Spanghero, M. Chiaravalli, S. Colombini, C. Fabro, F. Froldi, F. Mason, M. Moschini, C.Sarnataro, S. Schiavon, F. Tagliapietra, Animals 9, 1-14 (2019)

AUTHOR INDEX